Ar

The Metaphysical Found

editiones scholasticae

Edward Feser

Aristotle's Revenge

The Metaphysical Foundations of Physical and Biological Science

Bibliographic information published by Deutsche Nationalbibliothek
The Deutsche Nationalbibliothek lists this publication in the Deutsche Nationalbibliographie; detailed bibliographic data is available in the Internet at http://dnb.ddb.de

53819 Neunkirchen-Seelscheid - GERMANY
www.editiones-scholasticae.de

ISBN 978-3-86838-200-6

2019

Printed on acid-free paper

Printed in Germany
by CPI buchbücher.de GmbH

Contents

4. Space, Time, and Motion

5. The philosophy of matter

6. Animate nature

0. Preface

The central argument of this book is that Aristotelian metaphysics is not only compatible with modern science, but is implicitly presupposed by modern science. Many readers will be relieved to hear some immediate clarifications and qualifications. First, I am not talking about Aristotle's ideas in *physics*, as that discipline is understood today. For example, I am not going to be defending the claim that the sublunary and superlunary realms are governed by different laws, or the doctrine of natural place. I am talking about the philosophical ideas that can be disentangled from this outdated scientific framework, such as the theory of actuality and potentiality and the doctrine of the four causes. These are, again, *metaphysical* ideas rather than scientific ones. Or to be more precise, they are ideas in the *philosophy of nature*, which I regard as a sub-discipline within metaphysics, for reasons I will explain in chapter 1.

Second, I am not arguing that working scientists in general explicitly employ or ought to employ these philosophical ideas in their everyday research. I am arguing that the practice of science, and the results of science at least in their broad outlines, *implicitly* presuppose the truth of these ideas, even if for most practical purposes the scientist can in his ordinary work bracket them off. I am primarily addressing the question of how to *interpret* the practice and results of science, not the question of how to carry out that practice or generate those results.

Third, even then my remarks about those results will be very general. To be sure, I will have a lot to say about why relativity theory, quantum mechanics, chemistry, evolution, and neuroscience in no way undermine the central ideas of Aristotelian philosophy of nature, and even presuppose those ideas in a very general way. But there are in each case a number of different ways the Aristotelian might work out the details, and I make no pretense of having done more than scratch the surface here. An Aristotelian philosophy of physics, an Aristotelian philosophy of chemistry, an Aristotelian philosophy of biology, and an Aristotelian philosophy of neuroscience would each require a book of its own adequately to work out. If the relatively cursory treatments I provide encourage others to carry out these jobs more thoroughly, I will not be displeased.

This work is a sequel to my book *Scholastic Metaphysics: A Contemporary Introduction*, and builds on the main ideas and arguments developed and defended there. To be sure, this new book can be read independently of the older one, since I summarize in chapter 1 the most crucial points from the earlier book. But the skeptical reader who suspects that I have in the present work begged some question or insufficiently defended some background assumption is advised to keep in mind that he will find the full-dress defense in the earlier book.

I borrow the title *Aristotle's Revenge* from an article by the late James Ross (1990), from whose work I have learned so much. One of Ross's key insights concerns the ways that contemporary analytic philosophers have rediscovered and vindicated Aristotelian ideas and arguments, albeit often without realizing it. *Scholastic Metaphysics* developed that theme at length, and this new book continues in that vein. My subtitle is an homage to E. A. Burtt's classic book *The Metaphysical Foundations of Modern Physical Science*. Burtt's book is essential reading, but it raises more questions than it answers. The point of my book is to answer them.

This is the fruit of many years of work, and as always, I owe my wife and children an enormous debt of gratitude for patiently bearing with my absence through the hours I spend chained to the desk in my study. So, I give my love and thanks to Rachel, Benedict, Gemma, Kilian, Helena, John, and Gwendolyn. I also thank my publisher Rafael Hüntelmann for the superhuman patience he showed in awaiting delivery of the book, which came long after the original deadline.

In the years during which I worked on this book, I lost my sister Kelly Eells to pancreatic cancer, and then my father Edward A. Feser to Alzheimer's disease. I dedicate this book to their memory, and to my mother Linda Feser, whose example of self-sacrifice and dignity in the face of great suffering has been heroic and inspirational. Mom, Dad, and Kelly, I love you.

1. Two philosophies of nature

1.1 What is the philosophy of nature?

The nature of the *philosophy of nature* is best understood by way of contrast with *natural science* on the one hand and *metaphysics* on the other, between which the philosophy of nature stands as a middle ground field of study.

The nature of natural science is itself a topic about which I will have much to say in this book, but for present purposes we can note that the natural sciences are concerned with the study of the actually existing empirical world of material objects and processes. For example, biology investigates actually existing living things – the structure and function of their various organs, the taxa into which they fall, their origins, etc. Chemistry investigates the actually existing elements and the processes by which they get organized into more complex forms of matter. Astronomy investigates the actually existing stars and their satellites and the galaxies into which these solar systems are organized. And so forth.

Metaphysics, meanwhile, investigates the most general structure of reality and the ultimate causes of things. Its domain of study is not limited merely to what happens as a matter of contingent fact to be the case, but concerns also what *could have* been the case, what *necessarily must* be the case, what *cannot possibly* have been the case, and what exactly it is that *grounds* these possibilities, necessities, and impossibilities. Nor is it confined to the material and empirical world alone, but investigates also the question whether there are or could be immaterial entities of any sort – God, Platonic Forms, Cartesian *res cogitans*, angelic intellects, or what have you. In addition, metaphysics investigates the fundamental concepts that the natural sciences and other forms of inquiry all take for granted.

For example, whereas the natural sciences are concerned with various specific kinds of material substances – stone, water, trees, fish, stars, and so on – metaphysics is concerned with questions such as what it is to *be* a substance of any kind in the first place. (Is a substance a mere bundle of attributes, or a substratum in which attributes inhere? Are material substances the only possible sort? And so on.) Similarly, the natural sciences are concerned with various specific kinds of causal process –

combustion, gravitation, reproduction, and so on – whereas metaphysics is concerned with questions such as what it is to *be* a cause in the first place. (Is causation nothing more than a regular but contingent correlation between a cause and its effect? Or does it involve some sort of *power* in the cause by which it necessarily generates its effect? Is there only one kind of causality? Or are there four, as Aristotle held?) Whereas the natural sciences explain the phenomena with which they are concerned by tracing them to the operation of ever deeper laws of nature, metaphysics is concerned with issues such as what it is to *be* a law of nature and why such laws operate. (Is a law a mere *description* of a regular pattern in nature? If so, how could it *explain* such patterns? Why is the world governed by just the laws of nature that do in fact govern it, rather than some other laws or no laws at all?)

Of course, some philosophers and scientists deny that there is any reality other than material reality, and any method of studying reality other than science. That is to say, they defend materialism and scientism. But materialism and scientism are themselves metaphysical positions in the relevant sense. They too address the question whether reality extends beyond the natural world studied by empirical science, and simply answer in the negative.

Now, the philosophy of nature stands, as I say, in between natural science and metaphysics. It is more general or abstract than the former, but more specific or concrete than the latter. Metaphysics is concerned with all possible reality, not with empirical and material reality alone. The philosophy of nature is not like that. Like natural science, it is concerned only with empirical and material reality. Natural science, however, is concerned with the empirical and material world that happens as a matter of contingent fact to exist. The philosophy of nature is not so confined. It is concerned with what any possible empirical and material world would have to be like. What must be true of any possible material and empirical world in order for us to be able to acquire scientific knowledge of it? Are there general principles, deeper even than the fundamental laws of physics, which would have to govern any possible material and empirical world whatever those fundamental laws turned out to be? Those are the sorts of questions with which the philosophy of nature is concerned.

This is not to suggest that the boundaries between these three fields of study are always sharp. They are not, at least not in practice. The philosophy of nature might be thought of as a *branch* of metaphysics – as the metaphysics of any possible material and empirical reality, specifically. Alternatively, it might be thought of as the most philosophical end of natural science – natural science as it begins to stretch beyond what can be tested via observation and experiment and relies more on *a priori* considerations. As we will see, much work in contemporary analytic philosophy going under the "metaphysics" label in fact recapitulates traditional themes in the philosophy of nature. We will also see that many claims today put forward as "scientific" are in fact philosophical, or a mixture of the philosophical and the empirical. Nor are such confusions uniquely modern. Aristotle's *Physics* was as much a work of philosophy of nature as a work of physics, and the fact that there was no distinction drawn between these fields of study in Aristotle's day or for centuries thereafter led many erroneously to throw the Aristotelian philosophy of nature baby out with the Aristotelian physics bathwater. Precisely because of such errors, however, it is crucial to emphasize the difference in *principle* between the disciplines, occasional overlap notwithstanding.

How does the philosophy of nature relate to the *philosophy of science*? To a first approximation, it might be argued that it relates to it in something like the way metaphysics relates to epistemology. Epistemology is the theory of knowledge – the study of the nature of knowledge, of whether knowledge is possible, of the range of our knowledge, and of the ultimate bases of all knowledge. If metaphysics is concerned with the nature of reality itself, epistemology is concerned with how we *know* about reality. Similarly, it might be said that whereas the philosophy of nature is concerned with the character of the empirical and material reality studied by science, the philosophy of science is concerned with exactly how science gives us *knowledge* of that reality. It is a kind of applied epistemology, just as the philosophy of nature is a kind of applied metaphysics.

However, this analysis is an oversimplification. To be sure, it is not implausible if applied to the philosophy of science as it existed for much of the twentieth century. Logical positivist and logical empiricist philosophers of science like Rudolf Carnap, Hans Reichenbach, and Carl Hempel were hostile to metaphysics, rejected the scientific realist view that the theoretical entities posited by science exist independently of theory, and concerned themselves with the elucidation of the logic of scientific method. Falsificationist philosopher of science Karl Popper was not hostile to metaphysics or scientific realism, but also focused on questions of method, as did post-positivist philosophers of science like Thomas Kuhn, Imre Lakatos, and Paul Feyerabend. These major thinkers in the field were indeed largely concerned with epistemological matters.

Still, with the revival of scientific realism in the later twentieth century, philosophers of science once again took an interest in metaphysical questions – in the dispute between scientific realism and the various forms of anti-realism, of course, and also in issues such as whether we must attribute real causal powers to things in order to make sense of what science tells us about them, the status of laws of nature, how to interpret what modern physics tells us about the nature of time and space, and so on.

What this means, though, is that contemporary philosophers of science have essentially rediscovered the philosophy of nature, even if they don't always put it that way (though sometimes they do, as in the case of Brian Ellis (2002)). In practice, then, there is considerable overlap between the fields. But as with the distinction between the philosophy of nature, metaphysics, and natural science, it is nevertheless important to keep in mind the distinction in principle between the philosophy of nature and the philosophy of science. The emphasis in philosophy of nature is always on metaphysical questions, whereas the accent in the philosophy of science (at least where it isn't essentially just philosophy of nature under another name) is on epistemological and methodological issues.

What is the epistemology of the philosophy of nature itself? Is it an *a priori* discipline the way that mathematics and metaphysics are often claimed to be? Or are its claims subject to empirical falsification the way that those of natural science typically are? These alternatives are often thought to exhaust the possibilities, but they do not, and seeing that they do not is crucial to understanding how the philosophy of nature differs from natural science. There are propositions that are empirical rather than *a priori*, and yet which are not subject to empirical falsification. For example, the proposition that change occurs is one we know only through experience. But no experience could overturn that proposition, because any experience that purportedly did so would itself have to involve change. (More on this particular issue later on.) More generally, there might be very broad features of experience which, given that they are features of *experience*, are not known *a priori*, yet because they are so extremely *broad*, will feature in every possible experience, including any that might be brought forward to try to falsify claims about them.

Features like these are among those with which the philosophy of nature is especially concerned, because they are relevant to determining what any possible material and empirical world would have to be like. (Some Aristotelian philosophers of nature have labelled this particular kind of experiential basis for the philosophy of nature "pre-scientific experience." Cf. Koren 1962, pp. 8-10; Van Melsen 1954, pp. 12-15.) But the philosopher of nature is also bound to make use of empirical knowledge of the more usual sort, and in particular of knowledge gained from natural science itself. There is nothing in the nature of the subject that entails that we can determine *all* of the most fundamental features of some natural phenomenon from the armchair, as it were. For example, the Aristotelian philosopher of nature holds, as we will see, that all natural phenomena exhibit at least very rudimentary teleological properties. But exactly what the specific teleological properties of some particular natural phenomenon are, whether its apparent teleological properties are real or instead reducible to some more fundamental sort, and similar questions, can only be answered by bringing to bear what we know from chemistry, biology, and the other special sciences.

Brief comment must be made about a dispute in twentieth-century Aristotelian-Thomistic philosophy about the relationship between natural science, metaphysics, and the philosophy of nature. Thomas Aquinas (1986) drew a distinction between three ways in which the intellect abstracts from concrete reality (which have since come to be referred to as the three "degrees of abstraction") and three corresponding fields of inquiry. (Cf. Maritain 1995, chapter II; Smith 1958, chapter 1.) First, the intellect abstracts from the individualizing features of concrete material things, but still considers them in terms of the sensible characteristics that they have in common. Natural science and what would today be called the philosophy of nature (not distinguished by Aquinas himself, as they were not by Aristotle) correspond to this degree of abstraction. Second, the intellect abstracts from even the common sensible features of things and considers only their quantitative features. Mathematics is the field of inquiry corresponding to this degree of abstraction. Third, the intellect abstracts from even the quantitative features and considers only the most general ways in which a thing might be characterized – in terms of notions such as that of substance, attribute, essence, existence, etc. Metaphysics is the field of inquiry corresponding to this last degree of abstraction. For the Aristotelian philosopher of nature, the question naturally arises whether and how this way of carving up the conceptual territory has application today.

Three general views on this question were defended by twentieth-century Thomistic philosophers (Cf. Koren 1962, pp. 18-22). The first essentially endorses Aquinas's treatment of natural science and the philosophy of nature as continuous, amounting to the more concrete and the more abstract aspects, respectively, of a single species of knowledge distinct from metaphysics (Wallace 1982). This is the view associated with what is called either "Laval Thomism" (named for Laval University in Quebec, where its eminent proponent Charles De Koninck was a professor) or "River Forest Thomism" (named for a suburb of Chicago which was the location of the Albertus Magnus Lyceum for Natural Science, whose members were also associated with this view). The second view takes natural science and the philosophy of nature to be distinct species of knowledge, but nevertheless species in the same one genus, which is itself distinct from metaphysics. This view was associated with the Neo-Thomist philosopher Jacques Maritain (1951, pp. 89-98). The third view takes the philosophy of nature to amount not only to a distinct species of knowledge from natural science, but to be of a distinct genus as well, and in particular to amount to a branch of metaphysics. This view was presented in some of the manuals of Thomistic philosophy of the Neo-Scholastic era, such as Andrew Van Melsen's text on the philosophy of nature (1954, chapter 3).

As my characterization of the philosophy of nature above indicates, my sympathies are with this third approach to the subject. A powerful argument for its correctness is the fact that, at least in Aristotelian philosophy of nature, there is a considerable degree of overlap between the central concepts of metaphysics and the philosophy of nature that does not exist between either of these fields on the one hand and modern natural science on the other. For example, the theory of actuality and potentiality (to be expounded below) is not only central to Aristotelian philosophy of nature, but also to Aristotelian-Thomistic metaphysics. But it is utterly unknown to most contemporary physicists and inessential to dealing with the issues they are typically concerned with. Consequently, while it is fairly easy to transition from the study of Aristotelian philosophy of nature to that of Aristotelian-Thomistic metaphysics and vice versa, there is nothing in contemporary physics to orient one to the study of Aristotelian philosophy of nature and nothing in Aristotelian philosophy of nature to orient one to the study of contemporary physics. This state of affairs is easy to understand if one takes the philosophy of nature to be essentially a branch of metaphysics, but very difficult to understand if one supposes the philosophy of nature and natural science to be either of the same species, or two species in the same genus (Van Melsen 1954, pp. 98, 100-1; Koren 1962, pp. 21-22).

Insofar as these remarks entail my taking a side in this dispute, however, I would immediately add two qualifications. First, as I have said, the distinction between natural science and the philosophy of nature is not always observed in practice by either philosophers or scientists. Nor is it desirable that investigations in these areas be kept rigorously separate. Again, while very general concepts and lines of argument in the philosophy of nature (concerning the theory of actuality and potentiality, say, or teleology) can be developed without reference to the findings of natural science, their *application* to specific sorts of phenomena certainly requires attention to such findings, in ways we will be considering in the chapters to follow. Hence a fully adequate philosophy of nature will certainly have to be informed by natural science. And when it is so informed, it is hardly surprising if the relationship between the two fields of inquiry might seem as close as the Laval/River Forest school and Maritain took it to be.

Second, when each of the three main views in this dispute within Aristotelian-Thomistic philosophy has been suitably qualified, it is in my view not clear that much of substance really rides on the dispute. After all, all sides of the dispute would agree that the key concepts and arguments in question (concerning the theory of actuality and potentiality, hylemorphism, teleology, etc.) are sound and important, whether one classifies them as part of natural science, philosophy of nature, or metaphysics. All sides would agree that many standard criticisms of these concepts and arguments rest on a failure carefully to distinguish the ideas themselves (which remain valid) from their concrete application by thinkers of the past (which often rested on mistaken scientific assumptions). All sides agree that careful attention to the findings of modern natural science is crucial to the proper articulation and application of the ideas. Disagreement about whether to label the ideas in question "scientific" or "metaphysical" seems at the end of the day much less important than these matters about which there is agreement.

I ought also to comment on an important respect in which my characterization of the philosophy of nature differs from that of some other Aristotelian-Thomistic writers. I have said that the philosophy of nature is concerned with the most general features of *empirical and material* reality. Other expositions written from an Aristotelian-Thomistic point of view often characterize the field instead as concerned with *changeable* reality (though some earlier writers do characterize it the way I have, e.g. Bittle 1941, p. 13).

Now, I agree that the *correct, Aristotelian* approach to the philosophy of nature is indeed fundamentally concerned with changeable reality. Indeed, as we will see in this book, it is precisely by way of the analysis of change that the central concepts of Aristotelian philosophy of nature are introduced. However, I think my way of characterizing the philosophy of nature as a general field of inquiry is preferable, for two reasons. The first is that there are, as we will see, thinkers who do not deny the existence of the empirical and material world but do claim that modern physics (and in particular relativity) has shown that change is illusory. While I certainly think this view is false (for reasons to be set out in later chapters), it does reflect what might be called a philosophy of nature, albeit a rival to the Aristotelian philosophy of nature. Hence to characterize the very field of the philosophy of nature as essentially concerned with change might seem to beg the question in favor of the Aristotelian approach. Better to characterize it instead in terms of what both sides agree upon, viz. the existence of the empirical and material world.

A second and not unrelated consideration is that the Aristotelian position itself holds that change *presupposes* the reality of a material thing that undergoes the change. Hence this underlying material reality is plausibly the more fundamental subject matter of the philosophy of nature (Cf. McInerny 2001, p. 21). And one way the Aristotelian philosopher of nature might defend the reality of change against his rivals is precisely by appealing to the nature of the material reality that both sides affirm, and arguing that it entails the possibility of change.

Finally, some remarks about terminology are in order, because the very different senses attached to some of the same key terms by older and contemporary writers can and sometimes do lead to enormous but entirely needless misunderstandings. I have already mentioned that the domain of "physics," as Aristotle and Aquinas used that term, included both matters of the sort that contemporary physicists would be concerned with, and also matters that fall within the domain of what is today called the philosophy of nature. I have also mentioned that, for this reason, some modern writers fallaciously suppose that because Aristotle was mistaken about important matters of "physics" in the *modern* sense of the term, *everything* he said about the nature of the physical world (including what he said about matters of the philosophy of nature) is mistaken. Since this modern sense of the term is now the sense almost universally attached to it, it would be foolish, at least in most contexts, for a contemporary Aristotelian to insist on using the word "physics" in the older sense. So as to forestall misunderstandings of the sort in question, it is better to acquiesce to the modern usage of "physics" and apply instead the label "philosophy of nature" to those aspects of Aristotle's account of the nature of the physical world that are still defensible today (as most contemporary Aristotelians and Thomists in fact do).

Something similar can be said about the word "science." The older, Aristotelian use of this term is much broader than the standard contemporary usage. A "science," for the Aristotelian, is an organized body of demonstrated truths concerning the things falling within some domain and their causes. Hence, not only physics, chemistry, biology, and the like, but also metaphysics, ethics, natural theology, and indeed the philosophy of nature itself (since, for the Aristotelian-Thomistic thinker, these fields of inquiry rest on rational arguments and analysis no less than physics, chemistry, etc. do) count as sciences. But this broad usage of the term is so different from current usage (which confines the application of the term "scientific" to claims that are empirically falsifiable) that to insist on it would be to invite needless confusion. Better in most contexts (such as the present one) once again to acquiesce to standard contemporary usage and classify fields like metaphysics, ethics, natural theology, philosophy of nature, etc. as branches of philosophy rather than of "science." Nothing of substance is lost by this procedure.

Other potentially misleading terms include "cosmology" and "psychology." In older textbooks on the philosophy of nature, these are applied to the field's two main sub-disciplines. "Cosmology" is that branch of the philosophy of nature that is concerned with the most general features of inorganic phenomena, and "psychology" is the branch that is concerned with the most general features of living things. The trouble, of course, is that these terms are now generally used in very different ways. The term "cosmology" is now generally used as a label for the branch of natural science that studies the origins and development of the physical universe (which includes organic phenomena as well as inorganic phenomena). Obviously this overlaps somewhat with "cosmology" in the older sense, but modern cosmologists put a special emphasis on the *history* of the physical universe (e.g. tracing it back to the Big Bang) that "cosmology" in the older sense did not. "Psychology" is now generally used as a label for the empirical study of the mind and behavior, and would exclude (as "psychology" in the older sense would not) any concern with plants and other living things devoid of mental properties. The sorts of issues that older Aristotelian-Thomistic textbooks in "cosmology" concerned themselves with would in contemporary philosophy be classified as topics in the philosophy of physics and the philosophy of chemistry.

The sorts of issues that older Aristotelian-Thomistic textbooks in "psychology" concerned themselves with would in contemporary philosophy be classified as topics in the philosophy of biology and the philosophy of mind.

As with the terms "physics" and "science," so too with terms like "cosmology" and "psychology," the wisest policy is, in my view, not to quibble about contemporary usage but rather to use the best modern labels, qualify them as one sees fit, and then to get on with matters of substance. It seems to me that at least to some extent (*not* entirely, but, again, to *some* extent), the twentieth-century disputes between Aristotelian-Thomistic philosophers over the nature of the philosophy of nature may have reflected differing attitudes about how important it is to preserve older usage and to classify things the way Aristotle and Aquinas themselves classified them. This is in my view regrettable. Aristotelians and Thomists have in the past routinely been accused by their critics of being too deferential to authority and too concerned with merely semantic quibbles. Such accusations are, in general, unjust, and the cause of Aristotelian philosophy of nature is not well served by needlessly giving the critics ammunition.

1.2 Aristotelian philosophy of nature in outline

As I have indicated, the most fundamental concepts of Aristotelian-Thomistic philosophy of nature (the theory of actuality and potentiality, hylemorphism, and so forth) overlap with those of Aristotelian-Thomistic metaphysics. I have provided a detailed exposition and defense of those overlapping concepts elsewhere, in a book devoted precisely to Aristotelian-Thomistic metaphysics (Feser 2014b). The overlapping concepts most relevant to the specific topics in the philosophy of nature to be treated in this book will also be discussed in detail in the chapters to come, as each of these topics is treated successively. But it will be useful to provide at the outset a summary of the main concepts, of how they fit together, and of what sorts of considerations motivate the whole system. For one thing, this will give the reader a sense of the "big picture" that the various particular arguments in the chapters to follow are intended to uphold. For another, the "mechanistic" philosophy of nature that is the chief rival to Aristotelianism cannot be understood except by contrast with the latter.

Again, more detailed argumentation and responses to various objections will be developed later in the book. The aim in what follows is merely to provide an overview.

1.2.1 Actuality and potentiality

Aristotle's philosophy of nature was developed in reaction to the Pre-Socratic tradition, and aims for a middle ground position between the *dynamic monism* of Heraclitus and the *static monism* of Parmenides and Zeno (also called "Eleatic monism" after Elea, the town with which Parmenides and Zeno were associated).

Dynamic monism denies the reality of abiding objects. The man of common sense supposes that it is one and the same self that undergoes the bodily and psychological changes he experiences. For the Heraclitean, however, there is only the succession of stages, and no persisting *thing* that underlies them. There is the configuration of cells, molecules, atoms, and other particles that makes up your body now; the slightly different configuration that exists a few moments later as some of these particles drop away; the yet different configuration that exists later still as new cells grow to take the place of the lost ones; and so on. There is also the bundle of thoughts and sensations of which you are now aware; the somewhat different bundle of which you are aware a moment later as your attention turns to something else; the yet different bundle that takes its place when the course of your thoughts and sensations takes a new turn; and so forth. But there is no enduring self that underlies these constantly shifting collections of mental and physical constituents. What is true of human beings is, according to dynamic monism, no less true of everything else – of tables, chairs, rocks, trees, dogs, cats, stars, planets, and indeed molecules, atoms, and other particles themselves. In none of these cases are there really any abiding entities, but only the illusion thereof. The natural world just is this stream of *becoming* or *flux* and never coalesces into anything stable.

Static monism takes the opposite extreme position, and maintains that it is *change* that is illusory. For an ice cube to melt into a puddle of water, the puddle, which initially does not exist or lacks being, has to come into being. Parmenides argued that such a change would therefore entail being arising from non-being. But non-being is just nothing at all, and from nothing, nothing comes. Hence the puddle cannot come into being. But what is true of ice cubes melting into puddles is true of any other purported change. All of them would entail being arising from non-being, or something arising from nothing. Since this is impossible, change is impossible.

Parmenides' student Zeno reinforced such arguments with his famous paradoxes of motion. Consider, for example, the *dichotomy paradox*. Suppose a runner attempts to move from point *A* to point *B*. To get to *B*, he first has to get from *A* to the midpoint between *A* and *B*. But to get to that midpoint, he first has to get to the point which lies a quarter of the distance between *A* and *B*; and doing that in turn requires getting first to the point which lies an eighth of the distance between *A* and *B*; which in turn requires first getting to the point lying a sixteenth of the distance between *A* and *B*; and so on *ad infinitum*. Hence he cannot get to *B*. Indeed, he cannot so much as even *begin* the journey, for to get his foot even an inch off the ground would require first getting it half an inch off the ground, which would in turn require first getting it a quarter of an inch off the ground, and so on. But something similar could be said of any movement. Hence motion in general is impossible.

In response to dynamic and static monism, the Aristotelian argues, first, that both views are ultimately incoherent. If dynamic monism were true, then Heraclitus and every other dynamic monist would themselves be no less subject to the theory's analysis than anything else. That entails that there really is no abiding self associated with Heraclitus, for example, but only the constantly changing collection of thoughts, experiences, configurations of cells, and so on, associated with Heraclitus. In that case, the "Heraclitus" who entertains the first premise of an argument for dynamic monism is not the same as the "Heraclitus" who entertains any of the succeeding premises, who is in turn not the same as the "Heraclitus" who entertains the conclusion. There simply could not be any single mind which ever grasps all the steps of an argument for dynamic monism, and thus could not be any mind which could rationally be convinced of the view – or even irrationally convinced, for that matter, for in fact there would be no mind that persists long enough even to *formulate* dynamic monism or any premise in an argument for it. The truth of dynamic monism would thus be incompatible with the existence of people who affirm dynamic monism. Since there are such people, dynamic monism is false.

Static monism faces similar problems. For Parmenides to work through the steps of his argument, he has first to entertain its initial premise, then to entertain its succeeding premises, and then to entertain its conclusion. He will also thereby have gone from believing that change is real to wondering whether it is in fact real, and finally to being convinced that it is not real after all. But all of *that* entails the existence of change. If he considers such an objection, wonders how he might reply to it, and then finally puts forward a response, that too will involve change. The truth of static monism would thus be incompatible with the existence of static monists like Parmenides. Since there are such people, static monism is false.

But such arguments show at most *that* dynamic and static monism are false, not *where* exactly the flaws are in the various arguments given for such views. The second stage of the Aristotelian response is to identify these flaws. This brings us to the theory of *actuality and potentiality* (or *act and potency*, to use the traditional jargon), which is the core of Aristotelian philosophy of nature.

Again, Parmenides held that change entails being arising from non-being, which is impossible. The Aristotelian agrees that it is impossible for being to arise from non-being, but denies that that is what change involves. Among the things having being, we can distinguish actualities and potentialities. The water that makes up an ice cube is actually solid and actually cold, but it is potentially liquid and potentially lukewarm. When the sun melts the ice cube into a puddle, these potentials are actualized. What change in general involves is precisely that sort of thing, viz. the actualization of a potential. Accordingly, it is not a matter of being arising from non-being, because a potential has being. The potential to be liquid, for example, is something *really in* the water, in a way that a potential to be turned into gasoline is not. Its reality is what grounds the truth of counterfactual propositions such as the proposition that the ice *would have* melted *had* it been exposed to the sun, which is true even if the ice cube is not in fact so exposed. Given that a potential is something really in a thing, a change to the thing involves being of one sort (in the traditional jargon, *being-in-potency*) giving rise to being of another sort (*being-in-act*), rather than *non*-being or sheer nothingness giving rise to being.

The neglect of potentiality as a real feature of the world, and a middle ground between non-being on the one hand and actuality on the other, is at the root of Zeno's errors as well as those of Parmenides. The dichotomy paradox essentially supposes that each of the ever-smaller units of distance between *A* and *B* is actually present. The idea is that because, for every movement (even just the slight lifting of a foot), there is an infinite number of distances to traverse, the task cannot even be started much less completed. But in fact, the Aristotelian responds, the smaller distances, though not nothing or entirely lacking in being, are there only *potentially* rather than actually. Hence the *actual* distance that any movement would have to involve is finite, and the paradox disappears.

Now, though actuality and potentiality are distinct, the former is nevertheless more fundamental than the latter. For potentialities are *grounded in* actualities. It is because an ice cube is made of water that it has the potentiality to be melted by the sun. Had the cube been made instead of steel, it would not have had the potential to be melted in that way, but would require much higher temperatures if it is to melt. Water, steel, stone, flesh, etc. each have different potentials, and these differences reflect the differing actual features of these substances (such as their different chemical compositions).

Because potentiality is grounded in actuality, there cannot be something that is *purely* potential, *in no way* actual. This brings us to the Aristotelian response to Heraclitus. Heraclitus denies that there is any stable reality, only endless becoming. If static monism essentially affirms actuality while denying potentiality, dynamic monism essentially affirms potentiality while denying actuality. It is, on the Heraclitean picture of the world, as if every potentiality melts into another before it can ever be completely actualized. Whereas the Parmenidean universe is utterly frozen or rigidly locked in place, the Heraclitean universe is utterly protean and amorphous, never actually fixing on being anything in particular even for an instant. But if there were no stability of *any* sort, nothing *in any way* actual to ground the potentialities that manifest themselves in change, then there just could not be any potentiality or any change. There would be no melting in the sun, for example, if the ice cube were not actually water.

The opposed but equally bizarre conclusions that dynamic and static monism arrive at vis-à-vis change are only half the story. Equally notorious are their opposite extreme positions with regard to *multiplicity*. Parmenides famously held that there are no distinct things in reality, but only one thing – being itself, unique and undifferentiated. The reason is that for one being to be distinct from another, there would have to be something that distinguished them. Yet the only thing distinct from being that could distinguish them is non-being, and non-being, since it is nothing at all, does not exist. Hence there is nothing that can distinguish one being from another, and so there just is no more than one being. The multiplicity of things we encounter in everyday experience is in Parmenides' view as illusory as he takes change to be. The denial of change is what makes static monism *static*, and the denial of multiplicity is what makes it a kind of *monism*.

Zeno reinforced Parmenides' line of argument with his *paradox of parts*. Suppose there are distinct things in the world. Then, Zeno says, they would have to have some size or other, and of course, common sense takes things to have different sizes. But anything having size can be divided into parts of smaller size, and these parts can in turn be divided into yet smaller parts, *ad infinitum*. Hence things having size will have an infinite number of parts. But since something is larger the more parts it has, something with an infinite number of parts will be infinitely large. Hence if there are distinct things in the world they will all be of infinite size, and for that reason will all be the same size. But those conclusions are, needless to say, absurd. Hence the assumption that led us to these absurdities, namely the assumption that there are distinct things in the world, must be false.

Heraclitus, meanwhile, goes to the opposite extreme position in holding, as we have seen he does, that there is no *unity* to the stages of the objects that common sense supposes exist, but only the multiple stages themselves. There is the bundle of mental and bodily features we associate with you at this instant, the somewhat different bundle that exists at the next instant, and so on. But there isn't really any persisting *self* that underlies and unites these temporally separated bundles. Nor are there abiding objects of any other sort – tables, chairs, rocks, trees, dogs, cats, planets, molecules, and so on – but just various kinds of series of stages that we mistakenly suppose add up to persisting entities. Whereas Parmenides and Zeno hold that there is far *less* multiplicity than common sense supposes, Heraclitus holds that there is at least in one sense far *more*. Every one of the countless ephemeral stages of what we falsely suppose to be a single abiding entity is really *itself* a distinct entity – or would be, if it stuck around long enough to congeal into an entity, which for the dynamic monist it does not. To speak strictly, the only *thing* or entity there really is is just the whole world itself, understood as a vast river of becoming rather than a collection of discrete entities. Thus is dynamic monism too ultimately a kind of *monism*, but a *dynamic* rather than static kind because of its affirmation of radical change.

Once again the Aristotelian response comes in two stages. First, the Aristotelian points out that here too we have positions that cannot be made coherent. Even to make his case, the static monist has to work through the steps of an argument, and since these are *distinct* steps, we have exactly the multiplicity he denies. Indeed, even to formulate his position, he has to distinguish between the way things appear to common sense and the way they really are, and this too is an instance of multiplicity. Hence the truth of static monism is incompatible with the existence of static monists themselves. So, since there is at least one static monist (as any static monist himself would have to admit), static monism must be false. Similarly, if dynamic monism were true, then there would not be such a thing as a single abiding mind which holds together long enough to entertain an argument for dynamic monism, or indeed even to formulate the view. Hence there would be no such thing as a dynamic monist. Yet there are dynamic monists, as the dynamic monist himself would have to admit. So, dynamic monism must be false. Moreover, the dynamic monist has to appeal to certain universals in order to formulate his position. He has to say, for example, that there is the redness and roundness of a certain ball that we experience at one moment, the redness and roundness of the ball we experience at the next moment, and so on, but really no such thing as the ball itself in the sense of a single, persisting object that underlies these stages. It is, the dynamic monist claims, the similarity the stages exhibit insofar as they all instantiate these universals that leads us falsely to suppose that there is some persisting underlying entity. But then the universals *themselves* – the redness, the roundness, and so forth – will nevertheless persist. For it is the *same one thing*, namely redness, that we attribute to this stage of the ball, the next stage of the ball, and so on. Hence the dynamic monist has to admit one kind of persistence in the very act of denying another kind.

Secondly, the Aristotelian once again deploys the theory of actuality and potentiality to explain where static and dynamic monists go wrong in their arguments vis-à-vis multiplicity. Parmenides supposes that the only thing there could be to distinguish one being from another is non-being, which of course does not exist. But this is incorrect, for two *actual* beings can instead be distinguished in terms of a difference in their respective *potentialities*, and potentiality, though not the same as actuality, is nevertheless a kind of being rather than a kind of non-being. Furthermore, Zeno's paradox assumes that each of the infinite number of parts he attributes to a thing is present in the thing actually. But in fact, the Aristotelian holds, the parts are present only *potentially*. A thing with some particular size could be divided into parts of smaller size, but until this division actually occurs, the parts are not actually present. Hence things with size would not in fact actually have an infinite number of parts, and the paradox is blocked. Heraclitus, meanwhile, supposes that there is no single entity underlying and tying together the stages we associate with a thing because he is implicitly assuming that there is only ever potentiality that never congeals into actuality, and thus nothing with the kind of reality that could count as a stable object. But in fact, since all potentiality is grounded in actuality, there could not be change in the first place unless there *were* some actuality stable enough to ground the potentialities that change presupposes.

So, the sober, if pedestrian, truth is that there are multiple things which change in some respects (contra Parmenides and Zeno) while being stable in others (contra Heraclitus), because they are mixtures of potentiality and actuality. Once we make this distinction, what is correct in the static and dynamic monist pictures of the natural world can be affirmed while their excesses are avoided. But affirming the reality of both actuality and potentiality is not only key to resolving paradoxes raised by a few eccentric ancient thinkers. It is essential to the very possibility of natural science, because unless the natural world were a mixture of actuality and potentiality, it could not be the sort of thing science tells us it is, nor the sort of thing of which we could have scientific knowledge.

If Parmenides and Zeno were correct, then for one thing, we could not trust our senses, since the senses tell us that change occurs and that there are multiple things. Accordingly, the observational and experimental evidence upon which science rests could not be trusted. For another thing, much of what science tells us has to do with change – the developmental processes occurring within organisms, the origin of some life forms from others, the nature of processes like combustion and freezing, the motion of planetary bodies around stars, and so on. Much of it also has to do with multiplicity – the distinct kinds of fundamental particles that there are, the different elements in the periodic table, the different classes of organisms, and so forth. Hence if change and multiplicity are illusory, so too is what science tells us about these purported phenomena. Meanwhile, if Heraclitus were correct, then in that case too, we could not trust our senses or the observational and experimental evidence they provide, since the senses tell us that things are generally stable. Furthermore, much of what science tells us has to do with laws which hold unchangingly despite the changes occurring in the things governed by the laws, and with the universals in terms of which such laws are formulated (e.g. mass, force, energy, and so forth). If *nothing* were stable, then there would be no such laws and no such universals. (More on these points in later chapters.)

The theory of actuality and potentiality, then, is for the Aristotelian absolutely crucial to understanding what any empirical and material world would have to be like for scientific knowledge of it to be possible. Since it deals with the necessary metaphysical preconditions of any possible natural science, it is deeper than any finding of natural science – whether physics, chemistry, biology, or whatever – and thus cannot be overturned by any such finding. It is a theory of the philosophy of nature rather than of natural science, and indeed the foundation of Aristotelian philosophy of nature.

There is much more to the theory when worked out systematically. Particularly relevant to the philosophy of nature is the distinction within the domain of potency or potentiality between an *active potency* and a *passive potency*. An active potency is a capacity to bring about an effect of some sort. It is what in contemporary philosophy is typically referred to as a *causal power*. A passive potency is a capacity to be affected in some way. In contemporary philosophy it is sometimes called a *liability*. The debate in contemporary analytic metaphysics over categorical and dispositional properties in several respects recapitulates ancient debates about act and potency. (For discussion of the relationship of this recent debate to the theory of act and potency, and of other issues surrounding the theory, see Feser 2014b, Chapter 1.)

1.2.2 Hylemorphism

In change, there is, again, both the potential that is to be actualized and the actualization of that potential. Consider the ink in a dry-erase marker. While still in the pen it is actually liquid. But it has the potential to dry into a triangular shape on the surface of the marker board. When you use the pen to draw a triangle on the board, that potential is actualized. Having dried into that shape, the ink has yet other potentials, such as the potential to be removed from the board by an eraser and in the process to take on the form of dust particles. When you erase the triangle and the dried particles of ink fall from the board and/or get stuck in the eraser, those potentials are actualized.

Now, what we have in this scenario is, first of all, a determinable substratum that underlies the potentialities in question – namely, the ink.

We also have a series of determining patterns that that substratum takes on as the various potentials are actualized – patterns like *being liquid*, *being dry*, *being triangular*, and *being particle-like*. The determinable substratum of potentiality is what in Aristotelian philosophy of nature is meant by the term "matter," and a determining pattern that exists once the potential is actualized is called a "form." Since change is real, matter and form in these senses must be real. Matter is, essentially, that which needs actualizing in change; form is, essentially, that which results from the actualization. Note that *any* determining, actualizing pattern counts as a "form" in this sense. A form is not merely the shape of a thing, nor is it necessarily a spatial configuration of parts (though shape and spatial configuration are kinds of forms). *Being blue*, *being hot*, *being soft*, etc. are also forms in the relevant sense.

Change is not the only phenomenon that points to the distinction between matter and form. Note that a form or pattern like *triangularity* is universal rather than particular. It is the same pattern that one finds in green triangles and red ones, triangles drawn in ink and those drawn in pencil, triangles used as dinner bells and those used on a billiards table, and so forth. Triangularity is also perfect or exact rather than approximate. For example, being triangular in the strict sense involves having sides that are straight rather than wavy. Now, the triangle you draw on the marker board has straight sides, but only imperfectly or approximately. It is also a particular instance of triangularity rather than triangularity as such. Hence there must not only be something by virtue of which the thing you've drawn is triangular, but also something by virtue of which it is triangular in precisely the imperfect way that it is. There must also be something by virtue of which triangularity exists in *this particular* point in time and space.

Now if being triangular is a way of being *actual*, being triangular only in an imperfect way is a way of being *potential*. For insofar as the triangle's sides are only imperfectly straight, the ink in which you have drawn it has, you might say, only partially actualized the potential for triangularity. And insofar as the triangle has been drawn in some particular time and place, the potential in question is a potential at *that* time and place, rather than at another, that has been actualized. Now, that by virtue of which what you have drawn is actually triangular to the extent it is, is what Aristotelian philosophy of nature calls its *form*; while that by virtue of which it is limited, or remains merely potential, in the extent to which it is triangular, is its *matter*.

Insofar as form accounts for whatever permanence, unity, and perfection or full actuality there is in the natural world, it represents, as it were, the Eleatic side of things. The triangle drawn on the marker board persists to the extent that it retains its triangular form, is identical to other triangles insofar as it is an instance of the same form they instantiate, and is perfect or complete in its actuality to the degree that it approximates that form. Insofar as matter accounts for the changeability, diversity, and imperfection or mere potentiality that exists in the natural world, it represents the Heraclitean side of things. The triangle drawn on the board is impermanent insofar as its matter can lose its triangular form, is distinct from other things having the same form insofar as it is one parcel of matter among others which instantiate it, and is imperfect or potential to the extent that it *merely* approximates the form.

Matter is passive and indeterminate, form active and determining. The same bit of matter can take on different forms, and the same form can be received in different bits of matter. Hence matter and form are as distinct as potentiality and actuality. Still, just as potentiality is grounded in actuality, so too does matter always have *some* form or other. If the ink in our example is not in a liquid form, it is in a dry, triangular form, and if not that then in the form of particles. And if the particles are broken down further so that the ink is in no sense still present, then the form of the chemical constituents of the ink would remain. If matter lacked *all* form it would be nothing but the pure potentiality for receiving form; and if it were *purely* potential, there would be no actuality to ground it and it would not exist at all.

The distinction between form and matter is not, however, the *same* distinction as that between actuality and potentiality, but rather a special case of that distinction. Everything composed of form and matter is thereby composed of actuality and potentiality, but not everything composed of actuality and potentiality is composed of form and matter. An angelic intellect or a Cartesian *res cogitans*, being incorporeal, would not be a compound of form and matter, but it would still be a compound of actuality and potentiality (insofar as God would have to create it and thus actualize what would otherwise be its merely potential existence). The distinction between form and matter is an application of the distinction between actuality and potentiality to *corporeal* things, specifically – to the physical objects we know through experience. Hence, whereas the theory of actuality and potentiality has completely general metaphysical applicability, the proper application of the distinction between form and matter is within the philosophy of nature.

Now, several further distinctions are needed in order to set out the Aristotelian analysis of what it is to be a corporeal *substance*. First, there is the general distinction between any substance and its attributes. Consider a solid, gray, round, smooth stone of the sort you might pluck from a river bed. The solidity, grayness, roundness, and smoothness are *attributes* of the stone, and the stone itself is the *substance* which bears these attributes. The attributes exist *in* the stone whereas the stone does not exist *in* any other thing in the same sense. Substances, in general, just are the sorts of things which exist in themselves rather than inhering in anything else, and which are the subjects of the attributes which do of their nature inhere in something else. This is true of corporeal substances like stones, and it is true of incorporeal substances too, if such things exist.

Corporeal substances are, again, composed of form and matter, but here two further distinctions must be made. If we abstract from our notion of matter *all* form, leaving nothing but what I have called the pure potentiality to receive form, we arrive at the idea of *prime matter*. (More on this below.) Matter already having some form or other – that is to say, matter which is actually a stone, or wood, or water, or what have you, and is not merely potentially any of these things -- is *secondary matter*. There is a corresponding distinction between kinds of form. A form which makes of what would otherwise be utterly indeterminate prime matter some determinate concrete thing of a certain kind is a *substantial form*. A form which merely modifies some secondary matter – and in particular, which modifies matter which already has a substantial form – is an *accidental form*. A corporeal substance is, to state things more precisely, a composite of *prime matter* and *substantial form*.

The distinction between substantial form and accidental form is illuminated by comparison with the different but related Aristotelian distinction between *nature* and *art* – that is to say, between natural objects on the one hand, and everyday artifacts on the other. Hence, consider a *liana vine* – the kind of vine Tarzan likes to swing on – as an example of a natural object. A *hammock* that Tarzan might construct from living liana vines is a kind of artifact, and not a natural object. The parts of the liana vine have an inherent tendency to function together to allow the vine to exhibit the growth patterns it does, to take in water and nutrients, and so forth. By contrast, the parts of the hammock – the liana vines themselves – have no inherent tendency to function together as a hammock. Rather, they must be arranged by Tarzan to do so, and left to their own devices – that is to say, without pruning, occasional rearrangement, and the like –

they will tend to grow the way they otherwise would have had Tarzan not interfered with them, including in ways that will impede their performance as a hammock. Their natural tendency is to be liana-like and not hammock-like; the hammock-like function they perform after Tarzan ties them together is extrinsic or imposed from outside, while the liana-like functions are intrinsic to them.

Now the difference between that which has such an intrinsic principle of operation and that which does not is essentially the difference between something having a substantial form and something having a merely accidental form. Being a liana vine involves having a substantial form, while being a hammock of the sort we're discussing involves instead the imposition of an accidental form on components each of which already has a substantial form, namely the substantial form of a liana vine. A liana vine is, accordingly, a true *substance*, as Aristotelian philosophers understand substance. A hammock is not a true substance, precisely because it does not qua hammock have a substantial form – an *intrinsic* principle by which it operates as it characteristically does – but only an accidental form. In general, true substances are typically natural objects, whereas artifacts are typically not true substances. A dog, a tree, and water would be true substances, because each has a substantial form or intrinsic principle by which it behaves in the characteristic ways it does. A watch, a bed, or a computer would not be true substances, because each behaves in the characteristic ways it does only insofar as certain accidental forms have been imposed on them from outside. The true substances in these cases would be the raw materials (metal, wood, glass, etc.) out of which these artifacts are made.

It is important to emphasize, however, that the correlation between what occurs "in the wild" and what has a substantial form, and the correlation between what is man-made and has only an accidental form, are only rough correlations. For there are objects that occur in nature and apart from any human intervention and yet have only accidental forms rather than substantial forms, such as piles of stones that gradually form at the bottom of a hill, tangles of seaweed that wash up on the beach, and beaver dams. And there are man-made objects that have substantial forms rather than accidental forms, such as babies (which are in an obvious sense made by human beings), water synthesized in a lab, and breeds of dog. Of course, no one would be tempted in the first place to think of these as "artifacts" in the same sense in which watches and computers are artifacts. But even objects that are "artificial" in the sense that they not only never occur "in the wild" but require significant scientific knowledge and technological expertise to produce can count as having substantial forms rather than accidental forms. Styrofoam would be one possible example (Stump 2003, p. 44).

The basic idea is that it seems to be essential to a thing's having a substantial form that it has properties and causal powers that are irreducible to those of its parts (Stump 2006). Hence water has properties and causal powers that hydrogen and oxygen do not have, whereas the properties and causal powers of, say, an axe seem to amount to nothing over and above the sum of the properties and powers of the axe's wood and metal parts (Stump 2003, p. 44). When water is synthesized out of hydrogen and oxygen, what happens is that the prime matter underlying the hydrogen and oxygen loses the substantial forms of hydrogen and oxygen and takes on a new substantial form, namely that of water. By contrast, when an axe is made out of wood and metal, the matter underlying the wood and the matter underlying the metal do not lose their substantial forms. Rather, while maintaining their substantial forms, they take on a new accidental form, that of being an axe. The making of Styrofoam seems to be more like the synthesis of water out of hydrogen and oxygen than it is like the making of an axe. Styrofoam has properties and powers which are irreducible to those of the materials out of which it is made, which indicates the presence of a substantial form and thus a true substance.

There is a further complication to the story. On the Aristotelian-Thomistic account, among the attributes of a thing, we need to distinguish those that are *proper* to it from those which are not. It is the former alone which are labeled "properties" in Aristotelian-Thomistic philosophy, with the others referred to as "contingent" attributes. (This contrasts with the very loose way the term "property" is used in contemporary analytic philosophy, to refer to more or less any feature we might predicate of a thing.) The properties or proper attributes of a substance are those which "flow" or follow from its having the substantial form it does. Being four-legged, for example, flows or follows from having the substantial form of a dog. It is a natural concomitant of "dogness" as such, whereas being white (say) is not, but is merely a contingent attribute of any particular dog. Now this "flow" can, as it were, be blocked. For instance, a particular dog might, as a result of injury or genetic defect, be missing a leg. But it wouldn't follow from its missing that leg that being four-legged is not after all a true property of dogs, nor would it follow that this particular creature was not really a dog after all. Rather, it would be a *damaged or defective instance* of a dog. When determining the characteristic properties and causal powers of some kind of thing, then, we need to consider the *paradigm* case, what that kind of thing is like when it is in its mature and normal state.

So, a thing counts as a true substance when it has a substantial form rather than a merely accidental form, and the mark of its having the former is that in its mature and normal state, it exhibits certain properties and causal powers that are irreducible to those of its parts. A corporeal substance is a composite of a substantial form and prime matter, related to one another as actuality and potentiality; and once in existence, a corporeal substance or substances constitute the secondary matter that is the subject of an accidental form or forms. This is the Aristotelian doctrine of *hylemorphism* (or *hylomorphism*), the name of which derives from the Greek words *hyle* (or "matter") and *morphe* (or "form").

The Thomistic interpretation of hylemorphism insists on the doctrine of the *unicity* of substantial form, according to which a substance has only a single substantial form. Suppose *A* is a substance, and has *B* and *C* as parts. Since *A* is a substance, it has a substantial form. Do *B* and *C* have further substantial forms of their own? If they did, then they too would be substances. In that case, though, *A*'s form would relate to *B* and *C* as an accidental form relates to secondary matter. But then *A* wouldn't really have a substantial form after all, and thus not really be a substance. So, if *A* really is a substance, then its parts *B* and *C* must not themselves have substantial forms or amount to true substances in their own right. There is only the single substantial form, the form of *A*, which informs the prime matter of *A*. Another way to look at it is that if *B* and *C* had substantial forms, then *they* would be what actualizes the prime matter so that it constitutes a substance (or two substances in this case, namely *B* and *C*). In that case, the prime matter wouldn't *potentially* be a substance, but would already *actually* be a substance. That is to say, it would be secondary matter. But then there would be nothing left for the substantial form of *A* to do qua actualizer of prime matter. It would serve merely to modify an already existing substance and thus amount to an accidental form rather than substantial form. So, again, a substance *A* can really only have one substantial form.

To see the implications of this, consider a concrete example like water, which has hydrogen and oxygen as its parts. Since water is a substance, it has a substantial form. But since a substance can have only a single substantial form, it follows that the hydrogen and oxygen in water don't have substantial forms. That entails in turn that hydrogen and oxygen don't exist in water as substances. Now, this may seem odd, since hydrogen considered *by itself* and oxygen considered *by itself* each do seem to be substances. They have their own characteristic irreducible properties and causal powers, after all. But the lesson we should draw from these considerations, according to the Thomist, is that hydrogen and oxygen do not exist *actually* in water, but only *virtually*. Notice that the claim is *not* that they don't exist in water *at all*. It is rather that they don't exist in water *in the way* that they exist when they exist on their own. The situation is comparable to the Aristotelian's account of what is really going on in Zeno's paradox of parts scenario, in which the parts are present – they are not nothing or non-being – but only potentially rather than actually.

This too may sound odd, but it should sound less so upon reflection. Consider that if hydrogen and oxygen were actually present in the water, then they should possess their characteristic properties and powers. That means that we should be able to burn the hydrogen, and to boil the oxygen at -183°C. But we cannot do either. Hence the substantial forms of hydrogen and oxygen cannot be present, in which case the *substances* hydrogen and oxygen cannot be present. Furthermore, if hydrogen and oxygen were actually present, then for something to be water would be for it to have a merely accidental form, and properties and causal powers reducible to those of hydrogen and oxygen. But that is also not the case, since water has powers and properties that a mere aggregate of hydrogen and oxygen does not. Hydrogen and oxygen are present in water, then, in the sense that water has the *potentiality* to have hydrogen and oxygen drawn out of it – by electrolysis, say. (More on this issue in a later chapter. For discussion of the relationship of hylemorphism to contemporary debates over reductionism, and of other issues surrounding the theory, see Feser 2014b, chapter 3.)

1.2.3 Limitation and change

As indicated in the preceding section, two of the motivations for hylemorphism have to do with its application to the critique of static monism's denial of multiplicity and of change. These lines of argument for hylemorphism are sometimes labeled the *argument from limitation* and the *argument from change* (Cf. Koren 1962, chapter 2).

The basic idea of the first line of argument is, again, that a form is *of itself* universal, so that we need a principle to explain how it gets tied down, as it were, to a particular thing, time, and place. For example, *roundness* can be instantiated in multiple objects and at different times and spatial locations, and the geometrical truths pertaining to it remain true whether or not any particular round thing or group of round things comes into existence or remains in existence. *Roundness* is thus not *as such* limited, so that something needs to be added to it if we do in fact find it limited in some way. Matter – the matter of this individual bowling ball, of that individual wheel, and so forth – is what does this job. For example, it is the matter of some individual wheel that accounts for the fact that roundness is instantiated in some particular automobile, in a way it is not instantiated in (say) the tree next to the automobile or the road under it. Matter also accounts for limitation in another respect. The *roundness* of a circle as defined in geometry is perfect or exact, yet any particular triangle drawn on a chalkboard, in a book, or what have you, is always at least to some extent imperfect. Matter accounts for this kind of limitation too insofar as, qua the potentiality to receive form, it is never fixed or locked on to any one particular form, but always ready to take on another.

For the moment, however, it is the argument from change, especially, about which more must be said. On an Aristotelian analysis, a real change involves the gain or loss of some attribute, but also the persistence of that which gains or loses the attribute. For example, when a banana goes from being green to being yellow, the greenness is lost and the yellowness is gained, but the banana itself persists. If there were no such persistence, we would not have a *change* to the banana, but rather the annihilation of a green banana and the creation of a new, yellow one in its place.

Matter for the Aristotelian essentially *just is* that which not only limits form to a particular thing, time, and place, but also that which persists when an attribute is gained or lost. It is absolutely crucial to understand that the characteristics of matter identified so far – its correspondence to potentiality (as contrasted with form's correspondence to actuality), its status as the principle of the limitation of form, and its status as the principle of persistence through change – are *definitive* of matter as the Aristotelian understands it. That is to say, the Aristotelian is using the term "matter" in a *technical sense*. He is *not* saying that matter *as it has independently come to be understood in modern physics and chemistry* is what turns out to be the stuff that plays the roles of persisting through change, limiting form, and corresponding to potentiality. He is, so far, not saying anything about matter in the modern sense at all. Rather he is *defining* "matter" as used in Aristotelian philosophy of nature as *that which plays these roles.* (Nor is this some eccentric usage; in fact it is an *older* usage than that familiar from modern physics and chemistry.) Of course, how "matter" in this sense relates to "matter" in the modern sense is a good question, and it is one to be addressed in later chapters. The point for the moment is simply to forestall irrelevant objections and misunderstandings.

Now, one sort of change that takes place is change *to* a persisting substance. The subject of this sort of change is secondary matter, matter already having a substantial form. For the Aristotelian, we can identify three kinds of change falling into this class. There is, first of all, *qualitative* change, as when the banana in our example changes color, from green to yellow. Second, there is *quantitative* change, as when the banana, having begun to rot, shrinks in size. Third, there is *local motion* or change with respect to location or place, as when the banana flies through the air when you toss it toward the waste basket.

Another, more radical kind of change is change *of* a substance, *substantial* change. It is change that involves, not a substance gaining or losing some attribute while still persisting, but rather a substance going out of existence and being replaced by a new one. This is what happens when the banana is eaten, digested, and incorporated into the flesh of the animal that ate it, or when it is burned and reduced to ash. Because change requires some underlying persisting subject that does not change, there must be such a subject in the case of substantial change no less than in the case of the other kinds. But because it is the substance itself that goes out of existence in this case, it is a substantial form that is lost, not a merely accidental form. Hence it is not any kind of secondary matter that is the subject of this sort of change, but rather prime matter.

Now, since prime matter is that which underlies the loss of one substantial form and the gain of another, it does not of itself have a substantial form and is therefore is not any kind of substance. Nor, since the having of accidental forms and attributes in general presupposes being a substance, does it possess any attributes or accidental forms. It is not *actually* any *thing* at all. But that does not entail that it is nothing, for between actuality and nothingness or non-being, there is potentiality, which is a kind of being. *That* is what prime matter is – the pure potentiality to receive form.

Because potentiality cannot exist without actuality, prime matter does not exist without actuality. That is to say, it does not exist on its own, but only together with some substantial form or another. All matter as it exists in reality, outside the mind, is secondary matter. But that does not entail either that prime matter is not real or that it is not really distinct from the substantial forms with which it is conjoined. Being *trilateral* (having three straight sides) is a different geometrical feature from being *triangular* (having three angles) even though a closed plane figure cannot have the one without having the other. We can distinguish them in thought and what we thereby distinguish are features that are different in reality, even if outside the mind the one cannot be separated from the other. Similarly, prime matter and substantial form differ in reality even if they cannot be separated in reality, but only in thought. (For more on the idea of a real distinction in Aristotelian-Thomistic philosophy, see Feser 2014b, pp. 72-79.)

Without prime matter, there could be no substantial change, because there would be no subject of change that persists through the change. There would rather be the complete annihilation of one substance and the creation of another utterly novel substance in its place. That the world does not work like that is evident from the continuity that substantial change, no less than the other sorts of change, exhibits. For example, wood that is burned reliably turns to ash, not to water or cheese or rose petals. Why would this be the case if there were absolutely *nothing* that carries over from the wood to the ash, but rather the complete disappearance of the first and the appearance out of nothing of the second? Why wouldn't just any old thing appear in place of the wood?

It might seem that the ancient atomist account of change provides an alternative to prime matter and substantial form. Dogs, trees, stones, and all other physical substances are on this view ultimately just collections of fundamental particles in different configurations. Change involves the rearrangement of the particles. For example, when a tree is burned and turned to ash, what happens is that the particles that were once arranged so as to form a tree are now rearranged so as to form ash. But a problem with this view is that it entails that dogs, trees, stones, and the like are not really substances. The true substances are the fundamental particles, and to be a dog, a tree, or a stone is just for these particles to take on a certain kind of accidental form. Yet this seems clearly wrong insofar as these and other natural objects appear to have causal powers that are irreducible to the sum of the causal powers of fundamental particles. And again, such irreducibility is the mark of the presence of a substantial form rather than a merely accidental form. (More on this in later chapters.)

Another problem is that from the Aristotelian point of view, the atomist doesn't really get rid of substantial form and prime matter at all, but simply relocates them. Suppose that to be a dog, a tree, or a stone really is to have a merely accidental form, and that the only true substances are the fundamental particles. We would still have to regard *them* as composites of substantial form and prime matter, for the reasons given in the arguments from limitation and from change. For one thing, like any other form, the form of being a particle is universal, and so there must be something that ties that form down to some individual thing time, and place – to *this* particular particle at this particular time and place, that particular particle at that time and place, and so on. That is the job matter does, and since the particles in question are fundamental rather than composites of some more fundamental substances, it is only prime matter than can do the job rather than any kind of secondary matter. And only this prime matter together with the substantial form of a particle would give us an actual substance.

For another thing, as long as it is even in *principle* possible for a fundamental particle to come into existence or go out of existence, there will have to be something that underlies this substantial change, which brings us back to prime matter and substantial form. Of course, the ancient atomists held that the fundamental particles could be neither generated nor corrupted. But merely to assert this does not make it so, and it is hard to see how there could be such particles. Any particle is going to be limited in various ways – to being of this particular size and shape, at this particular location at any moment, and so on. But what is limited in such ways is a mixture of actuality and potentiality rather than pure actuality. It is actually of *this* shape and merely potentially of some other shape, actually at this location and only potentially at that one, and so on.

Now, only what is pure actuality – something which has no potentials that need to be or indeed could be actualized, but which is, as it were, always already actual – could exist in a necessary way. (The idea of pure actuality is in fact the philosophical core of the Aristotelian conception of God.) Anything less than that could exist only in a contingent way. But then the fundamental particles would have to be contingent rather than necessary, and thus the sorts of thing which could in principle either exist or not exist. This capacity either to exist or fail to exist must have an underlying basis, which brings us back to the conclusion that the particle is composed of prime matter and substantial form. (For more on prime matter and on atomism, see Feser 2014b, pp. 171-84.)

1.2.4 Efficient and final causality

The thesis that change involves the actualization of a potential tells us how change is *possible*, contra static monism. But how does change ever *actually* occur? That is to say, what is it that does the actualizing when a potential is in fact actualized? It can't be something merely potential that does it, precisely because it is merely potential. For example, potential heat cannot melt an ice cube. Only actual heat can do so. In general, *if some potential is actualized, there must be something already actual which actualizes it.*

This is the fundamental formulation of what is sometimes called the *principle of causality*. The principle is also sometimes formulated as the thesis that *whatever is contingent has a cause*, or the thesis that *whatever comes into being has a cause*. But these are really just applications of what I have called the more fundamental formulation, for a contingent thing or a thing that comes into being requires a cause precisely because its existence depends on certain potentialities being actualized. (Note that the claim that "*everything* has a cause" is *not* an application of the fundamental principle. That is a straw man that no Aristotelian or Thomist endorses, and indeed, Aristotelian-Thomistic metaphysics *denies* that everything has a cause. What is purely actual not only need not have a cause but cannot have one, precisely because it has no potentials which could be actualized.)

David Hume famously challenged the principle of causality by suggesting that it is conceivable, and therefore (he infers) possible, that something could come into being without any cause. What he has in mind is an event like a bowling ball (say) suddenly appearing at some spot which an instant before had been empty. But there are several serious problems with this argument. For one thing, in general, to conceive of *A* without conceiving of *B* simply doesn't entail that *A* could exist apart from *B* in reality. For example, we can conceive of something's being a triangle without conceiving of it as being a trilateral, but in reality any triangle will also be a trilateral. We can conceive of a man without conceiving of his height, but in reality any man must have some height. And so forth. By the same token, even if we can conceive of a bowling ball coming into existence without conceiving of its cause, it doesn't follow that it could exist in reality without a cause.

For another thing, the kind of scenario that is supposed to illustrate Hume's point is typically underdescribed. Simply to imagine a bowling ball suddenly appearing where before there had been nothing is not *by itself* to conceive of a bowling ball coming into being without a cause. For why would this not instead amount to its coming into being with an *unseen* cause, or to its being *transported* from somewhere else via teleportation? Hence we would need to add something to the example to get from it to the conclusion Hume wants. The trouble is that there seems to be no way to add anything to the example that won't lead instead to its *undermining* rather than supporting Hume's conclusion. For as Elizabeth Anscombe (1981) pointed out, the way we typically distinguish something's *coming into being* from its being *transported from somewhere else* is precisely in *causal* terms. We know that a drawing on a certain desk was caused to exist at noon rather than transported there at noon because we find out that it had a generating cause (a certain artist) rather than a transporting cause. We know that an apple on the desk was transported there rather than having come into existence there because we find out that someone put it there. And so on. But if we have to *bring in* the idea of a generating cause in order to know that the bowling ball came into being rather than being transported, then we've undermined Hume's argument, because the whole point of the example was to *get rid of* the idea of a cause. (For further discussion of the problems with Hume's argument, see Feser 2014b, pp. 109-14.)

A corollary of the principle of causality is the *principle of proportionate causality*. This is the thesis that whatever is in an effect must in *some* way preexist in the total cause of that effect. Otherwise there would be some potential in the effect that was actualized without something already actual doing the actualizing, contrary to the principle of causality.

There are several ways in which what is in the effect might preexist in the total cause, viz. *formally*, *virtually*, or *eminently*. Suppose the effect is your coming to possess a twenty dollar bill. If the reason you have it is that I had a twenty dollar bill and I gave it to you, then we have a case where what is in the effect was in the cause formally. That is to say, you come to have the form or fit the pattern of something possessing twenty dollars, because I myself, who caused you to have it, also had the form or fit the pattern of something possessing twenty dollars. Suppose instead that the reason you have it is that, though I did not have twenty dollars in cash in my possession, I did have at least twenty dollars in my bank account, and was able to go withdraw it to give to you. In that case, what is in the effect was initially in the cause, not actually, but virtually. Now suppose that I did not have twenty dollars in cash or even twenty dollars in the bank, but I was able to get access to a U.S. Federal Reserve printing press and print off a new twenty dollar bill to give you. In that case, what was in the effect was first in the cause eminently. That is to say, it was in the cause by virtue of the cause's having something even greater or more eminent than a twenty dollar bill, namely the power to *generate* twenty dollar bills. If what is in the effect is in no way first in the total cause, however (me together with my wallet, or my bank account, or some other collection of factors), then it would not be in the effect in the first place, for it would in that case not have had anywhere to come *from*.

Sometimes this principle is objected to on the grounds that there seem to be cases where the cause lacks what is in the effect, e.g. someone can get a black eye from a person who doesn't himself have one. But this objection simply ignores the fact that what is in the effect can be in the cause in several ways, not merely in a straightforward "formal" way. Sometimes it is suggested that evolution is a counterexample to the principle, but that this is not the case should be obvious from the fact that evolutionary changes are never treated in biology as if they simply arose from nowhere – which *would* violate the principle – but, on the contrary, are explained by reference to preceding factors such as genetic mutations, environmental changes, selection pressures, and the like. (More on this in a later chapter.) Or consider the debate between dualists and materialists in the philosophy of mind. Materialists often argue that there cannot be any immaterial aspect to the mind, on the grounds that no such aspect could have arisen out of purely material evolutionary processes. Many dualists argue that since, as they hold, there are immaterial aspects to the mind, there must be further factors to the mind's origin beyond the purely material ones allowed by materialists. Both sides implicitly suppose, however, that *if* there are immaterial aspects, *then* they would have to have come from something other than the purely material factors admitted by materialism. Hence both sides implicitly presuppose the principle of proportionate causality. (For more on the principle of proportionate causality, see Feser 2014b, pp. 154-59.)

The principle of causality and the principle of proportionate causality have to do with what Aristotelians call *efficient* causes, where an efficient cause is what brings something into being or alters it in some way. This is to be distinguished from a *final* cause, which is the end, goal, or outcome toward which something is directed or points. For example, an acorn "points to" or is "directed toward" becoming an oak. The phosphorus in the head of a match "points to" or is "directed toward" the generation of flame and heat. Ice "points to" or is "directed toward" being melted when heat is applied to it. And so forth.

The Aristotelian-Thomistic metaphysician holds that efficient causality is unintelligible without final causality. Efficient causality is manifest in causal *regularities.* Plant an acorn, and what will grow from it is an oak, not a rose bush, or a cat, or a Volkswagen. Strike a match, and it will generate flame and heat rather than turning into a snake or a bouquet of roses. Leave an ice cube out in the sun, and it will melt into a puddle of liquid water rather than turning into a stone or into gasoline. Of course, these effects might be blocked. The cause may be damaged in some way, as when an acorn is crushed underfoot or eaten by a squirrel or a match is soaked in water. Or a triggering factor that is needed if the cause is to produce its effect may be absent, as when a match is kept in a drawer instead of struck, or an ice cube is placed in the freezer rather than out in the sun. But it remains true that *had* the causes been undamaged and the relevant triggering factors been present, then the usual effect *would have* followed.

That an efficient cause *A* reliably produces a particular effect or range of effects *B*, rather than *C*, or *D*, or no effect at all, is intelligible only if generating *B* is the final cause of *A* – that is to say, if the generation of *B* is the end, goal, or outcome toward which *A* "points" or is "directed." Otherwise causes and effects would be "loose and separate" (as Hume would put it) and there would be no reason why *A* should not be associated with completely random and unpredictable effects rather than the regularity that we in fact observe. A Humean account of causality, on which there is no *objective* rhyme or reason to the causal order but only the regularity that the mind creates and projects onto the world, is inevitable if final causality is abandoned. Indeed, from the Aristotelian-Thomistic point of view, the early moderns' abandonment of final causality was a key factor in the development of thinking about causality that culminated in Hume.

Aristotelian-Thomistic philosophy thus affirms a third principle concerning causality, the *principle of finality*, which is traditionally formulated as the thesis that *every agent acts for an end*, an "agent" being an efficient cause. (Contemporary analytic metaphysicians who argue that dispositions or causal powers exhibit a kind of "physical intentionality" or "natural intentionality" insofar as they are directed toward certain characteristic manifestations have essentially rediscovered the principle of finality. Cf. Place 1996; Heil 2003, pp. 221-22; Molnar 2003, chapter 3.)

Final causality is also known as "teleology" (from the Greek *telos* or "end'), a term which in contemporary usage has several misleading connotations. For example, it is often assumed that to attribute teleology to something is *ipso facto* to think of it as a kind of artifact which has been "designed." That is not the case. Consider once again the examples of the liana vine and the hammock Tarzan makes out of liana vines. A hammock has a specific teleology, namely to function as a bed, and of course it is indeed an artifact which was designed by human beings to serve this function. But a liana vine also has a certain teleology insofar as it tends toward activities like taking in water and nutrients through its roots, growing in a specific way, and so on. Yet a liana vine is *not* an artifact but a natural substance. The reason it tends toward the activities it does is not because some human designer makes it do so (as Tarzan makes the liana vines serve the function of a hammock) but rather because that is simply what liana vines by nature do as long as nothing impedes them from doing it. Something that didn't do so just wouldn't be a liana vine.

In other words, liana vines and other natural substances have their teleological properties in an *intrinsic* or *built in* way, whereas artifacts like hammocks have their teleological properties in an *extrinsic* or *externally imposed* way. This reflects the fact that liana vines and other natural objects have *substantial forms*, whereas hammocks and other artifacts have only *accidental forms*. The liana vines that make up Tarzan's hammock have no tendency on their own to function as a bed. That end or final cause has, like the form of a hammock itself, to be imposed on them from the outside. By contrast, the vines do have a tendency on their own to take in nutrients, exhibit certain growth patterns, etc. That tendency, like the form of being a vine itself, is built into them.

So, from an Aristotelian-Thomistic point of view, to be a natural substance is precisely *not* to be an artifact, because to be an artifact is to have a merely accidental form and extrinsic teleology, whereas to be a natural substance is to have a substantial form and intrinsic teleology. Accordingly, if "design" involves the imposition on something of an accidental form and extrinsic teleology – after the fashion of a human artificer – then natural substances are precisely *not* the sorts of things that are "designed." Now, that does *not* mean that they are not designed *if* what we mean by "design" is merely that the divine intellect is the ultimate cause of their existing and having the natures, including the natural teleological features, that they have. On the contrary, the Fifth Way of proving God's existence put forward by Aquinas and developed by later Thomists argues precisely that even intrinsic teleology must have the divine intellect as its *ultimate* source. (Cf. Feser 2013b) But the Thomist nevertheless insists that the *proximate* source of a natural object's teleological features is its substantial form.

The need for a divine cause simply does not follow *straightaway* from the existence of teleology, then, but requires further argumentation. And that argumentation takes us beyond the philosophy of nature to the branch of metaphysics known as natural theology. For the specific purposes of the philosophy of nature, a thing's teleological features can be taken as simply a consequence of its having the substantial form it has, just as its efficient causal powers can be seen as a consequence of its substantial form. Just as we can determine what causal powers a natural substance like water, copper, or stone has by simply examining the substance itself without wondering what the divine First Cause intended in creating it, so too can we determine a natural substance's teleological features by simply examining the substance itself, without having to wonder what the divine Supreme Intelligence had in mind. That is why you can know that copper conducts electricity, that flowing water has the power to erode stone, etc. whether or not you believe in a divine First Cause. Similarly, you can know that an acorn is inherently "directed toward" becoming an oak, that eyes are inherently "directed toward" the function of allowing us to see, etc. whether or not you believe in a divine Supreme Intelligence.

The Aristotelian-Thomistic conception of teleology is therefore very different from that reflected in "design arguments" of the kind associated with William Paley and contemporary "Intelligent Design" theory. Such arguments tend to assimilate natural substances to artifacts, and also tend thereby to reduce all teleology to extrinsic teleology and all form to accidental form. They are in that respect simply incompatible with an Aristotelian philosophy of nature. Both "design argument" proponents and their atheistic critics tend to assume that to admit that there is real teleology in nature is *ipso facto* to commit oneself to an artificer who put it there. From the Aristotelian point of view, this is too quick and reflects too crude an understanding of teleology, for not all teleology is of the extrinsic or artifact-like kind that *by definition* entails a mind that put it there. The teleology found in nature is instead of the intrinsic kind. While that kind of teleology might also ultimately require a divine cause – and again, the Thomist agrees that it does – that conclusion does not follow *merely* from the existence of teleology itself but requires further metaphysical premises. Accordingly, the question whether teleology exists in nature can, for the purposes of the philosophy of nature, be bracketed off from the dispute between atheism and theism.

As these last remarks indicate, for the Aristotelian the existence of teleology does not by itself entail *conscious awareness* of the end toward which a thing is "directed." Acorns are "directed toward" becoming oaks and the phosphorus in the head of a match is "directed toward" the generation of flame and heat, but that is not because acorns consciously desire to become oaks or because phosphorus consciously desires to generate flame and heat. There is, of course, no conscious awareness here at all. Only in human beings and other animals is there such awareness. In the vast majority of cases in which teleology exists in nature, things are "directed" or "point" toward the ends they do in an entirely unconscious and unthinking way.

As my examples also indicate – and once again contrary to Paley and "Intelligent Design" theory – for the Aristotelian the question of whether teleology exists in nature has nothing especially to do with biology. The functions of biological organs are one *kind* of teleology, but by no means the only kind or the most prevalent kind. For one thing, most teleology in nature is not biological. For another, most of it does not involve anything like biological *function* in the sense of a part's serving to advance the good of a whole. Again, the phosphorus in the head of a match inherently "points to" or is "directed at" the generation of flame and heat. But phosphorus is inorganic, and to affirm its teleological features does not require us to see it as relating to the rest of the universe in anything like the way an eye, heart, or kidney relates to the organism of which it is a part.

Intrinsic teleology exists at at least five levels in the natural world. First, there is what the contemporary philosopher Paul Hoffman (2009) has called the "stripped-down core notion" of teleology, which is simply the bare pointing of an efficient cause towards its characteristic effect or range of effects. This is present even in the simplest inorganic phenomena. Second, there is the teleology manifest in complex inorganic processes such as the water cycle and the rock cycle, in which there are several successive stages to the causal process rather than the mere "pointing" of a cause toward a single immediate effect (Oderberg 2008). Third, there is the rudimentary sort of *organic* but still unconscious teleology exhibited by vegetative life. Fourth, there is the *conscious* organic teleology exhibited by animal life. And fifth, there is the organic, conscious, and *rational* teleology exhibited in human thought and action. More on these three kinds of organic teleology in a moment. (For further discussion of final causality in general, see Feser 2010 and Feser 2014b, pp. 88-105.)

When we combine what has been said in this section with what was said in the preceding sections, we have the famous Aristotelian doctrine of the *four causes*. The *formal* cause of a thing is its form (its substantial form, in the case of a natural substance). The *material* cause of a thing is the matter which has taken on the form (prime matter, in the case of a natural substance). The *efficient* cause of a thing is what brought it into being. The *final* cause of a thing is the end, goal, or outcome toward which it points (intrinsically, in the case of a natural substance). The causes are interdependent. In physical substances, form does not exist except in matter and matter never exists except with some form or other. A thing's inherent efficient causal powers and teleological features are grounded in its substantial form. Efficient causality presupposes final causality. But despite their interrelationships, these four aspects of an explanation are irreducible, and each is a necessary component of a *complete* account of any natural phenomenon.

1.2.5 Living substances

The mark of a substance, as I have said, is a thing's possession of causal powers which are irreducible to those of its parts. All natural substances exhibit *transeunt* (or "transient") causation, in which the effect is external to the agent. One boulder's knocking into another and causing it thereby to roll off of a cliff would be an example. The mark of a *living* or organic substance is that in addition to transeunt causation it exhibits *immanent* causation, in which the effect remains within the agent and perfects it. An animal's digestion of a meal would be an example insofar as it allows the animal to stay alive and grow (though there are also external or transeunt effects like the excretion of waste products).

It is insofar as they exhibit immanent causation that living substances are taken by the Aristotelian tradition to be capable of *changing* themselves in a sense that non-living things cannot. Machines that change themselves (such as a coffeemaker that turns itself on or a computer that periodically runs a malware scan) are not counterexamples, because they are not true *substances* in the first place, but rather collections of substances (the raw materials out of which they are made) on which a certain accidental form has been imposed. They carry out these activities only because we *make* them do so, not because the parts of which they are made have any intrinsic tendency to do so, any more than liana vines have any intrinsic tendency to function as a hammock. They are no more alive than the eye of a statue is capable of seeing.

The Aristotelian tradition draws a distinction between three basic types of living substance. These form a hierarchy in which each type incorporates the basic powers of the types below it but also adds something novel of its own to them. The most basic kind of life is *vegetative* life, which involves the capacities of a living thing to take in nutrients, to go through a growth cycle, and to reproduce itself. Plants are obvious examples, but other forms of life, such as fungi, are also vegetative in the relevant sense. The second kind of life is *animal* life, which includes the vegetative capacities of nutrition, growth, and reproduction, but in addition involves the capacities of a thing to take in information through specialized sense organs and to move itself around, where the sensory input and behavioral output is mediated by appetitive drives such as the desire to pursue something pleasant or to avoid something painful. These distinctively animal capacities are not only additional to and irreducible to the vegetative capacities, but also transform the latter. For example, nutrition in animals participates in their sensory, appetitive, and locomotive capacities insofar as they have to seek out food, take enjoyment in eating it, and so forth.

The third kind of life is the *rational* kind, which is the distinctively human form of life. This form of life incorporates both the vegetative and animal capacities, and adds to them the intellectual powers of forming abstract concepts, putting them together into propositions, and reasoning logically from one proposition to another, and also the volitional power to will or choose in light of what the intellect understands. These additional capacities are not only additional to and irreducible to the vegetative and animal capacities, but transform the latter. Given human rationality, a vegetative function like nutrition takes on the cultural significance we attach to the eating of meals; the reproductive capacity comes to be associated with romantic love and the institution of marriage; sensory experience comes to be infused with conceptual content; and so forth.

There are on the traditional Aristotelian view, then, three basic divides in the natural world: between the inorganic realm and the basic, vegetative form of life; between the merely vegetative and the animal forms of life; and between the merely animal and the rational forms of life. Echoes of these divides survive in three areas of contention in modern science and philosophy: the debate over the *origin of life*; the debate over the *qualia problem* (also known as the "hard problem of consciousness"); and the debate over the apparent *irreducibility of the propositional attitudes*. The first debate, of course, concerns the issue of how life could have arisen from inorganic processes. There is no generally accepted sketch of a theory of how this might have happened, much less a worked out account. The qualia problem has to do with the question of why any purely material states or processes, such as neurological states and processes, would be associated with any qualitative character of the sort conscious experiences possess (such as the way pain feels, or what it is like to perceive a color like red or green). (Cf. Feser 2006, chapters 4 and 5.) The debate over the propositional attitudes has to do with whether and how mental states like *believing that it is raining*, *desiring that the Lakers win the game*, and so forth can be reduced to or exhaustively explained in terms of neurological processes or some other purely corporeal phenomenon. (Cf. Feser 2006, Chapters 6 and 7.)

From the Aristotelian point of view, the difficulties notoriously facing modern origins of life research stem, not merely from any gap in current empirical knowledge, but from the irreducibility of even the simplest organic substances to purely inorganic phenomena. The intractability of the qualia problem stems from the irreducibility of sentient forms of life to merely vegetative forms of life. The difficulties facing materialist theories of the propositional attitudes stem from the irreducibility of the rational or human form of life to the merely sentient forms of life. In other words, the difficulties in question are essentially confirmation of the traditional Aristotelian position. Of course, most modern scientists and philosophers would disagree with this, and insist that each of the phenomena in question will eventually yield to a completely materialistic explanation given further scientific investigation. But their confidence stems, not from any actual findings of science, but rather from their explicit or implicit commitment to a philosophy of nature that is very different from the Aristotelian one. We will return to these controversies about the various kinds of living substances in later chapters, but for the moment let us turn to an overview of that rival philosophy of nature.

1.3 The mechanical world picture

Tim Crane has suggested that the Aristotelian conception of nature led to an essentially "organic world picture," on which even "the earth itself was thought of as a kind of organism" (Crane 2016, p. 2). In fact this is not true of Aristotle, Aquinas, or other mainstream Aristotelians, who certainly did not think that any natural substances other than the three kinds just described were alive. Still, Crane's characterization provides a helpful way of beginning to understand the difference between the Aristotelian philosophy of nature and what Crane calls the "mechanical world picture" that began to displace it in Western thought in the seventeenth century. It is in living things that the reality of intrinsic teleology and substantial form is most evident. Nothing could be more obvious than that eyes, ears, arms, legs, and other biological organs have final causes. That a whole is more fundamental than its parts is clearest in living things, whose various organs can properly be understood only by reference to the organism they serve. While most of nature is inorganic, every part of it nevertheless exhibits, in a more subtle and rudimentary way, these features that are most glaring in living things. Hence it is not surprising that the Aristotelian might *seem* committed to a kind of "organic world picture," even if that is a very misleading way of putting things. In any event, early modern critics of the Aristotelian philosophy of nature thought of themselves as replacing an "organic" conception of the world with what has been called a "mechanical philosophy," a "mechanistic" conception of nature, or a "mechanization" of the world. That is to say, they took the notion of the *machine* rather than that of the organism to be the best model for nature in general. (Cf. Dear 2006, pp. 15-16)

It is important to emphasize, again, that this was not in any way a scientific discovery. As we will see in this book, there is not a single empirical finding or successful scientific theory which strictly *must* be given a "mechanistic" rather than Aristotelian interpretation. The mechanical conception was rather a philosophical account of how best to carry out scientific investigation and/or to interpret its results. It is a methodological-cum-metaphysical theory *about* science, rather than strictly a part *of* science. That is to say, it is essentially a philosophy of science and a philosophy of nature. That it happens to be a philosophy of science and a philosophy of nature explicitly or at least implicitly accepted by most modern scientists themselves should not blind us to that fact. Nor should it lead us to accept it ourselves, or even to give it the benefit of the doubt. Scientists qua scientists are not experts on philosophical matters. Indeed (and as we will see), where a purportedly scientific claim embodies both empirical and philosophical assumptions, scientists who have no training in philosophy often fail to disentangle these components or even to see the difference between them, and commit philosophical errors as a result. While scientific knowledge is certainly relevant and necessary to evaluating the various specific areas of dispute between Aristotelianism and mechanism that we will be addressing in this book, ultimately the dispute is philosophical, and the mechanistic world picture must accordingly stand or fall on its philosophical merits.

1.3.1 Key elements of the mechanical philosophy

What exactly is the content of the mechanical world picture? What precisely does it mean to say that nature ought to be modeled on a machine? Part of the idea, of course, is that "if you [want] to understand how something work[s], you should try to take it apart and see how it runs," and in particular that "finding out the causal connection between the parts would reveal a lot about how the thing as a whole work[s]" (Churchland 1995, p. 24). Like a watch, a computer, or any other machine, a molecule, bodily organ, solar system, or other natural phenomenon can be understood by breaking it down into its components and determining the efficient causal relationships holding between them. But this is hardly sufficient to make a conception of the world "mechanistic" as opposed to Aristotelian. No Aristotelian has ever denied that natural objects are in *some* respects machine-like or that breaking them down into their parts and determining how those parts interact is *part* of a complete explanation.

The correct way to understand how the mechanical world picture differs from Aristotelianism is to recall the Aristotelian distinction between natural objects and human artifacts, or more precisely between, on the one hand, things having substantial forms and intrinsic teleology and, on the other, things having merely accidental forms and at most merely extrinsic teleology. A true natural substance is one having a substantial form and intrinsic teleology (as the liana vine of my earlier example does). Artifacts (including machines) and other objects with merely accidental forms and thus at most only extrinsic teleology (as in the case of the hammock) or no teleology at all (as in the case of a random pile of stones) are secondary kinds of reality, parasitic on the existence of natural substances. To be part of the "natural" order of things *just is* to have a substantial form and intrinsic teleology; and thus, to lack substantial form and intrinsic teleology *just is not* to be "natural" in this sense. Whatever else we say about nature, then, it is for the Aristotelian precisely *not* a machine or any other kind of artifact, despite the superficial similarities (such as the fact that both machines and natural substances have interacting parts).

The "mechanical world picture" is essentially a rejection of this fundamental conception of what is "natural," and of everything implicit in it. For the mechanical philosophy, a natural object *is* to be understood on the model of a machine or artifact, and therefore *not* in terms of substantial form or intrinsic teleology. Thus, of Aristotle's four causes, the mechanistic picture effectively rejects formal and final cause, and also radically redefines material and efficient cause (since for the Aristotelian these latter two kinds of cause were partially defined in terms of the former two). In turn, other elements of the Aristotelian philosophy of nature, such as the idea of a hierarchy of irreducibly different kinds of natural substance and the theory of actuality and potentiality, are explicitly or implicitly abandoned as well. There is also an emphasis on those aspects of nature which are *predictable and controllable* in the way the behavior of a machine (ideally) is. This entails a focus on the *quantifiable* aspects of nature (which are more susceptible of strict prediction and control), and thus on a mathematical description of physical systems as paradigmatic of scientific rigor.

Let's expand upon these points. First of all, the mechanical world picture abandons the idea of matter as the potentiality to take on form. It is committed instead to an essentially atomist model, even if, as the history of science has proceeded, that model has been very drastically modified. Matter is, to a first approximation, conceived of in terms of fundamental particles possessing only such "primary qualities" as size, shape, spatial position, and local motion, and devoid of "secondary qualities" like color, sound, odor, taste, heat, and cold, which are reinterpreted as mere projections of the mind rather than really inhering in material things as they are in themselves. In short, the essential properties of matter are taken to be those which are quantitative and to include none that are irreducibly qualitative.

For Descartes' plenum version of the mechanical picture, matter is pure extension in space, so that it is infinitely divisible (since whatever is extended can be divided into ever smaller units) and there is no void space through which bits of matter pass (since such purportedly empty space would itself be extended and thus filled with matter). Other early modern thinkers like Gassendi and Hobbes adhered instead to the traditional atomist picture of basic particles which are indivisible in principle and move through void space. Corpuscularians like Locke and Boyle modified atomism by taking the basic particles to be undivided merely in fact rather than in principle. Initially, mechanistic explanations sought to understand all causation on a push-pull model of the sort illustrated by the gears of a watch, a system of pulleys, or the like. With developments like the Newtonian theory of gravitation, Maxwell's theory of electromagnetism, and quantum mechanics, such early features of the mechanical philosophy came to be rejected as simplistic. For example, crude push-pull causation and the idea that particles ought to be thought of on the model of tiny marbles or BBs have long since been abandoned. Still, the idea that the natural world is essentially a vast sea of colorless, odorless, soundless, tasteless particles in motion is to this day taken to be at least a rough approximation to the truth.

This model has also always tended toward Parmenideanism, at least in some respects. The ancient atomists took the atoms roughly to correspond to Parmenides' absolutely static reality. Of course, they regarded the things that are *made up* of atoms as coming into being and passing away and in other ways too to be changeable. But the atoms *themselves* were taken to be neither created nor destroyed nor to be changeable in any way other than extrinsically, with respect to their spatial location. Similarly, as Dennis Des Chene notes:

> [T]he contrast between potential and actual... [is] banished in the Cartesian restriction of natural properties to figure, size, and motion... Cartesian matter... is, from an Aristotelian standpoint, at every instant entirely actual. (1996, pp. 5-6)

In early modern physics, the inertial motion of particles and of the things made up of particles was in turn taken to be a kind of *state* rather than a true change. Only a change in direction or speed was regarded as a true change. Thus were both the things that move and at least much of their movement itself assimilated to an essentially static conception of reality. Minkowski's interpretation of relativity in terms of a four-dimensional space-time manifold essentially takes this tendency to its logical conclusion, yielding a Parmenidean static block universe. (More on all this in later chapters.)

The radical transformation of the notion of material cause was, as I have indicated, a concomitant of the abandonment of formal cause, and in particular of substantial form. With the rejection of substantial form went also a rejection of the idea that there are any sharply demarcated and irreducibly different kinds of substance in nature. All natural objects were to be regarded instead as essentially the *same* one kind of thing, namely fundamental particles in different configurations. Whereas for the Aristotelian, the parts of natural substances are metaphysically secondary to the wholes of which they are parts, for the mechanical philosophy it is the parts which are metaphysically fundamental, and the wholes reflect, in effect, merely accidental forms. Just as Tarzan's hammock is really "nothing but" a collection of vines which would be just as they are even apart from their organization into a bed, so too are natural objects "nothing but" collections of particles which would be just as they are even apart from their organization into stones, trees, dogs, etc. (Though many contemporary philosophers sympathetic to the mechanistic tradition have, as we will see, moved beyond the radical reductionism this entails to embrace a "non-reductionist naturalism," their position is, as we will also see, unstable and threatens either to collapse back into reductionism or to give the game away to Aristotelianism.)

If their organization into stone-like, tree-like, dog-like, etc. configurations is nothing more than a superficial manifestation of what are really just fundamental particles in motion, than naturally the *tendency or directedness toward* distinctively stone-like, tree-like, dog-like, etc. ends or outcomes is also going to be regarded as illusory. That is to say, teleology or final cause is for the mechanical philosophy no more really a part of the objective natural world than substantial form is. Indeed, this is arguably *the* fundamental and non-negotiable component of the mechanical world picture's critique of Aristotelianism. (Cf. Koyré 1965, pp. 7-8; Hasker 1999, pp. 63-64; DeWitt 2004, p. 84; and the long list of references in Johnson 2005, p. 24, note 38.) Alex Rosenberg writes:

> Ever since physics hit its stride with Newton, it has excluded purposes, goals, ends, or designs in nature. It firmly bans all explanations that are *teleological*...
>
> There are several... outstanding problems that physics faces...
>
> No matter how physics eventually deals with these problems... [i]n solving them, physics will... not give up the ban on purpose or design. (2011, pp. 40-41)

Notice that it is not any actual empirical *finding* of physics that Rosenberg is or could be talking about here, but rather an *a priori* methodological *stipulation* about what is and will be *allowed to count* as a legitimate physical explanation. After all, to "firmly ban" something is not to *discover* that it doesn't exist, any more than to ban someone from your home is to discover that he is not there. Rather, just as to ban someone from your home is to stick to a policy of never letting him in in the first place, so too for physics to ban teleology is simply for physicists to stick to a policy of not letting themselves make use of the notion of final cause when giving explanations, *even if* it might seem to be called for in some situation. Similarly, to say that physics will stick to this ban "no matter how" it ends up dealing with the problems it faces can hardly be an empirical claim. How could Rosenberg or anyone else possibly know that empirical evidence for teleology will not turn up "*no matter how*" the outstanding problems of physics end up being resolved? It is rather the expression of a determination to continue sticking to a certain core element of the mechanical philosophy's anti-Aristotelian revolution that has persisted to the present day when other aspects (push-pull causation, reductionism, etc.) have fallen by the wayside. As David Hull notes:

> [M]echanistic explanation [is] a kind of explanation countenanced by views that range from the extreme position that all natural phenomena can be explained entirely in terms of masses in motion of the sort postulated in Newtonian mechanics, to little more than a commitment to naturalistic explanations. Mechanism in its extreme form is clearly false because numerous physical phenomena of the most ordinary sort cannot be explained entirely in terms of masses in motion... Historically, explanations were designated as mechanistic to indicate that they included no reference to final causes or vital forces. In this weak sense, all present-day scientific explanations are mechanistic. (1995, p. 476)

Now, the key early modern defenders of the mechanical world picture banished final cause from the *natural* world, but not from reality altogether. For thinkers like Descartes and Newton, while there is no teleology *intrinsic* to material things, there is certainly teleology *extrinsic* to them. For Descartes, purposes exist in the thoughts and volitions of the immaterial soul or *res cogitans*, and these purposes are reflected in speech, bodily behavior, and the things we make, even if the sounds we make, the motions of our limbs, and the objects we create would have no purposes *apart from* those thoughts and volitions. Descartes and Newton also took final cause to exist in God, who imparts purposes to the material world that it wouldn't otherwise have. This was a natural concomitant of their picture of the world as a kind of machine. If the world is like a hammock or a watch, then even though it is devoid of intrinsic teleology, it would still have an externally imposed teleology, just as these artifacts do.

Later thinkers, of course, would delete God from the mechanical world picture, and Darwinism was crucial to making this deletion seem plausible. As I have said, while biological phenomena are by no means the *only* teleological natural phenomena, they are the most *obviously* teleological natural phenomena. Hence when natural teleology was reinterpreted by the mechanical philosophy as entirely extrinsic rather than intrinsic, biological phenomena were regarded as the most obvious examples of teleology of the extrinsic sort – as natural "machines" which could only have come about by way of a divine machinist. Many adherents of the mechanical philosophy thus gravitated toward William Paley's "design argument" as clear evidence that a mechanical world was not *per se* an atheistic world. Darwinism, though, seems to make even the purportedly extrinsic teleology of living things illusory, and thus to undermine the inference from biological teleology to a divine machinist.

That still left the teleology or final causality associated with human thought and action, which for Descartes was to be located in an immaterial substance, and for later dualists at least in immaterial properties if not in an immaterial substance. But with God out of the picture, immaterial substances and properties came to seem extremely odd and unlikely things to have arisen in an otherwise completely material world. So, immaterial substances and properties were abandoned altogether by materialist adherents of the mechanical world picture. Now, if matter as understood by the mechanical philosophy is devoid of all intrinsic teleology, and there are also no immaterial entities having intrinsic teleology which could serve as the source of the apparent extrinsic teleology of some material things, then the teleology that exists in human thought and action – including the thoughts and actions of scientists and adherents of the mechanical world picture themselves – becomes highly problematic. Thus do we have what has come to be called the problem of "naturalizing" the intentionality or "directedness" of human thought and action (since this "directedness" would, on a mechanistic picture of nature, seem no less illusory than the "directedness" entailed by final causality). For the reductionist materialist, the way to do this is to show that while what we call the "directedness" or intentionality of thought is real, it is really nothing but a certain kind of efficient causality in disguise, falsely *appearing* to be irreducibly teleological. For the eliminative materialist, the intentionality of thought is *not* real in the first place, and thus needn't be reduced or otherwise explained. Either way, even the human mind can (so it is claimed) be assimilated to the picture of a world entirely free of any real "directedness" or "pointing" toward an end or outcome, any teleology or final causality.

Just as the rejection of formal causality necessarily went hand in hand with a redefinition of material cause, so too did the rejection of final causality entail a redefinition of efficient cause. Again, for the Aristotelian, efficient causation essentially involves the operation of causal powers, and different powers are inherently "directed toward" different outcomes or ranges of outcomes. The causal powers of the phosphorus in the head of a match are "directed toward" the generation of flame and heat, the causal powers of an acorn are "directed toward" the production of an oak tree, and so forth. Causal powers thus "point" a thing forward, as it were, toward its characteristic effect or range of effects. Meanwhile, given the principle of proportionate causality (according to which whatever is in an effect must in some way pre-exist in its cause), a thing is also "pointed" backward toward that which generated it. Intrinsic teleology thus cements causes and effects together.

Accordingly, the abandonment of teleology dissolved this cement, making causes and effects inherently "loose and separate," as Hume famously put it. In principle, any effect or none could follow upon any cause. Why, then, do we observe causal *regularities* in the world rather than the randomness that the abandonment of teleology would imply? The answer of the early proponents of the mechanical philosophy was that if nothing *intrinsic* to things could account for this regularity, something *extrinsic* to them still could, namely *laws of nature*. The notion of a law of nature was understood by thinkers like Descartes and Newton in explicitly theological terms, as a divine decree that things will behave in such-and-such a way. Whereas the Aristotelian position conceived of natural substances as inherently *active* and prone to operate in certain ways by virtue of their distinctive substantial forms and causal powers, the mechanical philosophy conceived of them as inherently *passive*, lacking any inherent tendency toward a characteristic way of acting (Ellis 2002, pp. 2-3). They are made to act in the regular ways they do only because God decided to impose on them *this* particular set of laws rather than that one. This too dovetails with the idea that the universe is a kind of machine or artifact with externally imposed teleology and accidental form. (Cf. Osler 1996)

As physicist Paul Davies has noted, the key assumptions modern scientists make about laws of nature have their origins in this theological picture. He writes:

> The orthodox view of the nature of the laws of physics contains a long list of tacitly assumed properties. The laws are regarded, for example, as immutable, eternal, infinitely precise mathematical relationships that transcend the physical universe, and were imprinted on it at the moment of its birth from "outside," like a maker's mark, and have remained unchanging ever since... In addition, it is assumed that the physical world is affected by the laws, but the laws are completely impervious to what happens in the universe... It is not hard to discover where this picture of physical laws comes from: it is inherited directly from monotheism, which asserts that a rational being designed the universe according to a set of perfect laws. And the asymmetry between immutable laws and contingent states mirrors the asymmetry between God and nature: the universe depends utterly on God for its existence, whereas God's existence does not depend on the universe...
>
> Clearly, then, the orthodox concept of laws of physics derives directly from theology. It is remarkable that this view has remained largely unchallenged after 300 years of secular science. Indeed, the "theological model" of the laws of physics is so ingrained in scientific thinking that it is taken for granted. The hidden assumptions behind the concept of physical laws, and their theological provenance, are simply ignored by almost all except historians of science and theologians. (2010, pp. 70-1)

Of course, when later adherents of the mechanical world picture deleted God from the story, this account of laws of nature was no longer available. But what could take its place? Those inspired by Hume have tended to regard laws as mere *regularities*. The idea here is that to say, for example, that it is a law of nature that events of type *A* are followed by events of type *B* is simply to assert that *in fact* events of type *A* are always followed by events of type *B*. It is merely to note that that is the way the world happens to work. One problem with this view is that if that is what laws are, then they don't seem to *explain* anything. To appeal to the notion of a law when giving a scientific account of the relationship between events of type *A* and events of type *B* ends up being nothing more than a *re-description* of that relationship in a new jargon, rather than a way of making it intelligible *why* that relationship holds. An alternative way to interpret laws is to see them as abstract objects analogous to Platonic Forms, in which the objects and events in the natural world participate. But that inevitably raises the question why the natural world participates in these particular laws rather than some other laws, and indeed why it participates in any laws at all. If the answer is like the view suggested in Plato's *Timaeus*, to the effect that a divine craftsman accounts for the world's being governed by just the laws that govern it, then theology will have brought back into the mechanical world picture, when the aim was to get rid of it. Yet another alternative view of laws of nature is to regard them as a kind of shorthand description of the way a natural substance will tend to behave given the essence and causal powers inherent to it. But this is essentially a return to the Aristotelian philosophy of nature that the mechanical philosophy was supposed to be replacing.

These difficulties, and the difficulties inherent in banishing immaterial substances (whether God or the soul) from the mechanical world picture, indicate that an atheistic version of the mechanical world picture is incoherent. If there is no "directedness" anywhere in reality, not even in immaterial substances external to the material world, how can we make sense of the intentionality of human thought and action, including the thought and action of scientists and adherents of the mechanical philosophy? If there is no God and no substantial forms either, how can we make sense of the operation of laws of nature?

But from the Aristotelian point of view, the mechanical world picture is incoherent even *if* it is supplemented with theism, as Descartes, Newton, and other early modern thinkers supplemented it. The notion of a machine or artifact presupposes that of a natural substance, and a natural substance, possessed as it is of intrinsic teleology and substantial form rather than extrinsic teleology and accidental form, is precisely not a machine or artifact. So to make machines or other artifacts the models for natural substances simply puts the cart before the horse. Attempts to banish intrinsic teleology and substantial form and replace them with fundamental particles in motion (as in atomism) doesn't really banish them at all, but simply relocates them. The fundamental particles become the true substances possessed of substantial rather than accidental forms, and the directedness of their causal powers toward certain characteristic effects constitute an ineliminable residue of intrinsic teleology. The Parmenidean tendency to redefine all local motion as a kind of stasis flirts with the incoherence that afflicted the views of Parmenides and Zeno themselves. The "non-reductionist naturalism" that the contemporary successors of the mechanical world picture have been forced into by the failure of reductionism essentially abandons a key element of the picture and gives the game away to the Aristotelian.

We will return to these various issues in later chapters. The point for the moment is simply to note how philosophically problematic the mechanical world picture can be seen to be even before its application to various specific areas of scientific study is considered. Why, then, did it succeed in pushing the Aristotelian philosophy of nature to the periphery in the early modern period?

1.3.2 Main arguments for the mechanical philosophy

Much of the motivation for the mechanical philosophy was *political* rather than philosophical or scientific. This political motivation reflected both negative and positive aims. The negative aim was to help undermine the authority of the Catholic Church, the intellectual foundations of which had by the late Middle Ages come to be standardly articulated in Aristotelian terms. As philosopher Pierre Manent observes, for the early modern philosophers, "in order to escape decisively from the power of the singular religious institution of the Church, one had to renounce thinking of human life in terms of its good or end" (1995, p. 114). Hence "it is the teaching of Aristotle, which was essentially adopted by Catholic doctrine, that Descartes, Hobbes, Spinoza, and Locke will implacably destroy" (Manent 1998, p. 113). In the same vein, intellectual historian Mark Lilla notes that Hobbes's materialism had a "political end," namely "the dismantling of Christendom's theological-political complex," toward which end "the whole of Aristotle would have to be scrapped, along with the shelves of medieval commentary on him" (2007, pp. 75 and 87).

The positive aim was to redirect Western thought away from the metaphysical and otherworldly orientation it had in the ancient and medieval periods and toward a more practical and this-worldly set of concerns. In particular, thinkers like Bacon and Descartes sought to understand nature in a way that would facilitate the control of natural processes and the development of new technologies. Bacon wrote in *The Great Instauration* of increasing "human utility and power" through the "mechanical arts," and Descartes speaks in the *Discourse on Method* of making us "masters and possessors of nature." Focusing on those aspects of nature which could be precisely quantified, and modeling nature on a finely tuned machine, were conducive to this end.

Needless to say, these political aims don't constitute arguments. The fact that general acceptance of a certain set of ideas would help promote (what is taken to be) a desirable goal gives the ideas no *logical* support, even if it contributes *psychologically* and *sociologically* to their acceptance.

It is also important to note, in this connection, how much the success of the mechanical philosophy owed to rhetoric rather than argumentation. As historian of science Peter Dear writes, Descartes' methods of responding to Aristotelian arguments included "ridicule" and a "pretended inability to understand the[ir] meaning" (2006, pp. 17-19). Historian of philosophy Dennis Des Chene notes that "Descartes and those who subscribed to his polemics exaggerated the sins of their opponents, ascribing to the Aristotelians views the Aristotelians would have repudiated" (1996, p. 169). Another historian of philosophy, Helen Hattab, adds:

> Descartes gives few philosophical arguments to directly support his rejection of forms in favor of mechanisms. Moreover, the scattered reasons he offers in his corpus are cryptic and hard to unpack. (2009, p. 1)

(As Hattab goes on to show, Descartes' arguments also often presuppose his own controversial metaphysics – thus begging the question against his Aristotelian opponents – or apply at most only to Francisco Suárez's conception of substantial form, not to Aquinas's.) Galileo too resorted to misrepresentation and "pillory" and got away with it "because the audience that mattered was already on his side" (Dear 2006, p. 21). Galileo's rhetorical tricks are famously recounted in Paul Feyerabend's *Against Method* (1993). Feyerabend even goes so far as to conclude that "the Church at the time of Galileo... kept closer to reason as defined then and, in part, even now" (p. 125). Historian of science E.A. Burtt speaks of the "wishful thinking" and "uncritical confidence" that often accompanied early modern defenses of the "mathematico-mechanical" world picture (1980, pp. 304-6).

All the same, there were and are also actual arguments for the mechanical world picture. They are of three general sorts:

1. Scientific objections to Aristotelianism: The first sort cites various scientific errors made by Aristotle and his medieval followers, and claims that these errors undermine the Aristotelian philosophy of nature. For example, Aristotle and medieval Aristotelians held that heavy objects naturally tend to fall to the earth, specifically. Of course, that is not correct, for there is nothing special about the gravitational pull of the earth *per se*. Now, the tendency to fall to the earth, specifically, was a purported example of intrinsic *teleology*, and that heavy objects had this tendency was supposed to reflect their *substantial forms*. Hence the falsity of the scientific assumption in question shows (so this sort of argument goes) that there is something suspect about the notions of intrinsic teleology and substantial form. Other scientifically erroneous illustrations of purported teleological features and substantial forms reinforce this conclusion.

Similarly, it is sometimes suggested that what medieval Aristotelians would have regarded as irreducible substances have been shown by modern science to be reducible, and that what Aristotelians would have described in terms of final causality has been shown by science to be describable in terms of efficient causality alone. Water, for example, is reducible to hydrogen and oxygen, and where the medieval Aristotelian might have described frozen water's tendency to cool down a surrounding liquid in terms of its directedness toward this outcome as its final cause, modern science would instead describe a complex causal interaction between the molecules making up the ice and the molecules making up the liquid water surrounding it.

To see the fallacy in such arguments, suppose you say that there are such things as murders and give as examples of murderers Dr. Sam Sheppard and Charles Manson. Suppose I reply: "Your claim is falsified by the fact that Sheppard was actually innocent, and Manson only gave orders to accomplices, who actually carried out the killings." Obviously this would be a silly reply. That a particular claim about a certain murder turns out to be false, and that certain other murders are more complicated than merely postulating a single murderer who directly kills the victim, in no way casts doubt on the reality of murder *per se*.

But it is no less silly to say: "Aristotle was wrong about the natural motion of sublunar bodies, therefore there are no final causes or substantial forms" or "What happens when ice is in water involves a complex exchange of energy among the molecules in the liquid water and the molecules in the ice, therefore there is no final causality here." For whatever the scientific details concerning gravitation, cooling, etc. turn out to be, they will (so the Aristotelian argues) involve patterns of efficient causation (gravitational attraction, molecular interaction, etc.); and *these patterns*, the Aristotelian holds for reasons described above, will necessarily presuppose teleology or directedness toward a certain outcome or range of outcomes. Similarly, whether or not a certain purported substance turns out to have a merely accidental rather than substantial form, the *bottom level* kinds of natural objects will, the Aristotelian argues for reasons described above, be true substances having substantial rather than accidental forms.

In short, what modern science shows is at most only that particular purported *examples* of intrinsic teleology or substantial form are not good ones, but it does not tell us that there are no such things as intrinsic teleology or substantial forms at all. General philosophical principles must be distinguished from concrete applications of those principles, and deficiencies in the latter do not necessarily entail deficiencies in the former. For all modern science has shown, the Aristotelian arguments to the effect that there must be intrinsic teleology and substantial forms at *some* level of nature still stand. The big picture remains untouched, even if some of the details turn out to have been mistaken. And as we will see in the chapters to follow, even where some of the details are concerned, the traditional Aristotelian view is in a much stronger position than its critics realize.

Similarly, appeals to various mechanisms uncovered by modern science (such as the molecular interactions between ice and liquid water that occur when the former cools down the latter) are in no way in *competition* with explanations in terms of teleology and substantial form. For the Aristotelian argues, of course, that there are formal and final causes in addition to material and efficient causes. To identify the latter is not by itself to cast doubt on the reality of the former, any more than identifying the chemistry of the paint used in the *Mona Lisa* casts doubt on, or is in any way in competition with, the claim that it represents a woman named Lisa del Giocondo.

As Peter Dear writes of the early modern dispute between Aristotelians and adherents of the mechanical world picture:

> [W]hat was at issue had nothing to do with disagreements over what phenomena there were in the world to be explained; empirical investigation would not settle matters. In that sense, this was a fundamentally philosophical debate, and specifically a natural-philosophical one. It deeply concerned the nature of the universe, rather than resting on the affirmation or denial of controversial physical phenomena. (2006, p. 17)

What was true in the seventeenth century remains true today. The dispute between modern Aristotelians and the successors of the mechanical philosophy ultimately has to do, not with the observational evidence or the reality of any specific physical phenomenon, but rather with the question of which metaphysics or philosophy of nature ought to inform our *interpretation* of the empirical evidence and the specific physical phenomena.

2. Philosophical objections to Aristotelianism: This brings us to the second general sort of argument for the mechanical philosophy, which challenges precisely the philosophical adequacy of Aristotelian philosophy of nature. The idea here is that, whatever one says about the empirical issues, notions like substantial form and intrinsic teleology are *conceptually* suspect. Molière's famous joke about opium's "dormitive virtue" illustrates one of the alleged problems. The claim that opium causes sleep because it has such a virtue or power is, so the objection goes, a mere tautology and says nothing informative. But *all* appeals to substantial forms, essences, powers, etc. are like this and thus do no explanatory work.

Common though this objection was and is, however, there are several problems with it. First, the claim that opium causes sleep because it possesses a dormitive power does in fact say something informative, even if it is only minimally informative. It tells us that the fact that sleep tends to follows the use of opium is not an accidental feature of this or that particular sample of opium or of the circumstances in which it is ingested, but rather reflects something in the nature of opium itself. True, the claim does not specify exactly what it is about the nature of opium that causes sleep. But that does not make the claim a tautology, any more than it is a tautology to claim that something in the chemical structure of opium causes sleep, without specifying exactly what that chemical structure is.

Second, even advocates of the mechanical philosophy don't really treat claims about powers, substantial forms, etc. as if they were mere tautologies. For example, they don't say: "Yes, opium has a power to cause sleep, but that's too trivially true to be worth saying." Rather, they *deny* that there are any such things as powers or substantial forms, on the grounds that any theory that posits them is metaphysically extravagant. But such a denial makes sense only if the affirmation of powers, substantial forms, etc. does in fact have some content, however minimal. The critic is conceding that for a thing to have powers *would* in fact be for it to have something over and above the features the critic is willing to acknowledge. He is treating talk of powers, substantial forms, etc. as substantive but false rather than true but uninformative.

Third, the substantive content is not in fact as minimal as the critic supposes. For example, part of the work that the notion of a causal power does in Aristotelian metaphysics is to explain how it can be true both that there is something in the very *nature* of a certain sort of cause *C* by which it tends to produce a certain kind of effect *E*, but also that *C* is not in fact always producing that effect. A power is something *C* can *have* without *exercising*. The *having* of it follows from the nature of *C*, but the *exercising* of it may depend on further conditions being satisfied. In the case of substantial form, I noted above how that notion is deployed in order to account for phenomena such as the persistence of a substance through change, and the difference between natural substances on the one hand and artifacts and mere aggregates on the other. In other ways too the Aristotelian metaphysical apparatus of substantial form, causal powers, intrinsic teleology, and all the rest is deployed in order to deal with a wide range of philosophical problems, something it could not do if it embodied mere tautologies.

Of course, it is true that modern science tells us a great deal that we would never learn merely from the application of this Aristotelian apparatus. But that is no objection to the latter, because it is not trying in the first place to answer the sorts of question that the physicist, chemist, or biologist is concerned with. Aristotelian philosophy of nature is, again, trying to tell us what the general structure of any possible natural world must be like. It claims that in any such world there must be a distinction between things having substantial forms and those having merely accidental forms, a distinction between the causal powers of a thing, the end or outcome toward which those powers point, the actual exercise of those powers, and so forth. But there are also questions about *what, specifically*, in the actual natural world we live in, are the things having substantial forms; about *what, specifically*, are the causal powers of various substances; about *what, specifically*, are the physical mechanisms underlying the operation of various powers; and so on. *Those* are the sorts of questions that the natural scientist deals with. To treat the answers the Aristotelian philosopher gives *his* questions and the answers the natural scientist gives *his own, different* questions as if they were in competition is simply to commit a category mistake.

It is no surprise then, that recent years have seen a revival of interest in causal powers, essences, and related Aristotelian notions in contemporary mainstream analytic metaphysics and philosophy of science. (Cf. Ellis 2002 for a general introduction to this "new essentialist" movement, Groff and Greco 2013 for a collection of papers, and Mumford 2009 and Feser 2014b, pp. 53-72, for surveys of the literature.) It would seem that it is not powers, essences, and the like, but rather Molière-style objections to them, that are passé. (For more detailed discussion of the Molière objection, see Feser 2014b, pp. 42-46.)

Another philosophical objection raised by early modern adherents of the mechanical philosophy was directed at the notion of intrinsic teleology. For the Aristotelian, such teleology or directedness toward an end exists even where there are no minds – in the directedness of the phosphorus in the head of a match toward the outcome of generating flame and heat, the directedness of an acorn toward becoming an oak, and so forth. But this (so the objection went) is unintelligible. There can be no directedness toward an end without a mind which grasps that end. (Cf. Des Chene 1996, pp. 393-94; Dear 2006, pp. 16-17; Ott 2009, pp. 30 and 41-43.)

As I indicated above, those Aristotelians who are also Thomists would distinguish between the *proximate* and *ultimate* sources of a natural substance's being directed toward an end. The proximate source is the substance's own nature. An acorn is directed toward the end of becoming an oak because being so directed is part of what it is to *be* an acorn as opposed to some other kind of thing. The ultimate source is the divine intellect, which "points" things toward their ends the way an archer points an arrow toward a target (where, again, the need for such an ultimate source is something Aquinas argues for in the Fifth Way). Now, if the early modern critics are saying that there could be no directedness toward an end in nature unless such directedness traced *ultimately* to a divine intellect, then this would hardly be an *objection* to the Aristotelian-Thomistic position, since this is just what that position itself says. So, the objection must instead be that, in things without minds, there cannot be any *proximate* source of directedness in their natures. Any teleology or directedness in such things would, according to the objection, have to be entirely extrinsic rather than intrinsic, derived *directly* and not merely ultimately from a divine intellect. That is to say, they could meaningfully be said to have teleology at all only if they were thought of as *artifacts*, as in Paley-style design arguments.

But what non-question-begging reason could be given for such a claim? It can't be maintained that it is somehow just intuitively obvious. After all, Aristotle evidently thought not only that there is intrinsic teleology or directedness toward an end in things which lack minds, but that this teleology was in *no way* connected to any intellect, not even *remotely*.

He thought that what I have been calling the *proximate* source of this directedness was the *entire* source of it. (It is true that Aristotle believed in a divine Unmoved Mover, but unlike Aquinas, he did not link the existence of teleology in natural things with the Unmoved Mover. He thought their teleological features simply followed from their natures.) Several contemporary philosophers also either endorse or at least treat as a live option the thesis that there is teleology in nature that is divorced from any mind. (Cf. the sympathetic discussions of Aristotle's conception of natural teleology in Johnson 2005, Ariew 2002 and 2007, and Nagel 2012, and the notion of "physical intentionality" or "natural intentionality" defended in Place 1996, Heil 2003, and Molnar 2003.) Evidently it is not intuitively obvious to these thinkers that there can be no teleology apart from some mind. So, to show that they are wrong would require an argument and not just an appeal to intuition.

Descartes perhaps implied such an argument when he suggested, in the Fourth of his *Meditations*, that we cannot know the final causes of things because we cannot presume to know God's intentions in creating them. The idea would be that it is only after we know that something is designed or directed by some mind that we can infer what its final cause is, or even that it has a final cause at all. The problem with such a claim, however, is that it is simply not plausible. It certainly *seems* quite obvious that eyes have the final cause of allowing us to see, that ears have the final cause of allowing us to hear, and so on, and these things seem obvious whether or not one supposes that there is an intellect which made them for these purposes. Of course, a theist might go on to argue that such final causes could be there only if there is such an intellect. Or a Darwinian naturalist might argue that since (as he supposes) there is no divine intellect, the appearance of final cause here is illusory. The point, though, is precisely that either conclusion would have to be *argued for*. There simply is no *manifest* connection between something's having a final cause and its being directed by some mind. We can perfectly well *understand* the idea that something has a final cause *whether or not* we think of it has having been designed. We don't *first* have to conceive of it as having been designed or directed by some mind in order to judge that it has a final cause, or to guess what that final cause is if it has one.

Then there is the fact that the Aristotelian has given *reasons* for concluding that there must be such a thing as intrinsic teleology, or a proximate ground of a thing's teleological features in its nature. For one thing, we need to suppose this in order to make sense of the difference between a natural substance like the liana vine of my earlier example, and a human artifact like a hammock. The difference between them is that the liana-like tendencies follow from the nature of the vine, whereas the hammock-like tendencies do not and have to be imposed from outside. To have tendencies of the former sort, though, just is to have intrinsic teleology. For another thing, we need to suppose that natural substances possess intrinsic teleology if we are to make sense of the fact that they reliably generate the specific sorts of effects they do.

Of course, the advocate of the mechanical philosophy will reject these claims, but the point is that he has actually to *answer the arguments* for the claims. He cannot, without begging the question, simply dismiss them out of hand on the grounds that teleology presupposes mind, so that there cannot be any such thing as intrinsic teleology.

Also question-begging is a further philosophical criticism some early modern thinkers raised against the Aristotelian philosophy of nature. Locke suggested in Book III of his *Essay Concerning Human Understanding* that the existence of severely deformed and mentally disabled human beings tells against the Aristotelian commitment to the reality of sharply demarcated and irreducibly different kinds of substance. In order to account for such people, Locke says, the Aristotelian will either have to posit some new species of thing in an *ad hoc* way, or to deny that they belong to any species. But this simply ignores, without answering, the Aristotelian point that the properties which flow or follow from the nature or essence of a thing might be prevented from manifesting. Again, dogs in their mature and normal state will have four legs, but a particular dog nevertheless might, because of injury or genetic defect, be missing a leg. That does not entail that it is not really a dog after all, but only that it is a defective instance of a dog. Similarly, injury or genetic defect might cause bodily deformity or mental impairment in a human being, but it simply doesn't follow that the people to whom this happens are not really human beings at all, and instead belong to some different species or to no species at all. Rather, this is merely a case where the properties which flow or follow from human nature, and which would be manifest in any normal and mature human being, are being prevented from manifesting. (Cf. Feser 2007, pp. 63-64; Feser 2014b, pp. 230-35.)

3. The appeal to Ockham's razor: The main argument for the mechanical world picture, though, has nothing to do with any specific scientific or philosophical errors made or allegedly made by Aristotelian philosophers of nature. The advocate of the mechanical philosophy might grant at least for the sake of argument that none of the outmoded scientific illustrations of the central Aristotelian theses are essential to the theses themselves and that none of the various specific philosophical objections raised by early modern critics has any force. He would nevertheless maintain that the central problem with the Aristotelian philosophical framework is that we simply have no need of it. Everything about the natural world that needs to be explained can be explained in terms of an exclusively mechanical conception of nature. Hence the central concepts of the Aristotelian philosophy of nature – actuality and potentiality, substantial form and prime matter, intrinsic teleology, and the rest – can be put aside on grounds of parsimony. To suppose that these concepts correspond to any real features of the world would be to multiply entities without necessity and thus violate Ockham's razor.

The trouble with this objection, however, is that it simply blatantly begs the question against the Aristotelian, who, in the case of every one of the key notions of the Aristotelian philosophy of nature, has put forward *arguments* which purport to show that we *do need* to affirm them in order to explain what needs to be explained. These arguments need to be engaged and successfully rebutted if the mechanical world picture is to be rationally justified. In the absence of such a rebuttal, any appeal to Ockham's razor is mere hand-waving.

It might be claimed that the success of modern natural science justifies the conclusion that recourse to the central notions of Aristotelian philosophy of nature is unnecessary. For example, Alex Rosenberg says of teleological explanations:

> Physics ruled out this sort of reasoning right at the start of its success. Ever since physics hit its stride with Newton, it has excluded purposes, goals, ends, or designs in nature. It firmly bans all explanations that are *teleological*... At each of the obstacles to its development, physics could have helped itself to purpose or design. No explanation of heat in Newton's laws? God must have added heat in separately. Why do electric fields produce magnetic fields and vice versa? God's clever design. Gravity is so mysterious, the way it moves through total vacuums at infinite speed and penetrates any barrier at all. How come? God made it that way to keep us from floating away from the ground.
>
> Theories about purposes at work in the universe could have gotten physics off the hook every time it faced a challenge. But physicists have always refused to shirk the hard work of crafting

> theory that increases explanatory precision and predictive application. Since Newton 350 years ago, it has always succeeded in providing a nonteleological theory to deal with each of the new explanatory and experimental challenges it has faced. That track record is tremendously strong evidence for concluding that its still-unsolved problems will submit to nonteleological theories...
>
> No matter how physics eventually deals with these problems... we can be sure... [that] physics will not give up... the ban on purpose or design... Physics' long track record of successes is the strongest argument for the exclusion of purpose or design from its account of reality. (2011, pp. 40-41)

Rosenberg then goes on in the rest of the book to argue that various phenomena that are evidently *not* susceptible of explanation in non-teleological or otherwise mechanistic terms – such as the intentionality of thought and the semantic content of language – ought to be eliminated from our ontology. If some phenomenon does not fit in with (Rosenberg's interpretation of) what physics tells us about reality, then in his view what that shows is not that there is after all more to reality than what (Rosenberg's interpretation of) physics tells us, but rather that the phenomenon in question simply must not be real.

But the problems with this argument are many and glaring. First, Rosenberg's characterization of teleological explanation doesn't even rise to the level of caricature, certainly not if it is the Aristotelian understanding of teleology that is in view. Proposing that "God did it" simply has nothing at all to do with arguments for intrinsic teleology of the sort summarized above. Second, attempts to eliminate intentionality, semantic content, and other intractable phenomena that Rosenberg would do away with are notoriously problematic and indeed incoherent. (We will have reason to revisit this particular issue in a later chapter.)

More relevant to the present point, Rosenberg's argument too is manifestly question-begging, in two respects. First, he speaks of physics "ruling out," "excluding," "banning," and "refusing" teleology. Naturally, if you simply *stipulate* that no appeal to teleology *will be allowed to count* as a genuinely scientific explanation, then it is no surprise if science so defined does not and never will find evidence of teleology. Second, Rosenberg insists that any phenomenon not susceptible of non-teleological explanation must be illusory. Naturally, if you simply refuse from the get-go to admit that there are any counterexamples to a claim, you will never find them. In both cases, Rosenberg is essentially committing a "No True Scotsman" fallacy, "proving" his thesis by simply *stipulating away* any possible evidence against it.

A further problem is that from the premise that physics has in the past successfully explained various phenomena in non-teleological terms, it simply does not follow that it will be able to explain *all* phenomena in such terms. This is like arguing that, since you have succeeded in getting rid of all the dirt in the different rooms of your house by sweeping it under a certain rug in the hallway, it follows that you're likely to succeed in getting rid of the dirt under the rug itself via the same method. In fact, of course, that method is guaranteed *not* to work in the case of the dirt under the rug. The analogy is apt because the way science has succeeded in getting rid of the apparent teleology or "directedness" of various natural phenomena is by treating it as a mere illusion or projection of the human mind. But the having of an illusion or act of projection themselves each involve a kind of "directedness," namely the directedness of the intentionality of thought and the directedness of human action toward an end. The mind is like the rug under which all other "directedness" is swept but which in the nature of the case cannot itself be emptied of it. (Again, more on this later.)

Finally, even if we were to accept Rosenberg's stipulation that physics and other sciences cannot make use of irreducibly teleological notions, this would not change anything of substance. It would only raise the question of whether there must in that case be something more to physical reality than what the methods of science, *so understood*, can capture. And if it turns out that there are good arguments for the reality of irreducible teleology (or substantial form, or the theory of actuality and potentiality, or whatever), then it will follow that there is indeed more to physical reality than what the methods of science, so understood, can capture. Whether one wants to call the arguments in question scientific arguments or arguments of the philosophy of nature is ultimately neither here nor there, and simply asserting that science makes no use of teleological notions settles nothing.

As I have said, I have elsewhere argued at book length and on general metaphysical grounds for the indispensability of the main Aristotelian theses – the theory of actuality and potentiality, hylemorphism, intrinsic teleology, and so forth (Feser 2014b). I have not attempted to recapitulate all those arguments here, but have merely summarized some of the key points with a view to setting out the main differences between the Aristotelian philosophy of nature and the mechanical philosophy. The aim in the chapters to follow is to show how both the practice and the results of natural science are not only in no way incompatible with these Aristotelian theses, but in many cases presuppose their truth. The mechanical world picture in the name of which modern science arose in fact cannot account either for the success of science's methods or the truth of its findings.

2. The scientist and scientific method

2.1 The arch of knowledge and its "empiriometric" core

The chapters to follow will argue that the *results* of modern science not only in no way conflict with the central claims of Aristotelian philosophy of nature, but in some respects even vindicate those claims. The present chapter argues that the very *methods* of modern science vindicate those claims -- and in an even more decisive way. For while the results of science might change (as currently accepted theories are abandoned and replaced by new ones), at least the core elements of scientific method will not.

Naturally, this raises the question of exactly what the scientific method *is* – something which has, of course, been a matter of great controversy in modern philosophy of science. I am not suggesting that that controversy is susceptible of easy resolution, nor will I try to resolve it here. My point is that although various details of scientific method are matters of dispute, there are some basic assumptions that all sides to the debate tend to agree on, and it is these which presuppose an essentially Aristotelian conception of nature (whether most philosophers of science realize this or not).

To see what these core assumptions are it will be useful to consider the history of what philosopher of science David Oldroyd (1989) has called the notion of "the arch of knowledge," and in particular the ways that notion was developed by early modern thinkers like Bacon, Galileo, Descartes, Newton, Boyle and others, and modified by more recent philosophers of science. As Oldroyd notes, the basic idea of the "arch" in fact goes back at least to Plato. But it is the construal of the "arch" associated with the fathers of the scientific revolution that has in modern times come to define what constitutes "science."

Bacon famously put heavy emphasis on observation as the evidential foundation of science. That was not by itself in any way novel. The thesis that *nothing is in the intellect that was not first in the senses* had been a commonplace of medieval Aristotelianism. Where Bacon took himself to be departing from his Aristotelian predecessors was in his *application* of this principle. For Bacon, the Aristotelians were too uncritical in their appeal to empirical evidence, in two respects. First, they were in his view too quick to draw general conclusions from that evidence. What was needed was patience, and in particular the slow and painstaking assembly of as many observations as possible of the phenomenon under investigation, under as wide a variety of circumstances as possible. Only after this was done could one be confident of the general conclusions one might draw about the nature of that phenomenon.

Second, in Bacon's view the Aristotelians had an insufficient appreciation of the biases that can infect individual observations. These biases were enshrined in what Bacon's *Novum Organum* characterizes as the "Idols of the Mind," of which there are four. The first are the *Idols of the Tribe*, by which Bacon means the biases inherent in human nature, such as our tendency to take it for granted that things really are as they appear to the senses. The second are the *Idols of the Cave*, or the biases that derive from a person's individual temperament, education, experiences, social setting, and so forth. The third are the *Idols of the Marketplace*, our tendency uncritically to suppose that the way language carves the world up corresponds to the way things really are objectively. The fourth are the *Idols of the Theatre*, our tendency to suppose that reality must conform to some philosophical or scientific theory to which we are especially attached. This can lead us to think that we are reading the truth of the theory *off from* what we observe, when in fact we are reading it *into* what we observe.

Once we have made a sufficient number of careful observations and corrected as far as we can for biases of these sorts, then in Bacon's view we can begin to reason inductively to general conclusions. The canons of inductive reasoning emphasized by Bacon were the sort refined and expanded upon by John Stuart Mill in the nineteenth century, in the famous "Mill's Methods" of establishing causal relationships between phenomena. In Bacon's view this kind of reasoning had been insufficiently appreciated by the Aristotelians, who, as he saw it, were impatient to reason *deductively*, *from* the general conclusions to which they had too hastily arrived *to* conclusions about what the empirical world must be like. Of course, Bacon was not opposed to such deductive reasoning, but his emphasis was on what he took to be the long neglected inductive aspect of science. With this we have one of classic expressions of the idea of the "arch of knowledge." At the foot of the left side of the arch is the body of empirical evidence painstakingly assembled. The left leg of the arch is constituted by the inductive reasoning that takes us from this evidence up to general conclusions, which form the apex of the arch. The right leg of the arch is constituted by deductive reasoning from these general conclusions, down to the specific empirical predictions that lie at the foot of the right leg.

Bacon represents, in effect, the empiricist side of the scientific revolution (to construe "empiricism" very broadly – a Baconian need not be committed to the desiccated conception of experience associated with the modern British empiricists). The mathematization of nature championed by Galileo, Descartes, and Newton represents the rationalist side (to construe "rationalism" very broadly too – naturally I am not attributing to Galileo or Newton all the epistemological commitments of Descartes and other continental rationalists). Galileo's *The Assayer* famously declared that mathematics is the language in which the book of nature is written, and attributed to matter only quantifiable primary qualities, relegating irreducibly qualitative secondary qualities to the mind. Galileo modeled the physical world in terms of mathematical abstractions like frictionless planes. Descartes reduced matter to the geometrical attribute of extension. Newton brought to fruition Galileo's and Descartes' project of tracing the behavior of all bodies down to the operation of a set of fundamental mathematically formulated laws.

Descartes qua rationalist tended to emphasize deductive reasoning *down from* an abstract mathematical model of nature, in contrast to Bacon's emphasis on inductive reasoning *up from* painstaking empirical observation. However, the greatest impact on modern scientific method of the mathematization of nature had to do neither with the legs nor the feet of the "arch of knowledge," but rather with the character of the general description of the world that would form its apex. The idea was that, at least ideally (if not always in practice, especially in special sciences far removed from physics), the general theories toward which scientific inquiry worked ought to be formulated in terms of quantifiable properties and mathematically expressible laws.

A further component of the early modern conception of the apex of the arch was the mechanical philosophy's program of analyzing observable bodies and their behavior in terms of unobservable particles in motion. There was, initially, some disagreement on how this program ought to be fleshed out. Descartes advocated a *plenum theory* on which matter is infinitely divisible – so that there is no *fundamental* level of particles – and on which there is no void or empty space between particles.

Gassendi and Hobbes advocated the ancient *atomist* view that there *is* a fundamental level of particles which are indivisible in principle, and empty space through which the particles pass. What won out, eventually, was the *corpuscularian* position of Boyle and Locke, which also affirmed a level of fundamental particles passing through void space, but held that they are merely undivided in fact rather than indivisible in principle.

What all of these variations on the mechanical philosophy agreed on, however, was a commitment to an essentially quantitative and mathematical conception of the particles and their properties and behavior. Secondary qualities like color, odor, sound, taste, heat, and cold were relegated to the mind. Final causes, the actualization of potentialities, substantial forms and the like were denied or at least ignored. For purposes of scientific description, quantifiable primary qualities alone were affirmed of the particles and of the objects they composed, and their changes were analyzed in terms of mathematically describable movements through space.

As physics progressed, however, commitment even to the corpuscles of Boyle and Locke disappeared. In place of discrete particles changing their positions in space over time, relativity theory speaks of four-dimensional space-time worms extending through a static block universe, and quantum mechanics speaks of wave functions. (More on these notions in later chapters.) What has remained, however, is commitment to an essentially quantitative and mathematical mode of describing nature.

In the *Opticks*, Newton gave classical expression to the settled early modern conception of the "arch of knowledge" with his *method of analysis and composition*. Analysis constitutes the left leg of the arch. Through observation and experiment, we work from compounds to the ingredients that make them up, from motions to the forces responsible for them, and from effects to their causes. The reasoning is inductive, both in the sense that we are working from particular cases to general principles and in the sense that it is probabilistic rather than demonstrative. The general principles should be modified only when further observation and experiment require it. Composition constitutes the right leg of the arch, and involves working from the general principles back down to the particular phenomena, showing how the former provide *explanations* of the latter.

Newton's *Principia* added further detail to this methodological story with his four "rules of reasoning" in natural philosophy or physics. The first rule is essentially a formulation of Ockham's razor, to the effect that we ought to admit only those general principles required for the explanation of phenomena. The second, to the effect that we ought to assign to the same sorts of effects the same sorts of causes, affirms the principle of the *uniformity of nature*. The third also affirms this uniformity of nature and states that what is true of bodies within the reach of our investigations is true of all bodies everywhere (contra the ancient and medieval supposition that the principles governing sublunary phenomena differ from those governing superlunary phenomena). The fourth states that we should regard as true those general principles arrived at by analysis of particular phenomena, until such time as further observation and experiment warrant revising them.

Now, every aspect of the "arch of knowledge" as hammered out by the early moderns has been the subject of debate in modern philosophy of science. There is, for instance, the question of whether we ought to favor a *realist* or an *instrumentalist* interpretation of the mathematical models of nature constituting the apex of the arch. Indeed, this question was central to the early modern controversy over Copernican astronomy, and it remains a live issue today. There is the question of *reductionism*. Are the various theories at the apex of the arch all ultimately reducible to physics? Or are the various special sciences autonomous domains of inquiry revealing aspects of the world just as real as those the physicist uncovers, even if irreducible to the latter? (We will have reason to address all of these particular topics later on.)

The empirical claims at the feet of the arch also raise a number of philosophical questions. Following the logical positivists (Carnap 1947, pp. 207-8; Neurath 1983, pp. 54-55), we can distinguish between "physicalistic" and "phenomenalistic" interpretations of these claims. On a physicalistic interpretation of an empirical proposition, the proposition is about public or intersubjectively accessible objects – tables, chairs, rocks, plants, animals, etc. On a phenomenalistic interpretation, an empirical proposition is about private or subjective entities accessible only to an individual, namely his own sense data. Which interpretation ought the scientist to favor? Positivists like Ernst Mach (1984) favored a phenomenalistic interpretation. Carnap in one place favors a physicalistic interpretation on the grounds that its intersubjectivity made it more suitable given that science is a cooperative activity (1959, p. 166), but in another place suggests that the choice between physicalistic and phenomenalistic language is pragmatic (1947, pp. 207-8).

There is also the issue of whether observations really can be as neutral between competing scientific and philosophical assumptions as Bacon supposes. Wilfrid Sellars (1956) famously attacked what he called the "myth of the given," and philosophers of science like N. R. Hanson (1958), Thomas Kuhn (1962), and others have argued that observation is essentially "theory-laden." According to such arguments, the background theoretical assumptions we bring to bear in an observational situation inevitably determine which aspects of the situation we judge to be relevant, how we interpret what we observe, why we take the observational instruments we use to be reliable, and so forth. There are always alternative assumptions that could be made, and what we take ourselves to have observed would differ with these different background assumptions.

Nelson Goodman's "new riddle of induction" (1983) raises problems both about theory-ladenness and about the inductive ascent from the observation of particular cases to general conclusions about unobserved cases. To take Goodman's famous example, suppose we say that something is "grue" if it is observed prior to 2020 and is green, or is observed after 2020 and is blue. Then all of the observations of emeralds made so far will support the claim that *all emeralds are grue* no less than they support the claim that *all emeralds are green*. Of course, we don't really consider "grue" to be what Goodman calls a "projectable predicate," while we do consider "green" projectable. But we cannot distinguish between projectable and non-projectable predicates, and thus decide how to characterize observed phenomena, apart from some background theory concerning the nature of the phenomena.

The inductive ascent to the general theories that form the apex of the arch was of course also famously challenged by Hume, who argued that induction could not be given a non-circular justification. Karl Popper's (1992) response was to concede that induction cannot be rationally justified but then argue that science doesn't depend on it anyway. What matters is not inductive reasoning *to* a theory, but deductive reasoning *from* it. In particular, science, for Popper, is about deducing observational predictions from a theory and then trying to falsify them. Among the problems raised against Popper's positon was the objection from the Duhem-Quine thesis, according to which a theory can always be saved from falsification if we make adjustments to the auxiliary hypotheses in conjunction with which we test the theory. Kuhn, Paul Feyerabend (1993), and others also argue that various extra-scientific and even non-rational factors inevitably determine how scientists decide between theories.

The explanatory, right leg of the arch of knowledge also raises difficulties. The classic modern account is Carl Hempel's (1962) "covering law" model of explanation, on which the explanation of a particular phenomenon involves showing how it falls under a general law. This mode of explanation can take either a "deductive-nomological" form (in which the occurrence of the phenomenon to be explained follows deductively from general laws taken together with particular circumstances), or a "probabilistic-statistical" form (in which the phenomenon follows only in a probabilistic way). One problem with this analysis is that it does not account for the *directionality* of explanation. To take a stock example, you can deduce the length of the shadow of a flagpole from the flagpole's length together with the laws of optics, etc. and you can equally well deduce the flagpole's length from the length of the shadow together with the laws of optics, etc. But while the former deduction counts as a good explanation, the latter does not. Another problem is that good scientific explanations can in some cases (and contra Hempel) involve neither a deductive nor even a probabilistic connection between the *explanans* and the *explanandum*. To take another stock example, that someone had syphilis can be a good explanation of why he later died from paresis, even if syphilis only rarely leads to paresis.

Such are some of the main issues that have arisen in modern philosophy of science. I will not attempt to deal with all of them here (especially those that are more epistemological than metaphysical in nature), though we will have reason to revisit some of them later on. The point to emphasize for the moment is that there are core elements to the "arch of knowledge" that all sides to all the various debates just summarized more or less agree on. The first is that science has to have *some* basis in observation and experiment, however the difficulties surrounding these are resolved. The second is that mathematical models and mathematically formulated laws of nature are essential to scientific theory, especially physics, whether we take a realist or instrumentalist interpretation of these models and laws and whether or not we accept reductionism. The third is that canons of formal reasoning play *some* role in science, whether they are purely deductive or a mixture of inductive and deductive, and whether there are other factors (social, aesthetic, or whatever) that influence theory choice.

Some notions introduced by Jacques Maritain and often deployed by other twentieth-century Aristotelian-Thomistic philosophers are useful in elucidating the point. (Cf. Maritain 1951, Chapter III and 1995, Chapter IV. See also Smith 1950, Chapter 5; Wallace 1996, pp. 224-27; and Rizzi 2004, pp. 152f. It is worth emphasizing that I do not mean to endorse everything Maritain and others had to say when deploying the notions in question.) As we saw in chapter 1, for the Aristotelian, the full story about any natural substance will make reference to its essence and the properties that flow from its essence; to its substantial form and prime matter as constitutive of this essence; to the teleological properties that follow from its having the substantial form that it does; and so forth. Now, our knowledge of these aspects of a thing is *ultimately* grounded in experience. But at least some of them are not themselves *directly* knowable via experience, not even in principle. For example, you cannot see, hear, taste, touch, or smell the prime matter or substantial form of a thing. You can see, hear, taste, etc. the thing *itself*, but its substantial form and prime matter do not exist in abstraction from it, as separable entities which might be directly empirically detected. Unlike a molecule or an atom, you could not perceive them no matter how small you were, no matter what sorts of special scientific equipment you might deploy, etc. These aspects of a thing are knowable only by way of philosophical analysis of what must be true of it in order for it to exhibit the changeability, multiplicity, etc. that empirical objects do. That is why concepts like substantial form, prime matter, essence, etc. fall within the domains of philosophy of nature and metaphysics.

Now, modern science from Bacon, Galileo, Descartes, et al. onward has deliberately refrained from deploying these or any other philosophical concepts. Some scientists have dismissed them altogether as unnecessary and even an obstacle to understanding. Others have not gone that far but have merely regarded them as inessential to the specific task to which empirical science has set itself. Either way, modern science has thereby confined itself to what Maritain calls a purely "empiriological" as opposed to "ontological" mode of investigation of the natural world. An "ontological" approach would be one which seeks a complete description of the nature of a concrete empirical substance, including those aspects of its nature which go beyond what is directly empirically detectable – which, for the Aristotelian, would include the essence underlying its properties, its substantial form and prime matter, etc. By contrast, an "empiriological" approach is one that confines itself to those aspects of a thing that are at least in principle directly empirically detectable, and organizes what it discovers about these aspects in terms of the abstractions of some formal theory. For example, Newtonian physics describes the observable motions of bodies by way of concepts like *force*, *mass*, *acceleration*, etc. defined in terms of mathematical equations. The theory does not attempt to explain what force or mass *is* in some deep metaphysical sense. As long as the empirical predictions of the theory are confirmed in observation and experiment, the physicist has essentially done his job.

Within the category of empiriological forms of inquiry, Maritain draws a further distinction between "empiriometric" and "empirioschematic" sciences. An empiriometric science is an empiriological one that organizes the directly empirically detectable phenomena it investigates in *mathematical* terms, specifically. Modern physics is the empiriometric science *par excellence*. Now, one could take the instrumentalist position that the mathematically characterized entities and properties to which physical theory makes reference (mass, force, space-time, wave function, etc.) are mere abstractions that are useful for making predictions, but correspond to nothing in concrete reality. One could take the realist position that there really are concrete entities in the world that correspond to these concepts, and that physics essentially reveals to us the entire nature of these entities. Or one could take the middle ground structural realist position that there really are concrete entities corresponding to these concepts, but that physics reveals only their abstract mathematical properties and does not tell us their entire nature. (Again, more on this issue later.) But whatever one thinks about the metaphysical status of the mathematical description of nature afforded by modern physics, that physics organizes empirically detectable phenomena by way of such a description makes it empiriometric in Maritain's sense.

Other modern sciences, however, have not been able to achieve such a purely mathematical description of the phenomena they investigate. They may try to approximate it in certain respects, but certain key concepts that are not susceptible of a purely mathematical analysis remain essential to the practice of the science. To this extent these sciences point back to the "ontological" or philosophical mode of investigation that modern science eschews, but like empiriometric science they refuse to countenance what is not at least in principle directly empirically detectable. Hence if they make use of philosophical concepts it will only be in what Kant called a "regulative" way – useful for directing inquiry, but not reflective of any deep metaphysical reality. Maritain labels these the "empirioschematic" kinds of empiriological science. Biology would be an example, and the use biology makes of the notion of teleology would be an example of how a philosophical concept might function in a "regulative" way in empirioschematic investigation.

Now, for the scientific reductionist, what Maritain calls the "empirioschematic" sciences are ultimately reducible to the "empiriometric" science of physics. Contemporary anti-reductionist philosophers of science would deny that this is possible, and some would deny that it is even desirable. The tendency within modern science and modern philosophy of science has, however, been to regard physics as the gold standard of science, and an approximation of its methods as that to which all genuine science at least ought to aspire. The empiriometric mode of investigation therefore constitutes a commonly accepted core of the "arch of knowledge." However, all the various controversies about scientific method get resolved, all sides are essentially in agreement that science ought to be grounded in what is directly empirically detectable, that the mathematical descriptions of physics are paradigms of scientific theory, and that the formal reasoning required to spell out and test such theories are a key part of scientific practice.

Now, it goes without saying that an Aristotelian would deny that what can be captured by way of the empiriometric (or, more generally, empiriological) modes of investigation is all there is to the natural world. But put aside for the moment the question of whether these methods capture all there is to some specific phenomenon studied by physics, chemistry, biology, etc. The point I want to emphasize for present purposes is that *the very practice of the empiriometric method itself presupposes the truth of all the fundamental claims of Aristotelian philosophy of nature* – about actuality and potentiality, substantial form and prime matter, efficient and final causality, and so forth. The empiriometric method could not possibly be deployed unless the natural world is more or less just as the Aristotelian says it is, and thus unless there is more to the natural world than can be captured by that method. The burden of the rest of this chapter will be to defend this thesis.

2.2. The intelligibility of nature

Even a scientist or philosopher of science who reduces the natural world to what can be captured by way of empiriometric methods takes that world to be *intelligible*. No scientist or philosopher of science would think it scientifically respectable to treat some observation as an unintelligible brute fact, beyond the range of scientific investigation and explanation. Of course, he may regard a report describing the observation as mistaken in some way, or he may judge that we simply lack the further evidence we would need in order to find out what the correct explanation of the observation is. But he would not treat any observation correctly reported as *intrinsically* without rhyme or reason, or as something that a scientist may ignore. At least in principle, a theory of some domain ought to account for every observation relevant to that domain. This is true even if a theory is interpreted in positivist or instrumentalist rather than realist terms. Even a scientist or philosopher of science who regards a theory merely as a useful fiction for making predictions supposes that every observation ought to be covered by the theory, and thus that every observation concerns something that is in principle predictable and thus to that extent intelligible.

Science is in this way committed to a version of the *principle of sufficient reason* (PSR). Indeed, a characteristic Aristotelian-Thomistic formulation of PSR is: "Everything is intelligible" (Garrigou-Lagrange 1939, p. 181). Another is: "There is a sufficient reason or adequate necessary objective explanation for the being of whatever is and for all attributes of any being" (Wuellner 1956, p. 15). The scientist is certainly committed to the truth of this proposition where observed phenomena are concerned. That such a phenomenon exists and has the attributes it has are taken to have explanations that it is the business of science to uncover.

Note that I said that science is committed to a *version* of PSR. I am not attributing to science a commitment to everything that has been defended in the name of PSR. Nor would Aristotelian-Thomistic philosophers endorse everything that has been defended in the name of PSR. For example, some philosophers take *propositions* to be among the entities requiring an explanation given PSR, and some have supposed that PSR requires that an explanans must *logically entail* the explanandum. Aristotelian-Thomistic advocates of PSR would reject these assumptions. It is only concrete entities rather than abstract objects (such as propositions) that are covered by PSR, and an explanans need not logically entail the explanandum, but need only make it intelligible in some way (e.g. probabilistically). Some objections to PSR presuppose interpretations of the principle like the ones I am rejecting, and thus have no force against the version of PSR I am attributing to science. (See Feser 2014b, pp. 137-42 for discussion of the differences between the Thomistic understanding of PSR and that of rationalists and other non-Thomists.)

Why should we believe PSR? An empirical argument for the principle would be that, considered as an inductive generalization, PSR is as well-supported as any other. For one thing, we do in fact tend to find explanations when we look for them, and even when we don't we tend to have reason to think there is an explanation but just one to which, for whatever reason (e.g. missing evidence), we don't have access. For another thing, the world simply doesn't behave the way we would expect it to if PSR were false (Pruss 2009, p. 32). Events without any evident explanation would surely be occurring constantly and the world would simply not have the intelligibility that makes science and everyday common sense as successful as they are. That the world is as orderly and intelligible as it is would be a miracle if PSR were not true.

But PSR is more certain than a mere empirical hypothesis can be. If the principle seems difficult to prove, that is not because it is doubtful, but on the contrary because it is more obviously true than anything that could be said either for or against it. As Reginald Garrigou-Lagrange writes, "though it cannot be directly demonstrated, it can be indirectly demonstrated by the indirect method of proof known as *reductio ad absurdum*" (1939, p. 181). One way in which this might go is suggested by some remarks made by Alexander Pruss, who was in turn developing a point made by Robert Koons (Pruss 2009, p. 28; Koons 200, p. 110). Denying PSR, Pruss notes, entails radical skepticism about perception. For if PSR is false, then there might be no reason whatsoever for our having the perceptual experiences we have. In particular, there might be no connection at all between our perceptual experiences and the external objects and events we suppose cause them. Nor would we have any grounds for claiming even that such a radical disconnect between our perceptions and external reality is improbable. For objective probabilities depend on the objective tendencies of things, and if PSR is false then events might occur in a way that has nothing to do with any objective tendencies of things. Hence one cannot consistently deny PSR and be justified in trusting the evidence of sensory perception, nor the empirical science grounded in perception.

But the Pruss/Koons line of argument can be pushed further than they push it. Consider that whenever we accept a claim that we take to be rationally justified – as scientists do when they judge a theory to be well-supported by the available evidence, consider a hypothesis to have been falsified experimentally, and so forth – we suppose not only that we have a reason for accepting it (in the sense of a rational justification) but also that this reason is the reason *why* we accept it (in the sense of being the cause or explanation of our accepting it). We suppose that it is *because* the rational considerations in favor of the claim are good ones that we are moved to assent to the claim. We also suppose that our cognitive faculties track truth and standards of rational argumentation, rather than leading us to embrace conclusions in a way that has no connection to truth or logic. But if PSR is false, we could have no reason for thinking that any of this is really the case. For all we know, what moves or causes us to assent to a claim might have absolutely nothing to do with the deliverances of our cognitive faculties, and our cognitive faculties themselves might in turn have the deliverances they do in a way that has nothing to do with truth or standards of logic. We might believe what we do for no reason whatsoever, and yet it might also falsely *seem*, once again for no reason whatsoever, that we do believe what we do on good rational grounds. Now, this would apply to any grounds we might have for doubting PSR as much as it does to any other conclusion we might draw. Hence to doubt or deny PSR undercuts any grounds we could *have* for doubting or denying PSR. The rejection of PSR is therefore self-undermining. Indeed, to reject PSR is to undermine the possibility of *any* rational inquiry.

There is another way in which science implicitly presupposes PSR. Some philosophers have taken the view that there can be genuine explanations, including scientific explanations, even if PSR is false. One finds such a view in J. L. Mackie and Bertrand Russell (Mackie 1982, pp. 84-87; Russell and Copleston 1964, pp. 168-78). The idea is that we can explain at least some phenomena in terms of laws of nature, those laws in terms of more fundamental laws, and perhaps these in turn in terms of some most fundamental level of laws. The most fundamental laws would, however, lack any explanation. That the world is governed by them would just be an unintelligible "brute fact."

But this is incoherent. Suppose I told you that the fact that a certain book has not fallen to the ground is explained by the fact that it is resting on a certain shelf, but that the fact that the shelf itself has not fallen to the ground has no explanation at all but is an unintelligible brute fact. Have I really explained the position of the book? It is hard to see how. For the shelf has in itself no tendency to stay aloft – it is, by hypothesis, just a brute fact that it does so. But if it has no such tendency, it cannot impart such a tendency to the book. The "explanation" the shelf provides in such a case would be completely illusory. (Nor would it help to impute to the book some such tendency, if the having of the tendency is *itself* just an unintelligible brute fact. The illusion will just have been relocated, not eliminated.)

By the same token, it is no good to say: "The operation of law of nature C is explained by the operation of law of nature B, and the operation of B by the operation of law of nature A, but the operation of A has no explanation whatsoever and is just an unintelligible brute fact." The appearance of having "explained" C and B is completely illusory if A is a brute fact, because if there is neither anything about A itself that can explain A's own operation nor anything beyond A that can explain it, then A has nothing to impart to B or C that could possibly explain their operation. The notion of an explanatory nomological regress terminating in a brute fact is, when carefully examined, no more coherent than the notion of an effect being produced by an instrument that is not the instrument of anything.

So, rational inquiry in general, and scientific inquiry in particular, presuppose PSR. A further argument which supports this judgment has been put forward by philosopher Michael Della Rocca (2010). Della Rocca notes that even among philosophers who reject PSR, philosophical theses are often defended by recourse to what he calls "explicability arguments." An explicability argument (I'll use the abbreviation EA from here on out) is an argument to the effect that we have grounds for denying that a certain state of affairs obtains if it would be inexplicable or a "brute fact." Della Rocca offers a number of examples of this strategy. When materialist philosophers of mind defend some reductionist account of consciousness on the grounds that consciousness would (they say) otherwise be inexplicable, they are deploying an EA. When early modern philosophers rejected the Aristotelian notion of substantial form (or what Aristotelians would regard as a *caricature* of that notion, anyway), they did so on the grounds that the notion was insufficiently explanatory. When philosophers employ inductive reasoning they are essentially rejecting the claim that the future will not be relevantly like the past nor the unobserved like the observed, on the grounds that this would make future and otherwise unobserved phenomena inexplicable. And so forth.

Now, Della Rocca allows that to appeal to an EA does not by itself commit one to PSR. But suppose we apply the EA approach to the question of *why things exist*. Whatever we end up thinking the correct answer to this question is – it doesn't matter for purposes of Della Rocca's argument – if we deploy an EA in defense of it we *will* implicitly be committing ourselves to PSR, he says, because PSR just is the claim that the existence of anything must have an explanation.

In responding to these different examples of EAs, one could, says Della Rocca, take one of three options:

(1) Hold that some EAs are legitimate kinds of argument, while others – in particular, any EA for some claim about why things exist at all – are not legitimate.

(2) Hold that no EA for any conclusion is legitimate.

(3) Hold that all EAs, including any EA for a claim about the sheer existence of things, are legitimate kinds of argument.

Now, the critic of PSR cannot take option (3), because that would, in effect, be to accept PSR. Nor could any critic of PSR who applies EAs in defense of other claims – and the EA approach is, as Della Rocca notes, a standard move in contemporary philosophy (and indeed, in science) – take option (2).

So that leaves option (1). The trouble, though, is that there doesn't seem to be any non-question-begging way for the critic of PSR to defend option (1). For why should we believe that EAs are legitimate in other cases, but not when giving some account of the sheer existence of things? It seems arbitrary to allow the one sort of EA but not the other sort. The critic of PSR cannot respond by saying that it is just a brute fact that some kinds of EAs are legitimate and others are not, because this would beg the question against PSR, which denies that there are any brute facts. Nor would it do for the critic to say that it is just *intuitively* plausible to hold that EAs are illegitimate in the case of explaining the sheer existence of things, since Della Rocca's point is that the critic's acceptance of EAs in other domains casts doubt on the reliability of this particular intuition. Hence to appeal to intuition would also be to beg the question.

So, Della Rocca concludes that there seems no cogent way to accept EAs at all without accepting PSR. The implication is that we can have no good reason to think *anything* is explicable unless we also admit that *everything* is.

Della Rocca's argument can, in my view, be pushed even further than he pushes it. Della Rocca allows that while it would be "extremely problematic" for someone to bite the bullet and take option (2), it may not be strictly "logically incoherent" to do so. However, I think this is too generous to the critic of PSR. Even if the critic decides to reject the various specific examples of EAs cited by Della Rocca – EAs concerning various claims about consciousness, substantial forms, etc. – the critic will still make use of various patterns of reasoning he considers formally valid or inductively strong, will reject patterns of reasoning he considers fallacious, etc. And he will do so precisely because these principles of logic embody standards of intelligibility or explanatory adequacy.

To be sure, it is a commonplace in logic that not all explanations are arguments, and it is also sometimes claimed (less plausibly, I think) that not all arguments are explanations. But certainly *many* arguments are explanations. What Aristotelian philosophers call "explanatory demonstrations" (e.g. a syllogism like *All rational animals are capable of language, all men are rational animals*, so *all men are capable of language*) are explanations. Arguments to the best explanation are (obviously) explanations, and as Della Rocca notes, inductive reasoning in general seems to presuppose that things have explanations.

So, to give up EAs of *any* sort (option (2)) would seem to be to give up the very practice of argumentation itself, or at least much of it. Needless to say, it is hard to see how doing *that* could fail to be logically incoherent, at least if one tries to defend one's rejection of PSR with arguments. Hence, to accept the general practice of giving arguments while nevertheless rejecting EAs of the specific sorts Della Rocca gives as examples would really be to take Della Rocca's option (1) rather than option (2). And as we have seen, there is no non-question-begging reason to accept (1).

Now, some of the defenses of PSR I have set out here are *retorsion* (sometimes spelled "retortion") arguments, viz. arguments that attempt to refute a claim by showing that anyone making it is led thereby into a performative self-contradiction. Some of the arguments against static monism and dynamic monism set out in chapter 1 were also retorsion arguments. But some have questioned the probative force of such arguments. For example, it is sometimes suggested that retorsion arguments essentially commit an *ad hominem* fallacy of the *tu quoque* sort. If Bob tells Fred that he should not drink to excess and in response Fred points out that Bob is a drunkard himself, Fred has not thereby refuted Bob's claim. That a person is a hypocrite does not entail that what he is saying is false. But don't retorsion arguments commit this same *tu quoque* fallacy of rejecting a claim merely because those who make it are hypocrites?

No, they do not. As every logic teacher knows, one of the problems one encounters in teaching about the logical fallacies is that students often settle into too crude an understanding of what a fallacy involves, and thus tend to see fallacies where there are none. Not every use of language which has emotional connotations amounts to a fallacy of appeal to emotion. Not every attack on a person amounts to an *ad hominem* fallacy. Not every appeal to authority is a *fallacious* appeal to authority. A *reductio ad absurdum* argument should not be confused with a slippery slope fallacy. And so on.

In the same way, by no means does every reference to an opponent's inconsistency amount to a *tu quoque* fallacy. On the contrary, pointing out that a certain view leads to inconsistency is a standard technique of logical criticism. It is, for example, what a *reductio ad absurdum* objection involves, and no one can deny that *reductio* is a legitimate mode of argumentation. The problem with *tu quoque* arguments isn't an appeal to inconsistency *as such*. The problem is that the *specific kind* of inconsistency the arguer appeals to is not relevant to the *specific topic* at issue.

So, suppose Bob is indeed a drunkard but tells Fred that it is bad to be a drunkard, on the basis of the fact that being a drunkard is undignified, is damaging to one's health, prevents one from holding a job and providing for one's family, etc. Bob's hypocrisy is irrelevant to the truth of the claim he is making, because the proposition:

(1) Bob is a drunkard.

is perfectly compatible, logically speaking, with the proposition:

(2) It is bad to be a drunkard.

and perfectly compatible also with the proposition:

(3) Being a drunkard is undignified, is damaging to one's health, prevents one from holding a job and providing for one's family, etc.

Hence it is unreasonable to reject (2), or to reject the argument from (3) to (2), on the basis of (1). But that is what Fred does, which is why he is guilty of committing the *tu quoque* fallacy.

A retorsion argument is not like that at all. Consider the objection against static monism, raised in chapter 1, to the effect that the static monist cannot coherently deny that change occurs. The idea here is that the static monist is committed to the proposition:

(4) There is no such thing as change.

but at the same time, carries out an act – for example, the act of reasoning to that conclusion from such-and-such premises, where this very act itself involves change – which entails the proposition:

(5) There is such a thing as change.

Now (4) is *not* logically compatible with (5). What we have here is a performative self-contradiction in the sense that the very act of defending the position entails the falsity of the position. So, it is not mere *hypocrisy*, but rather *implicit logical inconsistency*, that is at issue.

Here's another way to think about it. Could being a drunkard still be a bad thing, even if Bob is in fact a drunkard himself? Of course. That's why it is a *tu quoque* fallacy to reject Bob's claim that being a drunkard is bad, merely because he is himself a drunkard. But could change really be an illusion, if Parmenides is in fact reasoning from the premises of his argument to the conclusion? No. That's why it is *not* a *tu quoque* fallacy to reject Parmenides' denial that change occurs on the basis of the fact that he has to undergo change himself in the very act of denying it.

Of course, Parmenides might respond: "Ah, but that assumes that I really *am* reasoning from premises to conclusion, and I would deny that I am doing so, precisely because that would be an instance of change! So, you are begging the question against me!" But there are two problems with this response. First and less seriously, even if the critic's retorsion argument against Parmenides did amount to begging the question, it still would not amount to a *tu quoque* fallacy. Second and more importantly, it does *not* in fact amount to begging the question. The critic can say to Parmenides: "Parmenides, *you* were the one who presented me with this argument against the reality of change. I merely pointed out that since the rehearsal of such an argument is itself an instance of change, *you are yourself* already *implicitly* committed to its reality, despite your explicit denial of it. I am pointing out a contradiction in *your own position*, not bringing in some question-begging premise from outside it. So, if you want to rebut my criticism, it is no good for you to accuse me of begging the question. Rather, you have to show how you can restate your position in a way that avoids the implicit contradiction." Of course, no such restatement is forthcoming, because the very act of trying to formulate it would involve Parmenides in exactly the sort of implicit contradiction he was trying to avoid. But that would be *his* problem, not his critic's problem.

Now, the same point applies to the retorsion arguments in defense of PSR that I have put forward. They neither commit a *tu quoque* fallacy nor beg the question, but simply point out an implicit self-contradiction in the position of anyone who would reject PSR while at the same time trusting the evidence of perceptual experience, or accepting the deliverances of his cognitive faculties, or appealing to laws of nature, or deploying explicability arguments. In this way (and as Garrigou-Lagrange pointed out) PSR can be defended by way of a kind of *reductio ad absurdum* strategy. A retorsion argument is essentially a *reductio* insofar as it refutes a claim by showing that the claim leads to a contradiction. What is distinctive of retorsion arguments is the specific way they derive the contradiction, namely by calling attention to an inconsistency in the position of anyone who makes the claim. But as I have argued, that does not make them any more fallacious than any other *reductio* argument.

Another objection to retorsion arguments comes from a surprising quarter. Retorsion arguments had a prominent place in the work of "transcendental Thomists" influenced by Kant, such as Joseph Maréchal, Karl Rahner, and Bernard Lonergan. Thomist philosopher John Knasas thinks this method is too beholden to that Kantian influence, and thus incompatible with the fundamental epistemological claims of Thomism (2003, Chapter 4). For Kant, categories like substance, causality, etc. reflect only the way the mind has to *think about* reality, but not how reality *is* in itself. Similarly, in Knasas's view, the transcendental Thomists' retorsion arguments can only ever tell us that we have to *think about* reality a certain way. They cannot tell us that reality really *is* that way. But in that case, they fail to establish the truth of the claims they are deployed in order to defend.

For present purposes, I put to one side questions about whether any of the specific retorsion arguments given by the transcendental Thomists is successful and whether those arguments really do depend on objectionably Kantian assumptions. I also put to one side any general treatment of issues in Thomist epistemology, which would require a book of its own. Suffice it to note that there is simply nothing in the retorsion strategy *as such* that presupposes a Kantian epistemology or any other non-Thomist epistemology, any more than the *reductio ad absurdum* mode of argumentation in general presupposes such an epistemology – or indeed, any more than *modus ponens, modus tollens,* or any other mode of inference presupposes a Kantian epistemology. Certainly none of the retorsion arguments I have defended presupposes such an epistemology. It would be absurd, and certainly contrary to a Thomist epistemology, to claim that *modus ponens* arguments (say) tell us only how we have to think about reality, and nothing about reality itself. This would utterly destroy the very possibility of the intellect's contact with reality, which is central to Thomist epistemology. It would be equally absurd to claim that *reductio* arguments in general tell us only how we have to think about reality and nothing about how reality really is. But by the same token, it is absurd to claim that the specific sorts of *reductio* arguments enshrined in the retorsion arguments I have defended tell us only about how we have to think about reality, and nothing about reality itself.

As Knasas acknowledges, in his commentary on Book IV of Aristotle's *Metaphysics*, Aquinas himself defends the principle of non-contradiction by pointing out the inconsistency of anyone who claims to deny it. And Aquinas, obviously, was not beholden to any Kantian or otherwise non-Thomistic epistemological assumptions. Yet Knasas denies that this is a retorsion argument, precisely on the grounds that Aquinas and the opponents he had in mind were not operating within the context of a Kantian epistemological problematic. But now the issue appears to be essentially semantic. Knasas seems to be using the term "retorsion" so that it just *means* an appeal to performative self-contradiction *within the context of a Kantian conception of knowledge*. But as the example of Aquinas's commentary shows, there is simply no necessary connection between pointing out a performative self-contradiction and endorsing Kantian epistemological assumptions. One can consistently take the former to have probative force while rejecting the latter. And that is what the arguments I have been defending do. Whether or not one wants to apply to them the "retorsion" label, they are essentially arguments of the kind Aquinas was deploying in the commentary.

In defending the retorsion style of argument, I have been answering objections raised against certain *arguments for* PSR. There are also objections against PSR *itself.* Some of these are directed against versions of PSR that Aristotelians and Thomists would not endorse in the first place, so naturally they have no force against the version of PSR I am defending here. (Again, see Feser 2014b, pp. 137-42 for discussion of the differences between Thomist and non-Thomist interpretations of PSR and of why objections that might have force against the latter have no force against the former.) An objection that might seem relevant to the interpretation I've been defending would be one that deploys against PSR the argument Hume deploys against the principle of causality. In particular, it might be suggested that it is at least conceivable that something might come into being without any explanation. But as we saw in chapter 1, Hume's argument is seriously problematic, and it is equally so whether deployed against the principle of causality or against PSR.

2.3 Subjects of experience

An angel, as Aquinas conceives of it, is a creature of pure intellect, a mind without a body. It has neither a brain nor sense organs. To know what it knows, it neither relies on sensory experience nor reasons from premises to conclusions. Its knowledge is built into it when it comes into being, and it knows what it knows all at once, grasping propositions and their implications in a single act rather than in successive mental acts.

Scientists, needless to say, are not like that. Like other human beings, they are rational animals and thus possess not only intellects, but also bodies and sense organs on which their intellects rely for information about the external material world. In particular, the observation and experiment that provide science with its evidential basis involve series of perceptual experiences. Scientists must also reason discursively, moving successively from one thought to another. For example, they reason inductively from the results of particular observations to general hypotheses, and deductively from general hypotheses to testable predictions concerning specific empirical consequences.

Recall from chapter 1 that on an Aristotelian analysis, change involves the gain or loss of some attribute, but also the persistence of that which gains or loses the attribute. For example, when a banana goes from being green to being yellow, the greenness is lost and the yellowness is gained, but the banana itself persists. If there were no such persistence, we would not have a *change* to the banana, but rather the annihilation of a green banana and the creation of a new, yellow one in its place. Now, the perceptual and cognitive activities involved in the practice of science entail the existence of change in precisely this sense.

Hence, consider even the simplest observational or experimental situation, such as watching for the movement of a needle on a dial. When the needle moves from its rest position it loses one attribute and gains another (namely a particular spatial location), and it is one and the same needle that loses and gains these attributes and one and the same dial of which the needle is a component. If there were no gain or loss of attributes, or if the needle or dial were not the same, the observation would be completely useless. For example, if what you are doing is testing a prediction about whether the needle which is at its rest position at t_1 will be at a different position at t_2, it would be completely irrelevant to such a test if the needle you observed at t_2 was a *different* needle from the one you observed at t_1, and there would be nothing to watch for if it were not possible for this same one needle to gain or lose an attribute.

Naturally this presupposes the realist assumption that the needle and dial are mind-independent objects, but the basic point would hold even on a phenomenalistic interpretation of science. Hence, suppose that all you are really observing when you read the dial are certain sense data rather than any mind-independent objects, and suppose we interpreted scientific theories as mere descriptions of the relationships between sense data. Observational and experimental situations like the one we are considering would somehow have to be interpreted in a way consistent with this. But however that would go, you would still have to suppose that the person who has the initial sense data at t_1 (namely you) is the *same* person who has the different sense data at t_2, and that this same one person is capable of gaining and losing attributes (namely the sense data in question). The observation would be completely irrelevant if the person who has the sense data at t_2 was a *different* person from the one who had them at t_1, or if that same one person were incapable of gaining of losing attributes like sense data.

Consider also even the simplest cognitive activity involved in science, such as reasoning from a premise to a conclusion via the inference rule *modus ponens*. When you reason from the premises *If p, then q* and *p* to the conclusion *q*, you lose one attribute (namely, the attribute of having the conscious thought that *If p then q, and p*) and gain another (namely, having the conscious thought that *q*). Moreover, it is one and the same person (you) who both loses the one attribute and gains another. If the person who had the second thought were not the same as the person who had the first one, there would not be any *reasoning* going on, any more than there would be if (say) Donald Trump had had the conscious thought that *If p, then q, and p*, and Hillary Clinton had, a moment later by sheer coincidence, the conscious thought that *q*. Nor would there be any reasoning going on if the same one person were incapable of losing one attribute (having the conscious thought that *If p, then q, and p*) and gaining another (having the conscious thought that *q*).

A skeptic might object that the changes apparently involved in perception and cognition could be merely *illusory*. But the trouble is that, for all the skeptic has shown, this skeptical scenario *itself* presupposes change. The skeptic initially thinks that he has perceptual experiences and cognitive processes that manifest change; then he entertains arguments to the effect that this may all be illusory; then he concludes that such changes don't really occur after all. But all of *that* evidently involved changes of various sorts – for example, the skeptic first having one belief and then giving it up and coming to have another. (Cf. Dummett 1960 and Zwart 1975)

Could the skeptic plausibly accuse such a response of merely begging the question against him? No, because the response is not a matter of simply dogmatically appealing to a premise that the skeptic denies (to the effect that change exists), and then pretending to refute him on that basis. Rather, it is a matter of pointing out that the skeptic *himself* in fact seems implicitly to *accept* the premise in question, even in the very act of denying it. Hence the only reply open to the skeptic is to show that he is *not* implicitly committed to the premise. That is to say, the skeptic needs to give some account of how it is possible for him to so much as entertain his skepticism given that change does not exist. In the absence of such an account it is the *skeptic*, and not his critic, who is being dogmatic. Yet no such account is forthcoming.

It is sometimes claimed that science, and in particular the physics of relativity, has shown that the change we think we see in nature is illusory. In reality (so the claim goes) the physical world is a static four-dimensional block universe, which we mistakenly perceive *as if* it were changing. Hence, according to philosopher of physics Michael Lockwood, space-time physics shows that "everything that ever exists, or ever happens, at any time or place... [is] just as real as the contents of the here and now," so that there is no "conferring [of] actuality on what are initially only potentialities" and instead "the world according to Minkowski is, at all times and places, actuality through and through" (2005, pp. 68-69).

Now, I will argue in a later chapter that in fact physics has shown no such thing. But for the moment it will suffice to point out that anyone who claims that science has shown change to be illusory faces a dilemma. If he acknowledges the existence of the cognitive and perceptual states of scientists themselves, then he is implicitly committed to there being at least *some* change after all, namely the change that exists within these thinking and experiencing conscious subjects. This not only merely relocates rather than eliminates change, but opens up a Cartesian divide between the conscious subject and the rest of reality, with all of its attendant problems (Mundle 1967). How does a changeable conscious subject arise within a changeless natural order? Could there be causal interaction between two realities so radically unlike? Are we left with epiphenomenalism?

If instead the denier of change takes an eliminativist line and *denies* the existence of the cognitive and perceptual states of scientists, then he will be throwing out the evidential basis of the scientific theory that led him to deny the reality of change in the first place. (Cf. Healey 2002) Into the bargain, he will face the further incoherence problems that notoriously afflict eliminativism about cognitive states, even apart from the problem of undermining the evidential basis of science. (More on that subject below.)

So, the very existence of scientists themselves, qua perceiving and thinking subjects, presupposes the reality of change. But the reality of change, the Aristotelian argues, in turn presupposes the distinction between actuality and potentiality. Hence the very existence of scientists themselves presupposes the distinction between actuality and potentiality. In particular, it presupposes that scientists qua subjects of experience possess potentialities to have various perceptual experiences and conscious thoughts, and that the change from one perceptual experience or thought to another in the course of conscious awareness involves the successive actualization of these potentials. Thus, the very existence of scientists qua subjects of experience presupposes the fundamental thesis of Aristotelian philosophy of nature. Aristotelianism begins at home, as it were.

The only way to avoid this conclusion would be to find an alternative to the theory of actuality and potentiality where the analysis of change is concerned. Now, it is sometimes claimed that change can be analyzed in terms of *temporal parts theory* or *four-dimensionalism* (Sider 2001). The basic idea is that just as a physical object has spatial parts at any particular moment of time, so too each stage of a physical object's existence through time ought to be regarded as a *temporal* part of the object. For example, just as Bob's arms and legs are distinct parts of him occupying different points in space, so too Bob as he is at exactly noon on Sunday and Bob as he is at exactly noon on Monday are distinct parts of him occupying different points in time. The view is "four-dimensionalist" insofar as it adds to the three spatial dimensions of a physical object a fourth temporal dimension of that object. A physical object is on this view essentially a collection of all of its temporal parts together with its spatial parts. (The view has obvious affinities with the Minkowskian interpretation of relativity, but they are independent of one another. One could defend temporal parts theory on philosophical grounds independently of the physics of relativity, and as Theodore Sider notes (2001, pp. 79-87), one could accept the Minkowskian interpretation of relativity and still reject temporal parts theory.)

Change, on this view, would be analyzable in terms of an object's temporal parts having different features. For example, that Bob once had a beard but lost it would be analyzable in terms of the fact that the temporal part of Bob that existed at noon on Sunday had a beard, and the fact that the temporal part of Bob that existed at noon on Monday lacked a beard. This could be regarded as an alternative to the Aristotelian analysis of change, since it makes no reference to the potentialities of a thing and their actualization. Rather, it speaks only of different actual features had by different actual temporal parts.

However, on closer inspection it is clear that temporal parts theory does not really provide an alternative analysis of change at all, but in fact implicitly *denies* the reality of change (Oderberg 2004 and 2009). For one thing, what the temporal parts analysis leaves us with seems to be, not a single thing that persists through change, but rather a series of ephemeral things, one after the other being created and annihilated. The temporal parts theorist may reply that this would be true only if we assume a *presentist* view of time, on which the present moment alone exists, whereas temporal parts theory is more naturally understood in terms of an *eternalist* view according to which every moment of time is equally real. (We will examine these two views of time in detail in a later chapter.) On this interpretation, the entire series of an object's temporal parts or stages exists "all at once," as it were, as a single four-dimensional object or space-time "worm" (to use the standard Minkowskian metaphor). But this is only to fall from the Heraclitean frying pan into the Parmenidean fire. That is to say, whereas the presentist interpretation of temporal parts theory leads to the denial that there is a changing *thing*, the eternalist interpretation leads to the denial that there is a *changing* thing.

Four-dimensionalism, after all, essentially conceives of time as analogous to space. Yet the fact that the different *spatial* parts of a single object have incompatible features at a particular moment of time does not entail change. For instance, that a person's hair is red while his hands are not does not entail change. But then how can the fact that the different *temporal* parts of a single object have incompatible features entail change, if temporal parts are supposed to be analogous to spatial parts? Your weighing 250 pounds on January 1, 2017 and 150 pounds on January 1, 2018 will amount to merely your 2017 temporal *part* weighing 250 pounds while your 2018 temporal *part* weighs 150 pounds. Why is this a case of *change* any more than your hair's being red while your hand is not amounts to change – if, again, temporal parts are like spatial parts (Oderberg 2004, pp. 706-7)?

Sider claims that we do sometimes speak of differences between spatial parts in terms of change; we might say, for example, that a certain road changes in the sense that it *becomes bumpier* the further along one travels down it (2001, p. 216). A difference between temporal parts can, he suggests, be understood as involving change in the same way. But this merely equivocates on the word "change," as is obvious from the sentence: "That road *hadn't changed* at all; it still *became bumpier* the further along it I traveled." This sentence is, of course, not self-contradictory, because when it is said that the road "became bumpier" what is meant is that while one (spatial) part of it is not bumpy, another (spatial) part of it is bumpy; while when it is said that the road "hadn't changed," what is in question is not a difference in its (spatial) parts but rather the fact that it still, at a later point in time, had a feature that it possessed at an earlier point in time.

We might call the road's still becoming bumpier an instance of "change" in the *spatial* sense, while the sense in which the road hadn't changed is a case of the absence of change in the *temporal* sense. Now, the objection on the table is essentially that four-dimensionalism fails to capture change in the *temporal* sense insofar as it models temporal parts on spatial parts. Sider's response is to illustrate how change is to be understood in light of four-dimensionalism by appealing to an example of "change" in the *spatial* sense rather than in the temporal sense. Far from *answering* the objection, then, Sider's response only *reinforces* it.

Nor can "change" in Sider's spatialized sense of the term do justice to the succession of perceptual and cognitive states of scientists. Suppose at some time *t* I hold in my left hand a piece of paper on which are written the sentences "All men are mortal" and "Socrates is a man," and in my right hand a piece of paper on which is written "Socrates is mortal." There is, we can allow for the sake of argument, a spatial "change" in sentences from left to right. But of course, it would be absurd to suggest that this "change" involves anything like an inference. Nor is the point affected if we add conscious subjects to the picture. Suppose at *t* Fred is standing to the left thinking "All men are mortal and Socrates is a man," while Bob is standing to the right thinking "Socrates is mortal." Again, there is a spatial "change" from left to right, and again, it would nevertheless be absurd to suggest that the change involves an inference. Now if we think instead of a temporal part of Fred at t_1 thinking "All men are mortal and Socrates is a man" and a temporal part of Fred at t_2 thinking "Socrates is mortal," and the "change" from Fred at t_1 to Fred at t_2 as a case of "change" in Sider's spatialized sense, then *this* "change" too will no more count as an inference than the "change" from Fred to Bob did.

By the same token, a spatialized "change" from a temporal part of Fred at t_1 formulating a prediction to a temporal part of Fred at t_2 performing an experimental test would no more count as testing a scientific theory than the previous example counted as an inference. Thus the cognitive tasks presupposed in any scientific theorizing – or philosophical theorizing for that matter – simply cannot be made sense of on a four-dimensionalist picture. The view is self-undermining.

So, temporal parts theory fails to provide an analysis of change that could serve as an alternative to the Aristotelian theory of actuality and potentiality. In particular, it fails to provide a way of avoiding the opposite extreme errors of dynamic monism and static monism. If interpreted in presentist terms, temporal parts theory falls into the former, Heraclitean error. If interpreted in eternalist terms, it falls into the latter, Parmenidean error. (For discussion of further problems with temporal parts theory, see Feser 2014b, pp. 201-208 and Oderberg 1993.) Hence temporal parts theory does not block the conclusion that the very existence of scientists qua subjects of experience presupposes the theory of actuality and potentiality.

We must, then, draw a distinction within the scientist qua subject of experience between the ways in which he is actual and the various perceptual and cognitive potentialities he possesses. Now, the mechanistic philosopher of nature who is also a Cartesian dualist might be willing to accept this much. But he might argue that this much does not really establish anything relevant to the philosophy of *nature*. What it establishes, so the argument might go, is only that the theory of actuality and potentiality has application within the *res cogitans* or immaterial thinking substance with which the Cartesian would identify the scientist qua subject of experience. It does not show that that theory has application to the physical world external to the conscious subject. In particular, it does not show that physical objects are composites of *substantial form* and *prime matter* – which, according to the Aristotelian, are the fundamental manifestations of actuality and potentiality in the world of physical objects.

There are several things to be said in response to this. First, there are serious problems with the Cartesian picture which keep it from being an acceptable alternative to the Aristotelian philosophy of nature, even if one agrees with the Cartesian that the human intellect is incorporeal (as I would, for reasons set out in Feser 2013a and elsewhere). For one thing, there is the notorious interaction problem facing Cartesian dualism, which does not afflict the Aristotelian-Thomistic form of dualism. (Cf. Feser 2006, chapter 8; Feser 2009, chapter 4; Feser 2018.) An aspect of that problem that is especially relevant to our subject is that it makes the evidential basis of science problematic. If the Cartesian cannot account for the causal relationship between the physical world and the mind, then he cannot account, more specifically, for the causal relationship between the physical world and the perceptual experiences that feature in observation and experiment. In that case, the Cartesian cannot account for how those experiences give the scientist any actual information about the physical world (Burtt 1980, p. 123).

For another thing, Descartes' account of matter as pure extension makes of a physical object something utterly passive, lacking any causal power by which it might effect changes in other physical objects. Hence, though he attributed real causal power to immaterial thinking substances, Descartes essentially took an occasionalist position vis-à-vis physical substances (Garber 1992, pp. 299-305). That is to say, he thought that it is really only ever God who causes things to happen in the material world. Now, among the problems with occasionalism is that it essentially negates the very idea of nature as an object of scientific study. If physical objects themselves don't really *do* anything, then there is no point in trying to *study* what they do or how they do it. God, who alone ever really does anything in the natural world, becomes the sole worthwhile object of scientific study, and natural science gives way to theology. But the situation is even stranger than that, at least if we factor in the Thomistic principle that *agere sequitur esse* (or "action follows being") – that is to say, that the way a thing *behaves* reflects what it *is*. If physical objects do nothing and only God acts, then it would follow that physical things don't have any *existence* distinct from God's existence. Occasionalism would collapse into pantheism, and the Cartesian philosophy of nature would thereby abolish nature altogether. (Cf. Feser 2017, pp. 232-38.)

A further problem with the imagined Cartesian dualist response to the argument of this section is that it begs the question against the Aristotelian insofar as it assumes that the perceptual and cognitive states of subjects of experience can entirely float free of the body. From the Aristotelian point of view, that is not the case, even given that the human intellect is incorporeal. For one thing, perceptual experience is corporeal, presupposing sense organs and brain activity. For another thing, even cognition requires, in the ordinary case, brain activity as a necessary condition, even if it is not a sufficient condition. For the intellect requires sensory images as a concomitant of its activity, and these are corporeal. This is precisely why human beings can acquire knowledge only gradually and why perception and cognition in us involve successive episodes spread out over time. If we were entirely incorporeal, we would essentially be angels, having our knowledge in a single act and without relying on perceptual experience. The Cartesian notion of *res cogitans* is really the notion of an angelic intellect, not a human one. Hence, from the Aristotelian point of view, to establish that there is a succession of perceptual and cognitive states in the subject of experience just is to establish that that subject is corporeal, and thus that the way in which it manifests actuality and potentiality is in part by being a composite of form and matter.

Demonstrating the essentially embodied and non-Cartesian character of human cognition and perception is something to which I will turn presently. The point to emphasize for the moment is simply that the imagined Cartesian objection is hardly very powerful. The Cartesian is on *defense*, not offense. He owes us an account of exactly how physical things as he conceives of them can have natures and causal powers for the scientist to investigate, and an account that does not collapse into an implicit Aristotelianism. (There is a reason why the recent revival of interest in essences and causal powers in analytic metaphysics has been called a "neo-*Aristotelian*" tendency.) He owes us an account of how mind and body interact, especially in such a way that perceptual experiences give us information about external physical reality. He owes us an account of how a *res cogitans* could have perceptual experiences and discursive cognition and still be a *human* mind rather than an angelic intellect. Until he does all this, he will not have given us a good reason to think a Cartesian philosophy of nature a serious rival to an Aristotelian one.

In any event, the chief rival to the Aristotelian philosophy of nature today is not the *Cartesian* version of the mechanical world picture, but the *naturalistic* version. Naturalists, of course, would be the last to deny that subjects of experience are corporeal. Naturalism insists that human beings, and thus scientists, are organisms which came into being by way of natural selection. Naturalism insists that the perceptual and cognitive processes of human beings, and thus of natural scientists, are grounded in sense organs and neural processes. But organisms and their sense organs and neural processes are corporeal. Hence if the very existence of scientists qua subjects of experience presupposes a distinction within them between actuality and potentiality, then the naturalist, unlike the Cartesian, will have to acknowledge that it presupposes within the world of *corporeal or physical* things a distinction between actuality and potentiality.

Now, for reasons which I summarized in chapter 1 and have set out at greater length elsewhere (Feser 2014b), when the distinction between actuality and potentiality is applied to corporeal things, it entails the further distinction between substantial form and prime matter. To have a certain kind of substantial form is just the fundamental way in which a corporeal thing manifests actuality. For that thing's prime matter to be capable of taking on or losing that substantial form, so that the thing is capable of coming into or going out of existence, is just the fundamental way in which a corporeal thing manifests potentiality. So, to establish that there is in the world of corporeal things a distinction between actuality and potentiality is ipso facto to establish that there is there a distinction between substantial form and prime matter.

Keep in mind that, as I emphasized in chapter 1, we have to distinguish the question of whether substantial forms exist at all from the question of whether some particular thing has a substantial form or merely an accidental form (where the hallmark of something's having a substantial form is its having properties and causal powers that are irreducible to those of the thing's parts). What I am claiming so far is only that the considerations adduced in this section show that there are substantial forms and prime matter in the natural world. Whether any particular corporeal things – such as scientists themselves qua subjects of experience – have substantial forms or merely accidental forms is a further question. Considerations to be adduced below and in later chapters will show that human beings do indeed have irreducible properties and powers and thus substantial forms rather than merely accidental forms. But even if the properties and powers of human beings were reducible to those of their physical parts, that would not affect the present point, which is that the distinction within the natural world of actuality and potentiality entails that composites of substantial form and prime matter have to exist at *some* level. As I argued in chapter 1, at the very least they would have to exist at whatever the most fundamental level of physical reality turns out to be (whether that turns out to be basic particles or even the entire physical universe considered as one big substance).

2.4 Being in the world

Most scientists tend to take for granted the commonsense belief in material objects. When a chemist analyzes a substance, or a biologist studies a specimen of an organism of some kind, or a neuroscientist studies a brain, he ordinarily does not think of the object of his investigation as a collection of sense data existing within his own mind. Like the man on the street, he supposes that he is dealing with physical entities that exist independently of his conscious awareness of them, and he also supposes that his own eyes, ears, hands, etc., of which he makes use in carrying out his investigations, are further physical objects that exist alongside of and causally interact with the physical things he is studying.

That is one obvious way in which scientists presuppose that human beings, including scientists themselves, are essentially corporeal. Of course, this would be of limited interest if it were merely a matter of scientists being as philosophically unreflective as most other people, and naively assuming the truth of some entirely contingent and challengeable claim. But the presupposition that human beings are corporeal goes far deeper than that, and deeper even than many scientists' commitment, in their more philosophical moods, to the naturalism referred to a moment ago. It has to do with the very nature of human cognition and perception (and thus of the cognition and perception of scientists), which are *essentially embodied*.

The various ways in which this is true were explored in depth by twentieth-century philosophers like Ludwig Wittgenstein (1968, 1972), Martin Heidegger (1962), Gilbert Ryle (1945-46, 1949), Maurice Merleau-Ponty (2012), and Michael Polanyi (1962, 1966). Needless to say, these are thinkers of diverse and sometimes conflicting commitments. As that fact indicates, one need not endorse everything said by any of them in order to see the force of the lines of thought they have in common. The recurring theme most relevant to our purposes is that of *tacit knowledge*. The idea is that the explicit content of all our cognitive and perceptual states presupposes a body of *in*explicit knowledge, where this knowledge is fundamentally a matter of *knowing how* to interact with the world, rather than a matter of *knowing that* such-and-such propositions are true. It is knowledge essentially embedded in *bodily capacities*.

This conception of human knowledge contrasts with what is sometimes called a *representationalist* conception. Representationalism was the epistemological side of the early modern intellectual revolution, of which the mechanical world picture was the metaphysical side. Hubert Dreyfus and Charles Taylor identify four key components of the view (2015, pp. 10-12). First, representationalism holds that our knowledge of objective reality is *mediated* by knowledge of representations of some sort – whether ideas in a Cartesian *res cogitans*, or patterns encoded in neural structures, or the formal symbols of a computer program, or the observation sentences of the Quinean naturalist, or whatever. Second, the content of these representations is taken to be clearly and explicitly defined rather than tacit. Third, it is held that the justification of all knowledge claims can never get beyond or below these explicitly formulated representations – especially the subset of foundational or "given" representations, if there is one (though on some versions of representationalism, there is not).

The fourth component of representationalism is what Dreyfus and Taylor call the "dualist sorting" of reality into the representations themselves on the one hand and the physical world they represent on the other, where the latter is conceived of in terms of the mechanical world picture. Descartes, of course, put this dualism forward as an ontological thesis, carving the world into the material and the immaterial, *res extensa* and *res cogitans*. Materialists reject this aspect of the Cartesian picture, holding that the representations ought to be identified instead with some subset of the denizens of the material world (such as brain processes), construed mechanistically. Since, on that mechanistic construal, matter is devoid of teleology and secondary qualities, this leaves the materialist with the problem of explaining *how* the intentionality and qualia that characterize these representations could be properties of matter so defined. In these ways, as Dreyfus and Taylor note (and to echo a point I made in chapter 1), representationalism generated the modern "mind-body problem."

Now, epistemology *per se* is not our concern here, but the metaphysical aspects of this epistemological picture are relevant to our subject. Materialists suppose the representations in question to be mere bits of matter alongside all the others, construed mechanistically. By contrast, the Cartesian takes the representations to be essentially disembodied, since he supposes that their conceptual and perceptual content would be entirely transparent to the mind even in the absence of a material world. For critics of representationalism like Wittgenstein, Heidegger, and the others named above, neither of these suppositions is correct, so that neither the materialist nor the Cartesian account of human nature can be correct either. These thinkers argue (contra the Cartesian) that there is content to human cognition and perception that is not transparent to the mind, but rather necessarily exists below the level of consciousness and in an essentially embodied form. But the way it exists there entails (contra the materialist) that the body cannot be understood mechanistically, as a clockwork-like aggregate of insentient and meaningless parts. What these thinkers are engaged in is, in effect, a rediscovery of the Aristotelian conception of human nature, even if they do not always think of themselves as doing this and even if some of them would resist this characterization.

2.4.1 Embodied cognition

The considerations indicating the embodied nature of intellectual activity and those indicating the embodied nature of perception are related but distinct. Let us begin with intellectual activity, which, as noted in chapter 1, involves three main capacities: first, the capacity to form abstract concepts; second, the capacity to combine concepts into a complete thought or proposition; and third, the capacity to reason from one proposition to another in accordance with canons of logical inference. Critics of representationalism like those mentioned above sometimes present their objections in the form of *regress arguments.* (Cf. Gascoigne and Thornton 2013 for a useful overview.) These arguments come in different versions, which emphasize different intellectual capacities among the three I just identified.

Consider first the regress entailed by the grasp of a concept, and John Searle's way of spelling it out (1983, Chapter 5; 1992, Chapter 8). In order for you to grasp any one concept, you need to grasp others. For example, to understand the concept of a *bachelor*, you need to understand the concept of a *man* and the concept of *being unmarried*; and understanding these further concepts requires grasping yet other concepts in turn. *Applying* a concept also presupposes background knowledge. To borrow an example from Searle, in order to have the intention of running for President of the United States, you have to know that in order to become the President one has to win an election, that to win one needs to run a successful campaign, and so forth. Our understanding and application of concepts thus takes place within what Searle calls a *Network* of beliefs, intentions, etc.

Now, when applying a concept, we obviously don't bring to consciousness all the other concepts and beliefs that it presupposes. When you have the conscious thought "Fred is a bachelor," you don't necessarily at the same time consciously think "That Fred is a bachelor entails that Fred is a man," etc. That is one sense in which the explicit content of our thoughts presupposes something inexplicit. But there is a deeper sense in which it does so. To borrow another example from Searle, suppose you go into a restaurant and say "Bring me a steak with fried potatoes." Even if both you and the waiter do bring to consciousness the concept of *steak*, the concept of *bringing something*, etc. and consciously relate these concepts to the further concepts in terms of which they are to be defined, the precise way to *apply* all of this explicit knowledge is still as yet undetermined. For there is nothing in the Network of concepts and beliefs that go into defining what it is to be steak, etc. that by itself determines that when the waiter brings you the steak, it will be on a plate rather than encased in concrete, that he will place it on the table rather than shoving it into your pocket, and so on. Of course, in ordinary circumstances we would never for a moment expect these bizarre things to happen. The point, though, is that the supposition that they won't happen is one which is usually not *explicit* or conscious. We simply take it for granted that the steak will be on a plate, will be placed on the table, and so forth.

Naturally, we could make these assumptions explicit if we wanted to. You could consciously think "When the waiter brings me the steak, it will be on a plate and he will place it on the table." The waiter could consciously think "When I take the steak to the customer, I will not shove it into his pocket and it will not be encased in concrete." But as Searle points out, even if this happens, there will always be yet *further* assumptions that are not conscious or explicit. Precisely because there will be, the regress through the Network is not infinite. It ends with a set of capacities, dispositions, and ways of acting which Searle calls the *Background* against which the Network operates. The Background involves our behaving *as if* we were explicitly and consciously affirming propositions, when in fact we are not doing so. The waiter does not consciously entertain the thought that he needs to put the steak on the table and not in your pocket. He is simply unconsciously *disposed to act* in that particular way rather than some other way. You do not consciously entertain the proposition that the waiter will put the steak on the table rather than try to put it into your pocket. You are simply *disposed to act* in a way that presupposes this. For example, when you see him coming, you clear the area of the table directly in front of you, and you do not pull your pocket open. Insofar as the Background involves the exercise of capacities, the manifestation of dispositions, and the like, rather than the conscious entertaining of propositions, its operation is a matter of our *knowing how* to act rather than *knowing that* such-and-such propositions are true.

Searle draws a further distinction, between the "local Background" and the "deep Background." The local Background has to do with those unconscious capacities, dispositions, and ways of acting which are culturally and historically contingent, and thus which at least in principle can change from time to time and place to place. The custom of placing a customer's steak on the table rather than in his pocket and not encasing it in concrete first would be an example. There could be cases (even if very odd ones) where these particular Background dispositions change. For example, imagine that someone opens a theme restaurant devoted to performance art or pranks, where customers are told that they should expect the unexpected.

The deep Background, by contrast, involves capacities, dispositions, and ways of acting that are hardwired into us. For example, they might reflect our specific biological constitution. Even if cultural and historical circumstances change, we are not going to form a Background disposition to fly by flapping our arms, because the physical and biological facts simply do not allow for that. Or the deep Background dispositions might go even deeper than that, as those which presuppose realism about the material world outside our minds do. Even a reader of Descartes' *Meditations* who starts to wonder whether tables, chairs, rocks, trees, and even his own body are hallucinations will unthinkingly exercise capacities that presuppose that they are real. For example, he might put the book down momentarily and rub his chin pensively. He might walk over to the refrigerator to grab a beer to drink before he reads any further. If his roommate throws a baseball at him while he is reading, he will duck. And so forth. The deep Background dispositions that presuppose that his chin, the floor under his feet, the refrigerator, the beer, the baseball, etc. are all real run far deeper than any doubts about them he might entertain in his philosophical moments.

Now, these Background capacities, dispositions, and ways of acting, and especially the deep Background ones, are essentially *bodily* capacities, dispositions, and ways of acting. For they involve ways of speaking, gesturing, walking, picking things up, eating, etc. all of which involve use of the body and its organs. In this way our grasp of abstract concepts presupposes embodiment. Notice that this is as true of scientists in their carrying out of their investigations as it is of anyone else and of everyday activities. In dealing with telescopes, microscopes, gauges, and other scientific instruments, in conversing and cooperating with other scientists, and in examining minerals, plants, animals, lungs, hearts, brains, planets, stars, etc., scientists are exercising deep Background capacities that presuppose realism about the external material world. In applying certain methods of analysis, publishing in certain journals, teaching their students, etc., scientists are exercising local Background capacities, dispositions, and ways of acting acquired in graduate school, picked up in the laboratory, and so forth.

A second sort of regress argument involves our assent to propositions and our deployment of canons of logical inference. Suppose I explicitly assent to the proposition that *Socrates is mortal* upon considering the proposition that *all men are mortal* and the proposition that *Socrates is a man*. I have reasoned through what logicians call an AAA-1 form categorical syllogism, but it may be that I am unaware of having done so. After all, most people would draw that conclusion from those premises even if they had never taken a logic class and know nothing about the standard classification of forms of reasoning. When they entertain the propositions that *all men are mortal,* and that *Socrates is a man*, it just strikes them as obvious that Socrates must therefore be mortal. Their explicit knowledge that *Socrates is mortal* rests on inexplicit or tacit knowledge of the validity of reasoning of the AAA-1 type.

Now, a person could, of course, become self-conscious about what sort of reasoning he is deploying in cases like these, as logic students do. The fact that he is conscious of it may even play a role in his justification for believing the conclusion. Whereas the untutored reasoner might say "Socrates is mortal, *because* all men are mortal, and Socrates is a man," the logic student might say "Socrates is mortal, *because* all men are mortal and Socrates is a man, *and* an AAA-1 form syllogism is always valid." But even if this knowledge becomes explicit, there will always be yet further knowledge that is not explicit. As Ryle points out, a very slow student may explicitly know that *Socrates is a man*, that *all men are mortal*, and that *AAA-1 form syllogisms are valid*, and still not put all this knowledge together in the right way. It might somehow just not "click" for him that *Socrates is mortal*. What this student lacks is the inexplicit knowledge that the normal student has. Suppose we try to solve the problem by making this knowledge explicit and teaching it to the slow student that way. We might formulate it as the proposition that *if AAA-1 form syllogisms are valid and all men are mortal and Socrates is a man, then Socrates is mortal*, and then add this new explicit proposition to the already explicit set of propositions that *AAA-1 form syllogisms are valid, all men are mortal,* and *Socrates is a man.* But even if the student now sees that this new proposition is true, if he is very, *very* slow he may *still* not see that the conclusion that *Socrates is mortal* follows. And so on for any further proposition we make explicit and add to the mix. (Cf. Carroll 1895)

So, adding further explicit propositions will not solve the problem, and it is not what solves the problem in the case of the normal student. If it were, then since there is always yet another further explicit proposition we could add, what the normal student would be doing is explicitly grasping an infinite series of explicitly formulated propositions, all at once, when he judges that *Socrates is mortal*. Obviously, that is not what is going on. What is going on, Ryle argues, is that the normal student's explicit *knowledge that* the propositions in question are true and the inference rule valid leads the student to draw the right conclusion only because he also possesses practical *knowledge how* to apply that theoretical knowledge. This *knowing how* cannot be a matter of grasping explicit propositions, on pain of infinite regress, but rather involves (as it does for Searle) the having of certain capacities, dispositions, and the like.

For Ryle, what is true of logical reasoning is true of all intelligent behavior – playing chess, driving a car, operating machinery, or whatever. It cannot *merely* involve knowledge of explicitly formulated propositions, such as propositions stating rules for action. For one thing, there is always a gap between knowing the rule and actually applying it. For another, a rule can always be applied either intelligently or unintelligently, and intelligent application cannot be a matter merely of applying yet further rules, on pain of the same sort of infinite regress just mentioned. All intelligent behavior thus ultimately rests instead on *knowing how* – again, on dispositions and the like. Since playing chess, driving a car, and for that matter carrying out a conversation with someone about whether Socrates is mortal are all *bodily* activities, the dispositions in question are *bodily dispositions*. And once again, the intelligent behavior of scientists no less than of anyone else involves this embodied *knowing how*, and the manifestation of behavioral dispositions. Examples would be operating scientific equipment, writing journal articles, conversing with other scientists, and so forth.

Now, in an influential article (2001), Jason Stanley and Timothy Williamson have challenged Ryle's argument. But it seems to me that their objections rest on a number of misunderstandings. For example, they suppose that Ryle regards *any* kind of bodily activity as a manifestation of "knowing how," including what they characterize as the "action of digesting food" (p. 414). They then suggest, quite correctly, that it is implausible to think of digestion as involving "know how," given that it is not something we do intentionally. (Why, in that case, they would characterize it as an "action" in the first place is not clear.) They go on to propose that Ryle should therefore have confined his analysis to behaviors that are intentional. But in fact, Ryle did confine his analysis to such behaviors; as Stanley and Williamson themselves note (without seeing that it undermines their interpretation), Ryle speaks of operations that are "intelligently executed." What Stanley and Williamson portray as a correction of Ryle is in fact what Ryle himself was already saying all along. (Cf. Gascoigne and Thornton 2013, p. 55)

Stanley and Williamson also claim that the way Ryle generates a vicious regress is by supposing that:

> [I]f knowledge-how were a species of knowledge-that, then, to engage in any action, one would have to contemplate a proposition. But, the contemplation of a proposition is itself an action, which presumably would itself have to be accompanied by a distinct contemplation of a proposition. (p. 413)

Hence, according to Stanley and Williamson, what Ryle is arguing is that knowing how to do some action *A* would on this analysis involve the contemplation of a proposition about *A*, which would in turn involve the contemplation of a proposition about the contemplation of a proposition about *A*, which would in turn involve the contemplation of a proposition about the contemplation of a proposition about the contemplation of a proposition about *A*, and so on *ad infinitum*. Knowing how to do *A*, on this interpretation of Ryle, accordingly "would require contemplating an infinite number of propositions of ever-increasing complexity" (p. 414).

Stanley and Williamson then counter this argument that they attribute to Ryle by proposing that "knowledge that" need not in fact involve contemplating a proposition in the first place. As an example, they cite a person's knowledge *that one can get through a door by turning the knob*, which is manifest in the person's actually turning the knob automatically, without consciously entertaining any proposition as he does so (as one might absent-mindedly turn a knob while absorbed in a conversation with someone). And if "knowledge that" need not involve contemplation of a proposition, then (Stanley and Williamson conclude) it need not generate a regress of the sort they describe in their reconstruction of Ryle's argument.

But this criticism of Ryle is a tissue of confusions. First, the way Stanley and Williamson present the example of turning the knob is simply tendentious. Ryle would agree that the turning of the knob may well not involve the conscious entertaining of a proposition. But he would deny that the turning is itself a direct manifestation of "knowledge *that*" in the first place. Rather, the turning is a direct manifestation of "knowledge *how*," and it is *through* this "knowledge how" that the associated "knowledge *that*" one gets through a door by turning a knob is also *in*directly manifested. Stanley and Williamson would presumably reject this characterization of the situation, but merely to *assume* that their own characterization is correct and Ryle's alternative wrong simply begs the question against Ryle.

Second, Stanley and Williamson misunderstand the nature of the regress described by Ryle. They paint a scenario in which the very act of contemplating a proposition itself involves a further act of contemplating yet another proposition, which itself in turn involves a further act of contemplating yet another proposition, and so on. But that is not what Ryle is talking about, and the kind of regress he *is* concerned with would exist even if the contemplation of a proposition did not involve the specific regress Stanley and Williamson describe. What Ryle is saying is that once one contemplates a proposition – which, let us stipulate, of itself involves no regress – there is still the further question of how to *interpret* that proposition, and how to *apply* one's knowledge of the proposition in a practical way. Now, suppose that either interpreting or applying the proposition itself involved the contemplation of some further proposition, but suppose also that the mere contemplation of this further proposition of itself involved no regress. There would still be the question of how to *interpret and apply* that further proposition that one is contemplating, and *that would* entail a regress. So, it is not that the mere *contemplation* of a proposition entails a regress. It is rather that the *interpretation* and *application* of a proposition that one contemplates entails a regress, and this regress can be terminated only if we think of interpretation and application in terms of "knowledge how."

Third, as Neil Gascoigne and Tim Thornton point out (2013, pp. 55-56), what is at issue for Ryle is, fundamentally, not whether "knowledge how" involves *conscious* contemplation of a proposition, but rather whether "knowledge how" exists in a *propositional* form in the first place, whether consciously or unconsciously. Merely to note that we can turn a doorknob without consciously entertaining a proposition doesn't address that deeper point.

Finally, Stanley and Williamson challenge the thesis that "knowledge how" involves abilities. As counterexamples, they offer the case of a ski instructor who knows how to perform a certain stunt without being able to perform it himself, and a pianist who knows how to play a certain piece even though she has lost her arms in an accident and thus lost the ability to play it (2001, p. 416). Another critic of Ryle, Paul Snowdon, claims that a person can have an ability without possessing "knowledge how." He offers as an example a man in a room who has not yet explored it and thus is not aware of a certain exit he could easily access. He is in fact *able* to get out, says Snowdon, but lacks "knowledge how" to do so (2004, p. 11).

But these objections all equivocate on the expression "knowing how." Sometimes when we speak of a person "knowing how" to do something, we just mean that he knows *that* it is done by way of such-and-such a procedure. That is the sense in which, in Stanley and Williamson's examples, the ski instructor knows how to do the stunt and the pianist knows how to play the piece. But sometimes when we say that a person "knows how" to do something, we mean instead that the person has a certain ability. For example, when we say that someone knows how to ride a bike or knows how to swim, we mean that the person has the ability in question, and not merely that he knows that these things are done in such-and-such a way. Now, this is the sort of "know how" that Ryle is concerned with. By Stanley and Williamson's own admission, the people in their examples lack the relevant abilities. Accordingly, they *lack* "knowledge how" in the relevant sense, and thus these are not counterexamples to Ryle's position at all. Similarly, the man in Snowdon's example *does* "know how" to get out of the room in the relevant sense, because by Snowdon's own admission he has the relevant ability. What he lacks is knowledge *that* the room contains an exit. (Cf. Gascoigne and Thornton 2013, pp. 64-68 and, for a somewhat different response to these sorts of examples, Noë 2005.)

In short, when the misunderstandings and other errors are cleared up, there does not seem to be anything in Stanley and Williamson's critique that poses any challenge to what Ryle actually said, much less to other regress arguments of the sort we are considering here.

Now, Searle's analysis is in part inspired by Wittgenstein's philosophical anthropology, to which Ryle's position also bears a family resemblance. Other thinkers arguing for the essentially embodied nature of human intellectual activity have been primarily inspired instead by phenomenology (e.g. Dreyfus 1992). In both cases the considerations marshaled are of a philosophical rather than scientific character. However, similar conclusions have been arrived at by writers motivated precisely by findings in empirical science. Andy Clark has usefully summarized some of the key points (1997, Chapters 1 and 2).

For example, consider the *action loop* phenomena studied by psychologists, in which bodily action plays a crucial role in the solving of a cognitive task. Clark gives the example of trying to figure out where a certain piece fits in a jigsaw puzzle. The way we typically do this is not merely by intellectually representing the shape of the piece and the shapes of the spaces into which it might fit and then deducing which of the latter is the correct place to put it, though of course we do this to some extent. Rather, we also *physically manipulate* the piece by rotating it and trying actually to fit it into a certain space, adjusting our intellectual representations accordingly if we cannot. Our thought processes not only guide our bodily behavior but are influenced in turn *by* that behavior.

Then there is what Clark calls the phenomenon of "*soft assembly*." A "hard-assembled" system is one whose behavior is determined in a top-down way by a centralized body of information and cognitive "blueprint" for action and is thus ill-equipped to deal with circumstances that are not included in the body of information or covered by the blueprint. A "soft-assembled" system, by contrast, is more decentralized, sensitive to information coming in from the periphery of the system and thus more flexible in its responses and adaptable to unforeseen circumstances. Scientific study of human behavior shows that it is largely soft-assembled. For example, the specific way we walk across a certain surface is not determined entirely by centralized neural processes or conscious thought, but is highly sensitive to such localized factors as leg mass, muscle strength, the kind of shoes one is wearing, the presence or absence of blisters, the specific physical characteristics of the surface, and so on. All of these factors "partner" together to generate a particular gait, with the centralized neural and cognitive processes adjusting themselves to the deliverances of the body.

In these ways, *bodily factors* provide a kind of "scaffolding" for cognition, and as Clark emphasizes, material phenomena outside the body provide further "*external* scaffolding." For example, books, notes, pictures, or even just the having of certain specific physical objects around us all function as aids to memory. The presence or absence of physical objects of certain kinds also provides a context that both facilitates and delimits the actions we might perform and thus the practical reasoning we might engage in. To cite an example from Clark, the presence in a kitchen of certain specific spices, oils, eating utensils, etc. determines the range of the sorts of cooking options one will entertain and the decisions one will make about what specifically to cook.

Of course, much of this is just common sense, but it is common sense confirmed by psychological and neuroscientific study of human action, and also by research in robotics, insofar as application of principles like action loops, soft assembly, and external scaffolding often turn out to provide the most efficient solutions to the problem of getting a machine to simulate human behavior.

2.4.2 Embodied perception

Let's turn now to the ways in which perceptual experience too is essentially embodied. Once again, Clark provides a useful summary of some of the considerations from contemporary psychology and neuroscience which support this conclusion. There is, for example, the phenomenon of *niche-dependent sensing*, by which a creature's sense organs are adapted to detecting a specific range of environmental features. For instance, a tick is sensitive to the butyric acid on the skin of mammals, the olfactory detection of which will cause it to drop from a tree onto a passing mammal. Contact with the skin then initiates in the tick heat-detecting behavior, and the actual detection of heat will in turn initiate burrowing into the skin. Within the larger physical world, there is only a specific subset of phenomena that constitute the tick's "*effective environment*," with the rest of the world being largely invisible to it. Now, like other creatures, human beings too have sense organs that are keyed to certain features of the world and not others, and which determine for them their own unique effective environment.

Another example involves what researchers call *animate vision*, or visual sensing of a sort which crucially involves bodily engagement with the world. Saccades are quick movements of the eyes back and forth between fixation points, and they play a key role in visual perception. In viewing a particular scene, frequent saccades allow us, in the view of some researchers, to avoid having to construct an enduring and detailed neural model of the immediate environment. Instead we simply access the needed information by returning to the environment repeatedly during the course of the visual experience, letting the things in the environment code for themselves, as it were. In this way it is the things themselves, rather than our internal representations of them, that we deal with in perception.

Then there is the fact that we typically do not take gaps in sensory information to correspond to gaps in the thing sensed. Clark gives the example of grasping a bottle without looking at it, in which the absence of information about the areas of the bottle's surface between one's fingers is not interpreted as indicating that there are holes in those areas. What such examples show, some researchers suggest, is that sensation is not the mere passive intake of information, but rather an active bodily engagement with the world. The sense organs are essentially used as *tools* for exploring objects which are experienced as independent of us, and as extended beyond what we immediately perceive of them. As psychologist James Gibson famously argued (1979), things in our environment are perceived as "affordances" for action. We feel a bottle as something that affords us the possibility of picking it up and drinking from it, we see a doorknob as something which affords us the possibility of turning it and leaving the room, and so forth. In short, we perceive things precisely *as accessible to the body*.

This scientific work recapitulates and reinforces lines of argument developed along phenomenological lines by philosophers like Heidegger, Merleau-Ponty, and Polanyi. Dreyfus (1992, Chapter 7) summarizes three respects in which, according to this tradition, perception presupposes our being embodied subjects within a larger world of physical objects. First, there is the figure-ground phenomenon, in which the thing perceived is always perceived *as* distinct from some surrounding context. Dreyfus gives the example of the Rubin's vase image familiar from Gestalt psychology, which can be seen either as a white vase against a black background or as two black faces in profile against a white background. What is taken to be the background constitutes what Edmund Husserl (2002, Part III) called the "outer horizon" of the thing perceived, by contrast with the "inner horizon" or those aspects of the thing which are not perceived but are nevertheless presupposed in our perception of the thing. To borrow another example from Dreyfus, when we perceive a house we perceive it *as* something having a back and an inside rather than taking the front of the house as a mere façade, even though the front is all we directly see. Now, all of this presupposes embodiment insofar as what we take to be a thing's inner and outer horizons depends on how we regard our bodies to be situated with respect to the things perceived, and how we take those things qua physical objects to be situated with respect to one another. I take what I see to be the front of a house rather than a mere façade because I see what I see *from this angle*, and I note that it is next to *this other house*, behind *this driveway*, and so forth. Perception involves a particular perspective on the rest of the physical world, taken by one thing among others situated within that world.

Second, perception involves *anticipation* of a larger whole, of which what is immediately sensed is only a part. A musical note is perceived *as* a part of a piece of music, a nose or eye *as* a part of the face, and so forth. Furthermore, as Heidegger famously emphasized, we perceive things fundamentally in terms of their "readiness-to-hand," i.e. the way they might be deployed by us as "equipment" by which we might realize our ends. Now, at least much of the anticipation such perception involves is of a bodily nature. It is primarily in the act of *grasping* and *using* a hammer, in *feeling* in one's hand its weight and solidity, that we anticipate what might be done with it. It is primarily in *feeling in one's body* the rhythm of a piece of music that we anticipate the beats and notes that will follow, and in *moving our fingers* across a piece of silk that we anticipate that the rest of the fabric will have a similarly smooth texture.

Third, this anticipation is transferable across the body, from one sensory modality or organ of action to another. What is first learned through touch comes to be knowable also by sight; what is seen or touched thereby becomes graspable and otherwise subject to possible manipulation; what is heard coming toward us can thereby be avoided or approached via bodily movement; and so on.

Now, it might seem that such phenomenological descriptions of perceptual experience in terms of embodied subjects acting within a world of other physical objects could be replaced by descriptions couched either in the entirely "first-personal" and "subjective" terms of a sense datum language (as a Cartesian might propose), or in the entirely "third-personal" and "objective" terms of the theoretical entities postulated by physical science (as a materialist might propose). But such a replacement could never be carried out consistently, because the re-descriptions in question are parasitic on the phenomenological description.

For example, suppose I am looking at a tomato that I am holding in my hand, and that I try to describe the experience in terms of sense data such as *a roundish red patch in the center of my field of vision*, etc. It is only because I *first* have the experience of what I take to be a tomato situated in such-and-such a way relative to other external physical objects, and to me as an embodied subject, that I can go on to identify the sense data in question in just the way I do. For example, I have to say such things as that the redness of the patch is specifically of the sort that is typically seen on a tomato-like surface; that the red patch is surrounded by other color patches of a shape and texture that are typical of what one would normally take to be part of a hand; that all these patches have the appearance that a tomato and hand would have if looked at from above; and so forth. The sense datum description involves *abstracting* certain features *from* the commonsense phenomenological description of ordinary physical objects, and then treating these abstracted features as if *they*, rather than the physical objects, were what one is really perceiving. But the commonsense phenomenological description always remains lurking in the background, as that by reference to which we identify the sense data we are supposedly replacing it with. (Cf. Dreyfus and Taylor 2015, p. 53; Sellars 1956, p. 274; Strawson 1979, pp. 43-44.)

Something similar is true of any attempt to replace the phenomenological description in terms of a description couched instead in the language of scientific theory. Hence, suppose I try to replace any reference to tomatoes, hands, etc. with references to collections of particles organized in such-and-such ways. There is no way to identify exactly *which* collections of particles I have in mind in any particular case except by reference to the ordinary objects they are supposed to be replacing. I have to refer to particles arranged specifically in a *tomato-like way*, or to particles which to a normal observer would be *perceived as a tomato*, etc., and to the relations that collections of particles so described have to further collections of particles organized in an *eye-like way*, a *hand-like way*, etc. Once again, the attempted re-description is parasitic on the commonsense phenomenological description. (Cf. Polanyi 1966, pp. 20-21; Elder 2004, pp. 50-58; Elder 2011, pp. 118-24)

But even if such purported alternative descriptions are parasitic on the commonsense phenomenological description, might the latter not still be false? Might the external physical world and indeed one's own body not be illusory? Yet this familiar skeptical proposal presupposes what Dreyfus and Taylor characterize as the first, "mediational" assumption of representationalism. It supposes that we can abstract human cognitive and perceptual activity out of its bodily context and then intelligibly reify it as a set of "representations" which may or may not match up with a physical reality external to them. And that is precisely what the arguments we have been considering deny. These arguments maintain that the very idea that human cognitive processes and perceptual experiences might have just the content they do, yet without there actually being a physical world in which we are embedded as embodied subjects, is *itself* an illusion. The skeptic presupposes the possibility of a gap between the thinking and perceiving subject on the one hand and the corporeal world on the other that is not in fact intelligible.

So, the "mediational" assumption on which the skeptical objection rests simply begs the question. Worse, when put forward in the name of science, the "mediational" conception of perceptual experience leads to incoherence, in a way suggested by Frederick Olafson (2001, Chapter 3). (The remarks to follow are inspired by Olafson, anyway, though I will not be stating things exactly the way he does.) Science crucially depends upon observation. But what exactly is presented to us or given in observation? The commonsense view – traditionally known as "direct realism" or "naïve realism," and called by Olafson the "natural attitude" – takes ordinary physical objects themselves to be what we are directly aware of. The "mediational" component of representationalism rejects this assumption, and traditionally held that sense data or the like are in fact what are presented or given to us in perception, with the physical world known at best only *indirectly*, *through* our direct knowledge of sense data.

Now, contemporary philosophers have largely abandoned the sense datum theory, in part because the very notion of a sense datum faces problems like those summarized above. What is purportedly "given" to us on a sense datum account turns out to be no less "theory-laden," and thus subject to challenge, than commonsense or naïve realism is according to the "mediational" picture. Furthermore, if the sense datum theory is interpreted in Cartesian dualist terms, then we have to add the interaction problem to the list of difficulties. But if sense data are not what is presented or given to us in perception, what is? A naturalist might propose substituting some materialistically respectable representations in place of the notion of a sense datum – neural processes, computational symbols, or what have you. But these can hardly be taken to be *given* or *presented* to us in perception, since most people have no idea what is going on in their brains, and have no idea what a computational symbol is. It takes a lot of sophisticated theorizing to come to the conclusion that what one is "really" aware of when introspecting one's conscious experiences are brain states of a certain type, or computational symbols. To postulate neural or computational representations is therefore to open up a further gap between appearance and reality, in addition to the initial gap opened up by the original sense datum theory. Just as the original theory posits inner representations through which we get at the external world, we would now have to posit second-order inner representations through which we get at the (neural or computational) representations.

This puts the "mediational" picture in a dilemma. What motivated the picture in the first place was the idea that everyday perceptual judgments are so riddled with challengeable assumptions (the assumption that one is not dreaming, that secondary qualities correspond to real features of physical objects, etc.) that we cannot take physical objects to be what are presented to us in perception, but must replace them with sense data. But sense data themselves, and alternatives such as neural or computational representations, turn out to be no less theory-laden and thus no less subject to challenge. Hence if the theory-ladenness of ordinary perceptual judgments is taken to undermine the commonsense view that physical objects are presented to us in perception, then the theory-ladenness of judgments framed instead in terms of sense data or neural or computational representations should lead us to conclude that *they* are not what is given to us in perception either.

This generates a regress in which what is given or presented to us in perception keeps getting pushed back. There are two ways the "mediational" picture can deal with this regress, but both are fatal to it. On the one hand, one could break the regress by simply postulating that there is a stage that terminates it, despite involving judgments that are theory-laden and challengeable. One could say, for example, that sense data really are what is given to us or presented in perception, even though sense datum judgments are theory-laden and open to challenge. The problem with this move, however, is that it makes the abandonment of direct realism entirely pointless and unjustified. If theory-ladenness and the possibility of error do not suffice to show that sense data are not given or presented to us in perception, then how could they suffice to show that *ordinary physical objects* are not what is given or presented to us in perception? If we are going to end up admitting at the end of the day that there is after all some level at which things are just given in perception despite the possibility of error, we might as well take those things to be what common sense has always taken them to be – tables, chairs, rocks, trees, etc. – rather than philosophically problematic and unmotivated theoretical entities like sense data, neural or computational representations, or what have you. This way of dealing with the regress simply undermines the whole point of the "mediational" picture.

The second, alternative way of dealing with the regress would be to deny that we ever get to *anything* that is given or presented to us in perception. There is just the regress of representations themselves, which either proceeds to infinity or loops back around in a circle, and we cannot get beyond it. But if that is the case and nothing is really given or presented to us in perceptual experience, then perception loses all contact with external reality and cannot serve as an evidential basis for science.

This *epistemological* dilemma for representationalism parallels a *metaphysical* dilemma that afflicts naturalist versions of the mechanical world picture – unsurprisingly, given that representationalism and mechanism have, as noted earlier, gone hand in hand in modern philosophy. As noted in the previous chapter, the mechanical philosophy banishes from the material world any features that are irreducibly qualitative and cannot be accommodated to a purely quantitative or "mathematicized" conception of nature. Hence color, sound, heat, cold, etc. as common sense understands them are taken to exist only as the qualia of conscious experience rather than as features of physical objects themselves, and "directedness" toward an object is taken to exist only as the intentionality of thought rather than as the teleology or final causality of physical processes.

Now, this seems straightaway to entail Cartesian dualism, since if irreducibly qualitative features and "directedness" do not exist in matter but do exist in the mind, then that implies that the mind must not be material. But the naturalist, of course, wants to resist this conclusion. This leaves him with two options. On the one hand, he could expand his notion of what counts as "natural" so as to include irreducibly qualitative features and "directedness." This is essentially the option taken by contemporary non-reductive naturalists and by property dualists and panpsychists who regard themselves as naturalists. The trouble with this position, though, is that it makes the original move in the direction of mechanism pointless. If you are going to have to put irreducibly qualitative features and "directedness" back into the natural world at the end of the day, why take them out in the first place? What, in that case, would justify resisting the commonsense and Aristotelian claim that they have *always* been there?

The other option would be simply to deny that irreducibly qualitative features and "directedness" really exist at all, even in the mind. This is the option taken by eliminativists who deny the existence of qualia and intentionality. But this is to deny the existence of perceptual experience and intellectual activity themselves, and thus to undermine the evidential basis of the science in the name of which eliminativists take this extreme position – thereby "immolating themselves on the altar of their theory," as Olafson puts it (2001, p. 51).

As Erwin Schrödinger said of the extrusion of the sensory qualities from the modern scientific picture of the natural world:

> We are thus facing the following strange situation. While all building stones for the [modern scientific] world-picture are furnished by the senses qua organs of the mind, while the world picture itself is and remains for everyone a construct of his mind and apart from it has no demonstrable existence, the mind itself remains a stranger in this picture, it has no place in it, it can nowhere be found in it. (1956, p. 216)

Thomas Nagel (1979) makes essentially the same point when he notes that modern science works with a conception of the physical world that excludes from it anything that reflects the first-person point of view of the conscious observer, so that that point of view cannot *itself* be fitted within, or explained in terms of, the physical. Though we arrive at what Sellars (1963) called the "scientific image" (the world as described in terms of the concepts of physical science) only from *within* the "manifest image" (the world as it appears to us in ordinary perceptual experience), the manifest image cannot in turn be reconstructed from the scientific image. Accordingly, the scientific image *cannot account for its own existence* and thus cannot possibly give us an exhaustive description of reality (Olafson 2001, pp. 20-21).

The way to avoid these paradoxes is to abandon the anti-Aristotelian representationalist and mechanistic assumptions that inevitably lead to them. Contra representationalism, we need to acknowledge that it is physical objects themselves that are presented or given to us in perception. Contra mechanism, we need to put the first-person point of view of the conscious subject back into the body.

2.4.3 The scientist as social animal

Naturally, what is true of thinking and perceiving subjects in general is true of scientists in particular. The scientist is essentially an embodied subject in a world of physical objects. When he entertains hypotheses, thinks through the implications of a theory, weighs evidence, and so forth, he is deploying concepts, rules of inference, etc. that are ultimately rooted in bodily capacities and dispositions. When he makes observations, conducts experiments, operates scientific equipment, reads books and journal articles, listens to lectures, and so on, he makes use of bodily sense organs, perceives things from a particular bodily perspective, manipulates the objects perceived using hands and other organs, etc.

Moreover, like other human beings, the scientist is someone for whom the corporeal world he occupies contains other embodied subjects. In particular, it contains other scientists. It is *essential* to the practice of the individual scientist that there be other scientists with whom he interacts. One reason for this is that the practice of science in part involves the mastery and deployment of an existing body of scientific knowledge, which includes not only book learning but also (as Polanyi emphasized) *tacit* knowledge that is embodied in ways of perceiving and acting, and becomes part of the scientist's "local Background" (once again to deploy Searle's expression). Neither the book learning nor the tacit embodied knowledge is spun out of whole cloth by the individual scientist himself, but rather is acquired from other scientists – in college and graduate school, in the laboratory, at academic meetings, etc.

One need not endorse the more extreme relativist conclusions of some sociologists of science to see that there really *is* such a thing as the sociology of science – that scientists, like members of any other profession, inhabit communities which inculcate certain assumptions, practices, and norms, and that these assumptions, practices, and norms and the nature of their social inculcation can be identified and studied. Thomas Kuhn's analysis (1962) famously deploys the notion of a "paradigm," i.e. a set of ruling assumptions and standards of inquiry associated with a scientific theory, reflected in standard textbooks, etc. The training of a scientist essentially involves his initiation into a dominant paradigm, and ordinary scientific practice or "normal science" essentially involves the application of a paradigm to new problems, the attempt to resolve its outstanding problems, and so forth. Even "revolutionary science," by which a dominant paradigm is criticized and finally overthrown, is essentially a social enterprise insofar as it involves a shared judgment among a critical mass of researchers that a dominant paradigm is deficient and needs to be replaced, the organization of resistance to it, etc. Again, whether or not one agrees with all the conclusions Kuhn drew from this analysis (and I do not), there is obviously much truth in it.

Another respect in which interaction with other scientists is essential to the practice of the individual scientist concerns *language*. Obviously, it is by way of language that the existing body of scientific knowledge is transmitted to scientists in books, lectures, etc. Everyday scientific practice involves the deployment of technical terminology, mathematical equations, stock lines of argument, lists of elements or species, etc. and all of this is embodied in language. Now, language is an essentially *social* phenomenon, with scientific language being no different from any other sort in this respect. Moreover, it presupposes an objective world which different language users together occupy.

In an influential analysis, Donald Davidson (2001) speaks of a "triangulation" between the language user, other language users, and objects in their common environment. For one speaker to interpret another's utterances requires, in the most fundamental case, noting what is going on in that common environment and attributing to the other speaker thoughts that would, given what is going on, be the sort that would naturally be expressed by way of such utterances. To take a trivial example, if someone says "That must be John" in a context in which there has just been a knock on the door, we would naturally interpret his utterance as an expression of the thought that John is the one who knocked on the door. Knowledge even of the meaning of one's *own* utterances is similarly grounded, insofar as one takes oneself to be expressing the sorts of thoughts people would normally express by way of the words one is using, given the way those words are typically used in one's linguistic community. In this way, the very practice of using language presupposes that one is a thinking subject in a world of commonly accessible objects that is also occupied by other thinking subjects.

Of course, Davidson's particular way of spelling out the social nature of language raises all sorts of questions, and the topic of language is in any case a large one. But one needn't be a Davidsonian in order to acknowledge that language is essentially social. The point for present purposes is just to note that science presupposes not just that the individual scientist is an embodied subject, and not just that he inhabits a world of physical objects, but also that among the objects in that world with which he deals are other embodied subjects.

2.5 Intentionality

Intentionality is the "aboutness" or "directedness" toward an object familiar from thought and language. A distinction is commonly drawn between *intrinsic intentionality* and *derived intentionality* (Searle 1992, p. 78). The word "cat" is about or directed toward a certain kind of animal, but there is nothing in the physical properties of the ink marks, sound waves, or pixels in which the word is expressed that gives it that particular significance. The intentionality of the written, spoken, or typed word derives entirely from the conventions of language users. A thought about a cat is also directed toward that animal, but in this case the intentionality is intrinsic or built into the thought. It is the intrinsic intentionality of the thoughts of language users that is the source of the derivative intentionality of words and other symbols. (Searle speaks also of what he calls *as-if intentionality*, which is not really a kind of intentionality at all but rather has to do with the fact that some things can usefully be described *as if* they had intentionality. For example, I might say "My thermostat thinks the house has gotten too cold," when in fact it does not literally think anything.)

Intentionality is manifest in several of the phenomena we have been discussing – in particular, in the thoughts and perceptual experiences of scientists, in their actions, and in the semantic content of the things they and the books and articles they write. The intentionality of these latter, linguistic phenomena is derivative from the intrinsic intentionality of the thoughts, perceptual experiences, and actions of scientists, so let's focus on those.

Human actions, including the actions of scientists, exhibit a kind of intentionality even apart from the conscious thoughts that often generate them. Heidegger notes that we relate to things as "equipment" (in his sense of the term) insofar as we take them to exist "in-order-to" realize some end (1962, pp. 68 and 97). We relate to a hammer as that which might be used *in order to* pound nails, we relate to a cup as that which might be used *in order to* drink, and so on. As Dreyfus suggests (1993), what Heidegger is talking about here is a kind of intentionality that is different from and more fundamental than the kind familiar from conscious thought. Our use of things is "directed toward" certain ends without our always being aware of the fact. The carpenter often just hammers, and the coffee drinker often just takes a sip, without thinking about it – say, when absorbed in a conversation, or while one's conscious thoughts are occupied by some other topic. Merleau-Ponty describes this as "motor intentionality" (2012, p. 112-13), which he characterizes as something that exists in between the intentionality of thought on the one hand, and sub-personal causal processes on the other. Action has a "directedness" that is not present in such causal processes, but that is nevertheless not necessarily conscious the way that the intentionality of a thought is. "To move one's body," Merleau-Ponty says, "is *to aim at* things through it" (1967, p. 153, emphasis added).

The intentionality of perceptual experiences and of propositional attitudes (*believing* that such-and-such is the case, *hoping* that such-and-such is the case, *fearing* that such-and-such is the case, etc.) is the sort that is usually in view in contemporary philosophical discussion of the subject. A visual experience of seeing a tree is *directed toward* or *about* the tree. The belief that the tree is an oak is also *about* the tree, and *represents* it as being of a certain kind. Both propositional attitudes and, in human beings, perceptual experiences too involve the application of *concepts.* In a visual experience, you perceive the object you see *as a tree.* You apply the concept *tree* to it. In believing that it is an oak, you apply the concept *oak* to it. And so on.

Now, the notion of a concept is certainly relevant to understanding the nature of science and of scientific practice, since scientists make observations, weigh hypotheses, draw inferences, write up their results, present arguments, publish books, give lectures, etc., and all of this involves the application of concepts. But what is most relevant to the point I want to make at the moment is, again, the *intentionality or directedness* of perceptual experience and of the propositional attitudes, and intentionality or directedness can exist even in the absence of concepts. (Non-human animals have perceptual experiences, and these experiences are *directed toward* the objects that the animals see, hear, feel, etc. But a non-human animal does not *conceptualize* the objects of its experience.)

The reason it is relevant is that *directedness toward* an object is something that the mechanical world picture claims does not really exist in the natural world. For the mechanistic picture denies the existence of *immanent teleology* or *final causes* in nature, and *directedness* is the core of the notion of teleology or final cause. Hence to say that there is no teleology immanent to the natural world is at least implicitly to say that there is no *intentionality* there either. That is why Cartesians relocate intentionality and teleology in general out of the natural world and into the immaterial *res cogitans* and the divine mind. And it is why the materialist philosopher Jerry Fodor writes:

> I suppose that sooner or later the physicists will complete the catalogue they've been compiling of the ultimate and irreducible properties of things. When they do, the likes of *spin*, *charm*, and *charge* will perhaps appear upon their list. But *aboutness* surely won't; intentionality simply doesn't go that deep... If the semantic and the intentional are real properties of things, it must be in virtue of their identity with (or maybe of their supervenience on?) properties that are themselves *neither* intentional *nor* semantic. If aboutness is real, it must be really something else. (1987, p. 97)

To be sure, Fodor doesn't explicitly frame the issue in terms of mechanism and its rejection of immanent final causes. But it is only because materialists like Fodor implicitly presuppose a mechanistic or non-teleological conception of nature that it seems obvious to them that intentionality cannot be a fundamental feature of nature – that what *appears* to be intentionality "must be really something else."

Now, if human thought and action, including the thought and action of scientists, entails the existence of intentionality, and the existence of intentionality entails the existence of (a kind of) teleology or final causality, then scientific practice itself entails the existence of (a kind of) teleology or final causality. If Cartesianism is rejected, this teleology must be immanent to the natural world – which is exactly what Aristotelian philosophy of nature maintains.

To avoid this result, the materialist will have to show either that the intentionality of thought and action can be analyzed without remainder in terms of notions that the mechanistic picture is willing to countenance (such as efficient causation), or that intentionality is simply illusory. The first strategy, which is the one Fodor endorses, is reductionist; the second, favored by philosophers like Paul Churchland (1981) and Alex Rosenberg (2011), is eliminativist.

But neither strategy can succeed. Consider first human action, which seems as goal-directed or teleological as anything could be. It is sometimes claimed that this appearance is deceptive, and that action can be analyzed in terms that make no reference to goals or ends but only to efficient causes. To take a stock example, it is held that an explanation like *Bob knocked over the glass of water for the purpose of distracting Fred* can be rephrased as *Bob had the intention of distracting Fred and this caused him to knock over the glass of water*, where the latter description replaces the reference to purpose with a reference to efficient causation instead. Now, one problem with this is that there is still a reference to Bob's *intention* in acting, and this entails a kind of *directedness toward an end* that remains to be analyzed away. But suppose for the sake of argument that that could be done. As Scott Sehon (2005, Chapter 7) has noted, the analysis would still fail. For consider the case where Bob's intention to knock over the glass makes him so nervous that his hand shakes uncontrollably and knocks over the glass before he otherwise would have. Then it is certainly true that *Bob had the intention of distracting Fred and this caused him to knock over the glass of water*, but it is *not* true that *Bob knocked over the glass of water for the purpose of distracting Fred*. For in this case he knocked over the glass not *for the purpose of* distracting Fred (even though he did want to do that at some point), but rather because he lost control of his hand. So the two descriptions are not equivalent at all. To salvage his reformulation, the reductionist would have to stipulate that the intention in question can cause the resulting action only via bodily motions that the agent has *guidance* of or *control* over, rather than by involuntary shaking and the like. But the trouble with this is that "guidance" and "control" are *themselves* teleological notions – guidance or control is always guidance or control *towards* an end or goal – so that the analysis will not have truly eliminated teleology at all.

Reductionist accounts of the intentionality of thought and perception are, notoriously, no less problematic. There are two main varieties, *causal theories* (e.g. Fodor 1987) and *biosemantic theories* (e.g. Millikan 1984). The basic idea of a causal theory is that a neural state will represent some feature of the world outside the brain if it stands in the right sort of causal relation to that feature. For example, a certain neural state will represent water if it is caused by the presence of water under the right conditions. The basic idea of a biosemantic theory is that a neural state will represent some feature of the world external to the brain if that neural state was hardwired into us by natural selection because it caused our ancestors to interact with that external feature in a way conducive to fitness. For example, a neural state will represent water if natural selection favored creatures manifesting that neural state because the neural state caused them to seek out water.

Several technical objections have been raised against the various versions of these theories. (Cf. Feser 2006, Chapter 7) But in my view there are two fundamental and insurmountable problems with them. The first is that, so long as they presuppose a mechanistic conception of the material world, the most these theories can afford the materialist is what Searle calls *as-if* intentionality, which is not really intentionality at all. No matter how many details such theories add to the causal or evolutionary stories they tell, the stories will always be consistent with our merely behaving *as if* we had intentionality without our really having it (just as a thermostat behaves *as if* it thought the house is too cold even though it does not in fact think anything). So, there must be some further aspect to our having intentionality over and above what is captured by the theories. What that aspect is is the *directedness toward* an object that is precisely what the reductionist wants to avoid any reference to. Reductionist theories really just leave out the phenomenon to be explained and change the subject, rather than truly explaining it. (Cf. Searle 1992, pp. 49-52)

The second fundamental and insurmountable problem is that causal chains, evolutionary histories, and anything else the mechanistic world picture could countenance as a ground for intentionality cannot account for the *determinacy of content* that some intentional states possess. Take Quine's famous example of a native speaker's utterance of "gavagai," and a field linguist who considers the possibility that the correct translation might be "Lo, a rabbit." Quine focuses on the question of what might be gleaned from the speaker's behavior, but as others have emphasized, no matter *what* physical facts the interpreter might consider – the speaker's neural processes, causal relations between the speaker's brain and the external environment, the way natural selection molded his ancestors, and so forth – those facts will be perfectly consistent with the alternative hypotheses that what the speaker meant was really "Lo, an undetached rabbit part" or "Lo, a temporal stage of a rabbit." This is not merely an *epistemological* point but a *metaphysical* point. It's not just that we couldn't *know* from the physical facts alone what was meant, but that there simply *would be no* objective fact of the matter about what was meant, if the physical facts alone (especially as conceived of within the mechanical world picture) could determine what was meant. The physical facts to which the reductionist would appeal are systematically indeterminate in content in a way utterances and thoughts (at least often) are not. (This is merely one example of several in the large literature on this subject. I have developed and defended indeterminacy objections to materialist theories of intentionality at length elsewhere. See Feser 2011b and 2013a.)

The materialist's alternative strategy – to deny that intentionality is real in the first place, so that we ought to eliminate it altogether from our picture of the world rather than bothering to develop a better reductionist account – is even more hopeless. For one thing, there are no good arguments for it. Churchland and Rosenberg take the failure of reductionist accounts itself to constitute a good reason to embrace eliminativism. Rosenberg, who thinks there is no more to reality than what is described by physics, specifically, takes the absence of intentionality from the list of properties countenanced by physics to be a decisive consideration in favor of eliminativism. But of course, only someone who already agrees that physics as the mechanist would interpret it gives us an *exhaustive* description of reality, or that anything real is going to be susceptible of materialist reduction, would be moved by such arguments. Critics of eliminativism, including Aristotelians, do not agree with that at all. Hence these arguments simply beg the question. They are also not applied consistently. As Stephen Stich (himself a former eliminativist) and Stephen Laurence emphasize (1996), there are all sorts of notions for which we have no good reductionist analysis (they give examples like *couch*, *car*, *war*, *famine*, *ownership*, *mating*, and *death*) but which few would propose we eliminate from our ontology.

This brings us to the deepest problem with eliminativism, which is that it simply cannot possibly be coherently spelled out. The simplistic or "pop" way of making this point is to say that eliminativists claim to believe that there are no beliefs, which is a performative self-contradiction. As eliminativists rightly point out, by itself this is not a very impressive objection, because the eliminativist can always avoid using locutions like "I believe that..." and make his point in other terms instead. However, the real question is whether the eliminativist can spell out his position in a way that *entirely* avoids terms that explicitly or implicitly presuppose intentionality. That *cannot* be done. Even if you get rid of intentionality in one area it will, like the proverbial whack-a-mole, always rear its head somewhere else. (Cf. Baker 1987, Chapter 7; Boghossian 1990 and 1991; Reppert 1992; Hasker 1999, Chapter 1; Menuge 2007, Chapter 2; Feser 2008, pp. 229-37)

Hence, consider the eliminativist's central claims – that intentionality is *illusory*, that descriptions of human beings as possessing intentionality are *false*, that it is a *mistake* to try to reduce rather than eliminate it, etc. All of these notions are as suffused with intentionality as any that the eliminativist wants to overthrow. They presuppose the *meaning* of a thought or of a statement that has failed to *represent* things accurately, or a *purpose* that one has failed to achieve or that one should not have been *aiming* to achieve in the first place. Yet we are told by the eliminativist that there are no purposes, meanings, representations, aims, etc. of any sort whatsoever. So, how can there be illusions, falsehoods, and mistakes?

For that matter, how can there be *truth* or *correctness*, including the truth and correctness the eliminativist would ascribe to science? For these concepts too presuppose the meaning of a thought or statement that has represented things accurately, or the realization of a purpose. Thus, "Water is composed of hydrogen and oxygen" is true, while "Water is composed of silicon" is false, and the reason has to do with the meanings we associate with these sentences. Had the sentences in question had different meanings, the truth values would not necessarily have been the same. By contrast, "Trghfhhe bgghajdfsa adsa" is neither true nor false, because it has no meaning at all. Yet if eliminativism is right, "Water is composed of hydrogen and oxygen" is as devoid of meaning as "Trghfhhe bgghajdfsa adsa" is – in which case it is also as devoid of a truth value as the latter is. Moreover, if eliminativism is right, every statement in the writings of elminativists themselves, and every statement in every book of science, is as devoid of meaning as "Trghfhhe bgghajdfsa adsa" is, and thus just as devoid of any truth value. But then, in what sense do either science or eliminative materialist philosophy give us the *truth* about things?

Logic is also suffused with intentionality, insofar as inferences *aim* at truth and insofar as the logical relationships between statements presuppose that they have certain specific *meanings*. "Socrates is mortal" follows from "All men are mortal" and "Socrates is a man" only because of the meanings we associate with these sets of symbols. If we associated different meanings with them, the one would *not* necessarily follow from the others. If each was as meaningless as "Trghfhhe bgghajdfsa adsa" is, then there would be no logical relationships between them at all – no such thing as the one set of symbols being *entailed* by, or *rationally justified* by, the others. But again, if eliminativism is right, every sentence, including every sentence in every work of eliminativist philosophy and every sentence in every book of science, is as meaningless as "Trghfhhe bgghajdfsa adsa" is. In that case there are no logical relations between any of the sentences in any of these writings, and thus no valid arguments (or indeed any arguments at all) to be found in them. So in what sense do either science or the assertions made by eliminativist philosophers constitute *rational defenses* of the claims they put forward?

Notions like "theory," "evidence," "observation," and the like are as suffused with intentionality as the notions of truth and logic are. Hence if there is no such thing as intentionality, then there is also no such thing as a scientific theory, as evidence for a scientific theory, as an observation which might confirm or disconfirm a theory, etc. Eliminativism makes of *all* statements and *all* arguments – scientific statements and arguments no less than metaphysical ones, and indeed every assertion of or argument for eliminativism itself – a meaningless string of ink marks or noises, no more true *or* false, rational *or* irrational than "Trghfhhe bgghajdfsa adsa" is. As M.R. Bennett and P.M.S. Hacker put it (2003, p. 377), the eliminativist "saws off the branch on which he is seated," undermining the very possibility of science in the name of science.

The eliminativist owes us an explanation, then, of how he can so much as *state* his position in a coherent way in the absence of all these notions his position requires him to jettison. The stock eliminativist move at this point is to claim that future neuroscience will provide new categories to replace these old ones, at which point the view can be stated in a more consistent way. But that is like someone asserting that 2 + 2 = 23 and then, when asked what exactly this claim can mean given what it is to add, what it is for numbers to be equal, etc., responding that he can't really say but that future mathematicians will come up with a way of making sense of it. Until we have such an explanation, we don't even know so much as *what the claim is* that we are being asked to consider, much less whether it is correct. The same can be said for eliminativism. Until we are given a coherent way of *formulating* the thesis, we don't really have anything that *amounts to* a thesis, much less one we have reason to take seriously.

Hilary Putnam reports that Churchland once acknowledged in conversation that he needs a "successor concept" to the notion of truth, and that he doesn't know what it will be (Putnam 1988, p. 60). This is doubly problematic. For one thing, if an eliminativist like Churchland is admitting both that he cannot claim that eliminativism is *true* (since a consistent eliminativist has to regard the notion of truth as illusory as that of intentionality) and that he has nothing to put in place of truth, then it is not clear exactly what he is trying to say about eliminativism or to convince *us* to say about it. (It cannot be "Eliminativism is true," but at best something like "Eliminativism is _____ ," with no explanation of how to fill in the blank. Until we have such an explanation, what is it that are we supposed to do with this utterance?) For another thing, the notion of a *concept* is as suffused with intentionality as the notions of truth, meaning, etc. are. So, if Churchland is consistent, he cannot say that he needs a "successor concept" for the notion of truth, but rather a "successor _____ ," where we now have a second blank to fill in with a term that does not entail the existence of intentionality, and where we are once again at a loss as to what that term might be.

We will have reason to revisit eliminativism later on, when we examine issues in the philosophy of neuroscience. For the moment it will suffice to note that the irreducible and ineliminable intentionality of the perceptual experiences, thought processes, and actions of embodied human subjects – including scientists as they engage in the practice of science – entails the existence in those subjects of *immanent finality or directedness* of the sort posited by Aristotelian philosophy of nature. The very existence of thinking, perceiving, and acting scientists themselves thus puts an absolute limit on how far teleology might be eliminated in the name of science.

2.6 Connections to the world

Like final causality, efficient causality is sometimes said to have been banished from the natural world by modern science. Bertrand Russell (2003) at one point in his career argued that causation cannot be real, because no such notion appears in the differential equations of physics. There are several problems with this argument, which I have detailed elsewhere (2014b, pp. 114-18) and merely summarize here. For one thing, and as I will argue in the next chapter, it is simply a mistake to suppose that if something is really there in the natural world, it will show up in the mathematical description afforded by physics. For another thing, if Russell's argument were applied consistently, we would also have to eliminate notions like "law" and "event," since they too do not appear in the equations themselves (Schaffer 2007). Yet these notions are indispensable in physical explanation. Nor, as we will see later on, is the notion of causation eliminable from other sciences, or even from physics itself at the end of the day.

In any event, Russell himself came later in his career not only to abandon this position, but to make crucial use of the notion of efficient causation in spelling out the epistemology of physics. (Cf. Eames 1989) In *The Analysis of Matter*, Russell says:

> Percepts are in our heads, [and] they come at the end of a causal chain of physical events leading, spatially, from the object to the brain of the percipient. (1927, p. 320)

And in *Human Knowledge*, he writes:

> Everything that we believe ourselves to know about the physical world depends entirely upon the assumption that there are causal laws. Sensations, and what we optimistically call "perceptions," are events in us. We do not actually see physical objects, any more than we hear electromagnetic waves when we listen to the wireless. What we directly experience might be all that exists, if we did not have reason to believe that our sensations have external causes. (1948, p. 311)

Now, Russell is here evincing a commitment to a representationalist theory of perception and knowledge, but that is not essential to his main point, which is that we cannot take perceptual experiences to give us knowledge of a mind-independent physical world unless we also take those experiences to be *causally connected* to such a world.

The thesis that a perceptual experience can give us knowledge of an object only if the experience is caused by that object has been a commonplace in contemporary philosophy at least since Paul Grice's influential essay on the subject (1961). Among the standard arguments for the thesis is the consideration that, in order for a perceiver genuinely to perceive an object, it is not enough that the object be present and that he have an experience *as of* seeing the object. For we can imagine a case in which he is merely having a hallucination and, by chance, an object whose appearance matches that of the hallucinated object happens to be present before him. For an experience to count as a genuine perception of the object, the presence of the object itself has to play some role in *generating* the experience. Then there is the fact that aberrant causal factors are implicated in cases where perception goes wrong. When we fail to perceive a thing that is there, that is typically because there is some disruption in the causal chain by which the thing would otherwise generate an experience of it (e.g. a physical barrier between a person and the thing he would otherwise see, or damage to the eye or optic nerve). When we think we perceive something that is not really there, that is often because there is some dysfunction in the proximate causal processes underlying experience (e.g. brain damage, or the presence of hallucinogens). Even in perceptual experiences that are not dysfunctional, changes in the phenomenal character of a perceptual experience tend to co-vary with changes in the object perceived or in the condition of the sense organs (e.g. the lighting in the object's surrounding environment, or the wearing of sunglasses or earplugs).

To be sure, there has been much controversy over whether the existence of a causal connection between an object and a certain experience is *sufficient* for the latter to count as a perception of the former (Lewis 1980; Snowdon 1980-81). For one thing, there are multiple causal factors involved in the perception of an object – the source of the electricity that powers the lights by which we see an object in an otherwise dark room, the neural processes involved in perception, etc. – and most of these are not perceived. An aberrant causal process, such as a malfunction in the brain, may cause a hallucination, but that process is not perceived. There are even more eccentric examples in the literature of cases in which an experience is caused by an object and the thing experienced resembles the object but where it seems implausible to say that the object is actually perceived. What matters for present purposes, however, is that a causal connection between an object and an experience is a *necessary* condition for the latter to count as a perception of the former, even if it is not a sufficient condition. (See Fish 2010, chapter 7, for an overview of the contemporary debate over perception and causation.)

Needless to say, physics and other empirical sciences rely for their evidential basis on observation and experiment, and thus on perceptual experience. Science also rests on perception in a more mundane way insofar as scientists can converse with one another, read each other's journal articles, purchase and operate lab equipment, etc. only if they can perceive fellow scientists and other material objects. Hence, given that perception presupposes causal connections with the things perceived, the very practice of science presupposes that external physical objects are among the efficient causes of our experiences of them. But the causation goes in the other direction as well, for a scientist has to *do* something in order to carry out observations and experiments, write up his results, and so forth. Thus he has to have a causal influence on the experimental setup, on lab equipment, other scientists, etc.

Moreover, the objects with which the scientist must be causally related in order to practice his science are objects *more or less as common sense conceives of them* rather than objects as re-described by physical theory. When making an observation, the scientist takes himself to be looking at a *gauge*, or at a *timer*, or looking through a *microscope*, or the like – as opposed, say, to looking at or through a collection of unobservable particles. It is only by perceiving something *as* a gauge or *as* a microscope or *as* a timer or whatever (and thus as something which has a certain specific function, which will operate properly under certain specific conditions, etc.) that he can *intelligibly* take it to be providing just the information he takes it to provide. Physical objects *qua* collections of particles do not register pressure, or measure the passage of time, or magnify small objects. It is only physical objects *qua* gauges, timers, microscopes, etc. which do so. By the same token, the scientist has to perceive his fellow scientists *as scientists* and thus *as human beings* rather than as collections of particles or the like. For it is only by doing so that he can take what they say or do as intelligent speech and action, the relating of arguments or evidence, etc. Physical objects *qua* collections of particles don't speak or act intelligently, relate arguments or evidence etc. Only physical objects *qua* human beings can do that.

In other words, at least many of the perceptual experiences on which the practice of science rests, and at least many of the objects which these experiences are experiences of, can intelligibly be described only in terms of the "folk ontology" of common sense. It is not merely that the perceptual experiences of scientists must be causally related to physical objects of some sort or other. It is that they must be causally related to *human beings*, *gauges*, *timers*, *microscopes*, etc., specifically. Nor do human beings and scientific instruments alone make up the folk ontology presupposed by the perceptual experiences of scientists. Natural substances in general are part of it too. The chemist perceives the substance he is analyzing *as* stone or water, the biologist perceives the organism he is studying *as* a frog or a tree, and so forth. A reductionist or eliminativist program for replacing such "folk" notions with descriptions couched exclusively in terms of fundamental particles or the like cannot be coherently carried out (Elder 2004, pp. 50-58; Feser 2014b, pp. 177-84). However, for present purposes it suffices to point out merely that at least *some* folk notions are implicated in the perceptual experiences of scientists.

Naturally, concepts like "scientist," "human being," "gauge," "timer," "microscope," and the like don't appear in the equations of physics any more than the notion of "cause" does, despite the fact that the very practice of science is unintelligible without them. This reinforces the point that if some notion does not appear in the ontology of some science (including physics), that does not by itself give us any reason to conclude that there is nothing in reality that corresponds to that notion. What it may tell us instead (as it does in this case) is that the scientific ontology in question is simply incomplete. There can be no more conclusive reason for judging that some science gives us only an incomplete description of reality than that it fails to account for the existence of scientists and scientific practice themselves!

A failure to see this difficulty is one of several problems with John Norton's (2007) attempt to revive the early Russell's skepticism about causation. Norton rehearses the ways in which the notion of causation has undergone transformation since the seventeenth century. First, he reminds us, final causality was banished by the mechanical philosophy. Early modern scientists held that efficient causes operate only locally, but in the centuries after Newton the judgment prevailed that such causes sometimes act at a distance. In the nineteenth century, Mill downplayed the distinction between active and passive causes and abandoned the thesis that the continued existence of a cause is needed for the persistence of an effect. What remained was merely the idea of an efficient cause as antecedent to an effect which followed upon it deterministically. But determinism was in turn overthrown by quantum mechanics. And so forth. It is therefore implausible, Norton concludes, to suppose that some interesting notion of causation will survive further advances in science.

Norton poses a dilemma for anyone who insists that there is some general notion of causation which each of the various special sciences are in the business of applying to their own domains of study. Either there is some factual content to this claim which makes it empirically confirmable by the findings of these sciences, or there is not. If the defender of causation takes the first horn of this dilemma, then he faces the problem that the historical development of the notion of causation just rehearsed makes it doubtful that there is any such content. The notion has over the course of the history of science become so stripped of its traditional content that very little now remains. That leaves the defender of causation with the second horn of the dilemma, which involves holding on to a thinned out residual notion of causation which persists despite the transformation just rehearsed, is untouched by the specific factual findings of the various special sciences, and can therefore still be applied to all of them. The trouble with this option, Norton says, is that it makes of causation a mere "empty honorific." The defender of causation will in this case be engaging in *a priori* armchair theorizing which we have no reason to think corresponds to anything in reality.

Hence, while in *practice* the various sciences make use of causal notions all the time (as Norton concedes), such notions are in his view not in fact fundamental to the scientific picture of reality. The most that can be said is that it is often useful to describe natural phenomena *as if* they manifested causality, but these descriptions are ultimately dispensable. To hammer home his point, Norton describes a mathematical model of a simple Newtonian system (never mind quantum mechanics) in which motion occurs without a cause.

Now, one problem with this line of argument is that it simply begs the question, in multiple ways. Norton presents his summary of the history of thinking about causation as if it were a rehearsal of various *findings of empirical science*. Of course, the Aristotelian would say that it is not that at all. Rather, it is a history of what are essentially *philosophical* ideas about causation that have influenced both the way philosophers and scientists have *interpreted* the findings of science, and what they are prepared to count as the proper *methodology* of science. What Norton describes, in other words, is in fact a history of ideas *within the philosophy of science and the philosophy of nature*, rather than a history of results in physics, chemistry, etc. (even if the distinction between the former and the latter is often blurred, and often blurred by scientists themselves). Moreover, it is a history of *mistaken* ideas about causation, or so the Aristotelian would argue. So it is no good for Norton glibly to assert that such-and-such elements of the traditional notion of causation have been refuted by science. For whether that is really the case is precisely part of what is at issue between him and his opponent.

Second, Norton's dilemma would be a false one even if he *had* accurately characterized the history he recounts. He supposes that if the reality of causation is not revealed by the findings of physics and the various special sciences, then the only approach left to its defender is *a priori* speculation whose deliverances will be "empty" of substantive content. But as we saw in chapter 1, Aristotelian philosophy of nature maintains that there are propositions that are empirical and substantive yet nevertheless not subject to empirical falsification the way that the claims of physics, chemistry, and the other empirical sciences are. The proposition that *change occurs* was offered as an example. It is knowable via sensory experience, but no sensory experience could overturn it, because any sensory experience that purportedly did so would itself involve change.

Now, it is in the context of analyzing change that Aristotelian philosophy of nature introduces the distinction between actuality and potentiality, which in turn grounds the analysis of efficient causation as the actualization of potential. The reality of causation is thus knowable *empirically*, since it follows from the fact of change, which is knowable empirically. At the same time, knowledge of its reality does not depend on the findings of physics or of any other empirical science, since the very existence of sensory experience itself (on which any possible empirical science must rest) suffices to reveal its reality. Hence the options are not limited to the ones Norton recognizes. Furthermore, the Aristotelian analysis of causation is by no means "empty" or without substantive content, for it does significant theoretical work. For example, it accounts for the possibility of change, contra Parmenides and Zeno; it has been developed by Scholastic writers into a nuanced and multi-layered account of diverse causal phenomena (Feser 2014b, Chapters 1 and 2); it grounds several important arguments in natural theology (Feser 2017); and so forth. Norton would no doubt be as critical of these applications of the Aristotelian account of causation as he would be of the account itself, and would also no doubt reject the claim that there is some third possible avenue for the defender of causation beyond grounding it either in the findings of the empirical sciences or in *a priori* speculation. The point, though, is that he gives no non-question-begging *reason* for such a position.

A further problem for Norton's argument is that his proposed mathematical model of a Newtonian system in which motion occurs without a cause simply does not do the work he thinks it does. In particular, it lends no plausibility to the claim that causation is absent from physical reality. Physicists find the notion of a frictionless plane very useful, but that is no reason at all to believe that there really are any frictionless planes in nature. The Newtonian idea of inertial motion is also very useful, though no such motion actually exists in nature (since every physical object is in fact always acted upon by outside forces). Examples could be multiplied of idealizations that possess great utility but do not strictly correspond to reality. Norton gives us no reason to believe that his imagined system is any different. Russell himself came in later years to emphasize that the mathematical picture of the natural world afforded us by physics is by no means an *exhaustive* description of that world. Hence the absence of some feature (such as causation) from that *picture* by itself gives us no reason to think that that feature is also absent from *nature*. (More on this subject in the next chapter.)

Additionally, and as already indicated, Norton seems completely blind (as, again, the later Russell was not) to the problem that to strip the natural world of efficient causation makes the epistemology of physics completely mysterious. It also undermines standard contemporary naturalistic accounts of knowledge in general, of mental content, and of other phenomena, which make crucial use of the notion of efficient causation. To the extent that thinkers like Norton deny the reality of causation in the name of a scientifically informed naturalism, then, they threaten to make their overall metaphysical picture of the world incoherent.

In any event, whether or not *naturalism* can explain how the scientist has perceptual knowledge of the material world external to the mind, there must *be* an explanation, given PSR. That explanation cannot be in terms of some *necessary* connection between perceptual experiences and material objects, for their relationship is manifestly *contingent* insofar as it is possible for a perceiver to have an experience as of some object even when the object is not really present (as in hallucination), and possible for the object to be present even when the perceiver has no experience of it (for example, when his perceptual apparatus is damaged). But the only candidate contingent connection there is is a *causal* connection of some sort. Even if we were to reject the commonsense supposition that external objects *themselves* cause our experiences of them, in favor of some eccentric theory like occasionalism or Leibniz's theory of pre-established harmony, this would not eliminate causation, but merely relocate it in God.

There is also the consideration that, even apart from their causal relations to perceivers, we cannot make sense of the notion of a world of mind-independent material objects at all unless we suppose that they bear causal relations to *each other*. As P. F. Strawson argues:

> [O]ur concepts of objects are linked with sets of conditional expectations about the things which we perceive as falling under them. For every kind of object, we can draw up lists of ways in which we shall expect it not to change unless..., lists of ways in which we shall expect it to change if..., and lists of ways in which we shall expect it to change unless...
>
> [C]oncepts of *objects* are always and necessarily compendia of causal laws or law-likeness, carry implications of causal power or

> dependence. Powers, as Locke remarked – and under "powers" he included passive liabilities, and dispositions generally – make up a great part of our idea of substances. More generally, they must make up a great part of our concepts of any persisting and re-identifiable objective items. And without some such concepts of these, no experience of an objective world is possible. (1989, pp. 145-46)

In short, belief in *causation* as a mind-independent feature of the world and belief in mind-independent *objects* stand or fall together. Barry Stroud points out that Hume's skepticism about causation avoids this problem because Hume was also a skeptic about material objects. But contemporary skeptics about causation (like Norton) tend not to share Hume's skepticism about such objects, and thus they owe us an explanation of how they can consistently deny causation while affirming the existence of mind-independent objects (Stroud 2011, pp. 23-24).

Even Hume is not consistent in another respect, however. Skeptics about causation also owe us an account of exactly how we come to *believe* that causation is a real feature of mind-independent reality if (as they claim) it isn't. But the trouble is that any such account is bound to be a *causal* account, and in particular a description of the psychological mechanisms by which the purportedly false belief is generated. Hume himself offers an account of the psychological process he thinks "produces" or "gives rise to" the notion of causal necessity in us – despite the fact that notions like *production* and *giving rise to* are themselves *causal* notions, appeal to which Hume is not entitled to, given his denial that such notions track objective reality (Stroud 2011, pp. 33, 56-57).

Like final causality, then, efficient causality cannot coherently be banished from the world altogether in the name of empirical science. On the contrary, the very practice of science – and indeed, the very attempt to banish efficient causation – presupposes its reality.

2.7 Aristotelianism begins at home

It is time to tie together the many threads of argument developed in this chapter. We began by noting that, notwithstanding the many disputed issues in modern philosophy of science, there is broad agreement that scientific method entails an "arch of knowledge," the feet of which comprise observation and experiment and at the apex of which are what Maritain calls "empiriological" theories, with the subclass of "empiriometric" theories regarded as the gold standard of empiriological inquiry (and thus as the gold standard of science).

Now, the empiriological description of nature is essentially what Sellars calls the "scientific image" of the world, as opposed to the "manifest image" of common sense and ordinary experience. Since the subclass of "empirioschematic" sciences make use of concepts that are widely regarded as merely regulative rather than corresponding to anything in mind-independent reality, there is a tendency to identify the scientific image, strictly construed, with the empiriometric description of the world, specifically – that is to say, with a mathematicized conception of nature of the kind toward which the "mechanical world picture" tended, and that has become definitive of modern physics. That is not to say that those who take the scientific image to exhaust reality would all hold that everything real can be *reduced* to entities within the ontology of physics. Some would say instead that everything real need only *supervene* on the latter. Either way, though, for those who take the scientific image to be an exhaustive picture of reality, the ontology of physics "wears the trousers," as it were.

So understood, the scientific image also essentially corresponds to what Bernard Williams calls the "absolute conception of reality" (1990, p. 65) and Thomas Nagel calls "the view from nowhere" (1986). The basic idea of this "absolute conception" is to construct a description of the world that is entirely free of any explicit or implicit reference to the point of view of any particular observer, or any particular type of observer. As Nagel emphasizes, the conception in question regards anything that depends on the point of view of particular observers as "subjective," and thus it takes itself to be by contrast an entirely "objective" description. (The deletion of the points of view of particular observers is what makes this a view "from nowhere.") The basic idea has antecedents in the Greek atomists, and took center stage in Western thought with Galileo, Descartes, Locke, and other early modern proponents of the mechanical world picture. The distinction between primary and secondary qualities became the standard way of expressing the idea, with secondary qualities regarded as reflecting the observer's subjective point of view and primary qualities alone constituting the truly objective features of reality.

As Lorraine Daston and Peter Galison argue (2007, chapter V), in the late nineteenth and early twentieth centuries it became common to hold that *structure* is the key to objectivity so understood. The idea was to get away from the *self* even of the scientist, with his individual physiology, the contingencies of whatever natural language he speaks, etc., and to formulate scientific claims in terms of what is *intersubjectively communicable* between scientists. This project was pursued in different ways by Poincaré, Frege, Carnap, Schlick, Russell, Weyl, et al. Some emphasized *relations* as what could be communicated; some emphasized *invariance under transformations*; some emphasized the construction of formal languages. But the general idea that communicable structure is the key to scientific objectivity reinforced the tendency toward a purely quantitative and mathematicized conception of nature.

Part of what this chapter has been concerned to show is that the manifest image, the world as it appears from the "subjective" point of view of the conscious subject, cannot coherently be eliminated and replaced entirely by the "objective" or "absolute" perspective of the scientific image. For the latter presupposes the former, in two fundamental respects. First, abandoning the manifest image while trying to maintain the scientific image is tantamount to attempting to keep the apex of the "arch of knowledge" aloft while destroying its feet and legs. As Colin McGinn writes, the scientific image "purchases [its] absoluteness at the cost of removing itself from the perceptual standpoint" (1983, p. 127). Hence, "to abandon the subjective view is to abandon the possibility of experience of the world" (p. 127), and thus to abandon the evidence of observation and experiment on the basis of which the claims of the scientific image are supposed to be justified. It is also to abandon the reasoning processes that take us from that empirical evidence up to the scientific image and then back down from it to testable predictions. For the subjective view includes the *cognitive* (as well as the perceptual) states and processes of the scientist.

Second, the scientific image even *in itself*, considered apart from the specific perceptual experiences and specific cognitive processes that historically lead to it and give it a rational justification, cannot be understood apart from the subjective or manifest image. We are not in a situation of having a scientific description of reality that is *intrinsically* free of any taint of specifically human modes of cognition, with only the *justification* of the thesis that that description is correct having to reflect human perceptual experiences and reasoning processes. Rather, human "subjectivity" reaches all the way *into the scientific description itself*. However apparently "absolute" or "objective" it appears to be, it is always constructed from the point of view of a specifically human mode of cognition and always reflects that point of view.

To see this, consider the ways in which, for Aquinas, an angelic intellect differs in its mode of knowledge from a human intellect. (Note that whether angelic intellects actually exist is neither here nor there for present purposes. The point is just to illustrate the idea that there can at least in principle be intellects of radically different sorts.) For Aquinas, an angel does not acquire its concepts from sensory experience but has them all "built in," as it were, when it comes into being. Its concepts represent not only the universal natures shared by individual things that fall under the concepts, but also all the individuals themselves. For example, an angel who has the concept *triangle* thereby grasps not only triangularity in the abstract but also all concrete individual triangles. An angel's knowledge is not discursive. It does not have to reason from premises to conclusion or compare and contrast concepts and propositions in a piecemeal way in order to determine their logical relationships. Rather, it grasps these logical relationships in a single act. The more powerful an angelic intellect, the more of reality it can grasp in a single concept. And so forth. (Cf. *Summa Theologiae* I.58)

Now, human intellects are not like this, and the scientific image, even at its most rarefied, always reflects the ways in which they are not like this. For example, its description of nature deploys a great many different concepts – *force, mass, acceleration, energy, spin, charge*, etc. – rather than conveying all of nature in a single concept. Those concepts abstract *from* individuals and capture only what is universal, rather than taking both all of the individuals and the universal together in one cognitive act. And so on. The scientist never really "gets outside of his own skin," as it were, to take in nature from a "view from nowhere." As a human being, he always inevitably cognitively carves up reality in a distinctively human way. In short, an "objective" description is itself an extension of the "subjective" point of view, and the scientific image is itself merely a component of the manifest image. (Cf. Thomasson 2007, pp. 147-50.) As Hilary Putnam concludes (quoting William James), "the trail of the human serpent is over all" (Putnam 1987, p. 16). Or as Polanyi writes:

> [A]s human beings, we must inevitably see the universe from a centre lying within ourselves and speak about it in terms of a human language shaped by the exigencies of human intercourse. Any attempt rigorously to eliminate our human perspective from our picture of the world must lead to absurdity. (1962, p. 3)

It might be objected that the practice of science could be turned over entirely to machines – computers making use of artificial sense organs and the like – and that if this were done, the point of view of the human observer would be eliminated entirely. The notion of "android epistemology" (Ford, Glymour, and Hayes 2006) might be deployed in developing such an objection. However, there are several problems with such a proposal. First, even if it were plausible to regard "android epistemology" as a genuine alternative epistemic perspective, why would it be any more "objective" or "absolute" than the human epistemic perspective? Why would it not be merely one further particular point of view from which a "view from nowhere" or "absolute conception" should try to escape? Second, why regard it as an *alternative* epistemic perspective in the first place? Computers are, after all, made by human beings, and reflect a human scientific understanding about how information might be acquired and processed by machines. So why wouldn't "android epistemology" be merely an extension of the human epistemic point of view rather than an alternative to it? Third, as we will see in a later chapter, any advocate of the "mechanical world picture" who attempts to deploy computational notions in an anti-Aristotelian argument faces a dilemma. For given that picture, computation must (in light of an argument developed by John Searle, which we will consider later) be considered an observer-relative rather than intrinsic feature of the physical world. That reinforces the point that computers are merely an extension of the human perspective rather than an alternative to it. On the other hand, if it could be shown that computation is an intrinsic feature of nature, that would implicitly reintroduce into our picture of nature the Aristotelian notions of formal and final cause (for reasons which will, again, be developed in a later chapter). That would defeat the whole purpose of deploying computational notions in arguing against the Aristotelian philosophy of nature.

(It is important to emphasize that the impossibility of eliminating the human perspective in favor of an entirely "objective" or "absolute" point of view in no way entails that the scientific description of nature is not *true* or that it does not correspond to mind-independent reality. That science reflects *our* point of view does not mean that it reflects a *mistaken* point of view, or that its deliverances reflect *only* that point of view and nothing about the world itself. (Cf. Stroud 2000, pp. 33-34.) To draw such a conclusion would be like concluding that the sentence "Snow is white" does not convey any truth about mind-independent reality, on the basis of the premise that the English language is merely one human language among others. In fact the English sentence "Snow is white," the German sentence "Schnee ist weiss," and parallel sentences in other languages are all true and all correspond to the same mind-independent reality, despite the fact that there are significant differences between English, German, and these other languages. By the same token, that human intellects, angelic intellects, and perhaps other possible intellects grasp reality in different *ways* does not entail that they don't grasp reality at all.)

One main lesson of this chapter, then, is that, as Nagel puts it, "any objective conception of reality must include an acknowledgement of its own incompleteness" (Nagel 1986, p. 26). In particular, it must acknowledge that the subjective point of view of the scientist himself, which the objective conception leaves out, is no less real than what that conception does capture.

The other main lesson of this chapter is that the reality of the subjective point of view of the scientist cannot be made sense of without deploying the central notions of Aristotelian philosophy of nature. In particular, we cannot coherently deny the existence of the scientist as a conscious and rational subject who both undergoes change and persists through change. We have to affirm within this changing yet persisting subject a distinction between potentiality and actuality. We also have to affirm that this subject is embodied, and thus that there is within it a further distinction to be made between substantial form and prime matter (given that the distinction between actuality and potentiality entails that further distinction when applied to corporeal things). We have to affirm that this subject is, in its perceptual and cognitive states and in its actions, *directed toward* various objects and ends, as toward a final cause. We have to affirm that it bears various efficient-causal relations to the world. In short, the theory of actuality and potentiality, hylemorphism, and the doctrine of the four causes are all implicit in the existence and activity of the scientist as a subject of thought and experience. Accordingly, they are implicit in the very practice of science.

If the advocate of the mechanical philosophy fails to see this, that is because, as Orwell famously put it, to see what is in front of one's nose needs a constant struggle. Even if the mechanical world picture could drive actuality and potentiality, the four causes, etc. out of our picture of the world, it cannot drive them out of *us*, any more than an artist could demonstrate the non-existence of painters, paintbrushes, and palettes merely by refraining from putting images of these things into the painting he paints. Just as the very existence of the painting, whatever it represents, in fact *points to* the reality of painters and their tools, so too does the very existence of science point to the reality of thinking conscious subjects and their efficient and final causal relations to the world, the actualization of their potentials, and so forth. The key elements of Aristotelian philosophy of nature inevitably remain implicit in the practice of science, like the dirt that remains beneath the rug when swept under it after being removed from the rest of the house, or like J. L. Austin's "frog at the bottom of the beer mug" which "just when we had thought some problem settled, grin[s] residually up at us" (1961, p. 179).

Naturally, that does not by itself suffice to show that these various components of Aristotelian philosophy of nature have all of the applications which Aristotelians traditionally have thought them to have in physics, chemistry, biology, and so forth. The extent to which they do will be treated in the chapters to follow. The point is just that it will not do for naturalists and other proponents of a mechanistic picture of nature to wave away appeals to teleology, potentiality, and other Aristotelian notions with the glib assurance that modern science has banished them once and for all. Science has done no such thing. The question is not *whether* these notions have application, for that they have application *at least* in the analysis of the thinking, conscious, embodied subject is unavoidable. What is in question can only be the extent to which they have application in other areas. And as we will see, the frog refuses to stay in the mug, for their continued applicability is extensive indeed.

3. Science and Reality

3.1 Verificationism and falsificationism

In the previous chapter it was argued that the scientific image of the world presupposes the existence of something that is not captured by it, namely the scientist himself qua conscious and thinking subject. It is like a painting which leaves the painter out but nevertheless could not exist unless the painter did. But does the scientific image at least capture everything *else* in nature, everything in the world *beyond* the conscious thinking subject?

Naturally, the Aristotelian philosopher of nature would answer in the negative. But it is important to emphasize that one needn't be an Aristotelian to do so. For one thing, and as we also saw in the previous chapter, to affirm the conscious thinking subject while at the same time denying to nature any attributes not countenanced by the mechanical world picture would be to commit oneself to a Cartesian bifurcation of the world, with all of its attendant problems. Needless to say, Aristotelians are not the only ones who would reject such a bifurcation. For another, there are, even apart from the question of Cartesianism, insuperable problems with the notion that science gives us an exhaustive picture of nature – problems that even many thinkers outside the Aristotelian orbit have identified.

There are two basic versions of the idea that science gives us an exhaustive picture of nature. One of them emphasizes the *apex* of the "arch of knowledge," maintaining that an empiriological description of the world, or even an empiriometric description specifically, is an exhaustive description. This is *scientism* of the kind one finds in writers like Alex Rosenberg (2011). We will address this version in the subsequent sections of this chapter. The other version emphasizes the *feet* of the "arch," maintaining that there is nothing that can be known about the world, and perhaps even nothing that can meaningfully be *said* about it, that goes beyond what we can experience – thereby essentially collapsing the arch of knowledge down to its foundations. This is *verificationism*, which will be addressed in this section.

Verificationism has its roots in early modern empiricism. It is anticipated in Hume's Fork, the thesis that "all the objects of human reason or inquiry may naturally be divided into two kinds, to wit, *Relations of Ideas*, and *Matters of Fact*" (David Hume, *Enquiry Concerning Human Understanding*, Section IV, Part I). On one interpretation, "relations of ideas" concern propositions that are *analytic*, or true or false by virtue of the meanings of their constituent terms. "All bachelors are unmarried" would be a stock example of an analytically true proposition. "Matters of fact," meanwhile, concern propositions that are *synthetic*, or true or false by virtue of something more than just the meanings of their constituent terms – in particular, by virtue of what happens to be the case in the world. "Many bachelors go to singles bars" would be an example. Propositions concerning relations of ideas are knowable *a priori* and with certainty and are true of necessity, but are also trivial, giving us no knowledge of mind-independent reality. Propositions concerning matters of fact are non-trivial, but are knowable only *a posteriori* and are contingent and never knowable with certainty. These latter propositions would be the kind with which science is concerned.

Now, propositions of metaphysics and philosophy of nature are supposed to give us substantive knowledge of mind-independent reality but also to be true of necessity and knowable with certainty. (For rationalists, they are also knowable *a priori*, though as we saw in chapter 1, Aristotelians and Thomists do not take this position.) But if Hume's Fork is correct, there can be no such propositions. For if a proposition really is true of necessity and knowable with certainty, then in Hume's view it would have to be a trivial truth concerning only the relations of our ideas and not mind-independent reality; and if it really does give us non-trivial knowledge of mind-independent reality, then it must be true only contingently and not knowable with certainty. Hence, if Hume's Fork is correct, the only substantive knowledge to be had of the world concerns "matters of fact," which are the domain of science. There is no substantive knowledge of nature to be gained from metaphysics or philosophy of nature.

Hume's Fork is famously problematic, however. One problem is that it cannot plausibly account for mathematics. Given their necessity, certainty, and *a priori* character, truths of mathematics have to be located by Hume on the "relations of ideas" tine of his Fork. Yet mathematical truth is something we *discover* rather than invent and thus has a substantive and mind-independent character which makes it impossible to assimilate to trivialities like "All bachelors are unmarried." Another problem is that Hume's Fork is self-refuting. For it is itself neither analytically true nor empirically verifiable. Hence Hume's Fork presupposes exactly the third, metaphysical sort of perspective on reality that the principle denies can be had.

Now, Georges Dicker proposes that Hume's Fork can be interpreted in an alternative way which sidesteps the self-refutation problem (1998, pp. 53-55). In particular, it is in Dicker's view a mistake to identify Hume's "relations of ideas" with analytic propositions simpliciter. Rather, analytic propositions are one of two kinds of proposition concerning the relations of ideas. The other kind are synthetic propositions that (unlike matters of fact) do not assert or imply the existence of anything and are knowable *a priori*. Now, Hume's Fork is not analytic, but it could still be claimed to concern relations of ideas if we interpret it as a synthetic proposition that does not assert or imply the existence of anything and is knowable *a priori*. In that case, it would not be self-refuting, since it would fall under one of the two classes of statement that it recognizes.

Let us put to one side the exegetical question of whether Dicker's proposal is correct as an interpretation of what Hume himself actually meant. For even if it is, it is hard to see how the proposal rescues Hume's Fork, for the simple reason that the principle is no more plausibly knowable *a priori* than it is analytic. Dicker characterizes an *a priori* proposition as "one that can be known just by thinking" and that is not falsifiable by experience (1998, p. 11), and propositions other than analytic ones might fall into this class. But why should anyone believe that *Hume's Fork*, specifically, falls into it? After all, many philosophers have doubted Hume's Fork even after thinking carefully about it. Nor is it easy to see how just thinking carefully about it *could* result in knowledge of its truth. For example, one could not argue for it by applying the method of retorsion, since that would involve showing that denying Hume's Fork entails a performative self-contradiction, and denying it does *not* entail such a self-contradiction. But short of actually giving some *a priori* argument for Hume's Fork, Dicker has not rescued it from the charge of self-refutation at all. For if it is not defensible *a priori*, then it will not after all fall into either the "matters of fact" class or the "relations of ideas" class even given Dicker's interpretation of the latter.

Moreover, even if Dicker could get around that problem, there is another. For it is not enough for the defender of Hume's Fork to show that it can be established *a priori*. He would have to show that it can be established *a priori* via a method that would *not also* establish *a priori* metaphysical propositions of the sort Hume's Fork is supposed to rule out. Consider, for example, Descartes' famous *Cogito, ergo sum* ("I think, therefore I am"). Descartes takes this to be knowable *a priori*, since he thinks the intellect could grasp its truth even if the senses turned out to be completely unreliable. But Hume would have to deny that it is knowable *a priori*, since it asserts the existence of something, and Hume claims that all statements asserting the existence of something fall into the "matters of fact" class. Yet how could it plausibly be shown that *Hume's Fork* is knowable *a priori* if even the *Cogito* is not?

No less problematic than Hume's Fork is the logical positivists' *verifiability criterion of meaning* (Ayer 1952; Carnap 1959b). Analytic and synthetic statements, the positivists held, exhaust not only what is *knowable* but also what is strictly *meaningful*. More precisely, they held that a meaningful statement was true or false either by virtue of its logical form, or by virtue of the empirical facts. The former sorts of statements are either tautologies or self-contradictions, and tell us nothing about reality but only about how we use language. The latter do tell us something about reality, and comprise the propositions of empirical science. Any utterance that purports to tell us something about reality must therefore be empirically verifiable. If it is not, then it is strictly meaningless or devoid of cognitive content. The positivists alleged that utterances of a metaphysical, theological, and ethical sort were meaningless in this sense.

Here too the self-refutation problem rears its head, since the principle of verifiability is itself neither analytic nor empirically verifiable. Like Hume's Fork, it presupposes precisely the third, metaphysical sort of standpoint that it claims to rule out. Some logical positivists tried to get around this problem by proposing that the verifiability principle be interpreted not as a proposition that is either true or false, but rather as a convention to be adopted on grounds of utility – for example, as "a norm proposed for the purpose of avoiding unanswerable questions" (Feigl 1981, p. 311). But there are several problems with this attempted solution (Misak 1995, pp. 79-80). For one thing, no non-verificationist has any reason to adopt this convention, and the claims that the principle has greater "utility" than any alternative and that the questions it rules out of bounds are "unanswerable" simply beg the question. For non-verificationist philosophers would argue that such questions *are* answerable, and they have in fact proposed answers. Though those answers are, of course, controversial, so too is the verifiability principle even when interpreted as a mere convention, so it can hardly claim an advantage on that score. Furthermore, to regard the principle of verifiability and alternative principles as mere rival conventions entails that meaningfulness, and thus truth and falsity, are matters of convention. This not only renders the principle ineffectual as a critique of metaphysics, but it opens the verificationist to the standard objections to cognitive relativism – one of which is, of course, that such relativism is self-refuting. Hence a move to conventionalism would save the principle of verifiability from one self-refutation problem only at the cost of landing it in another.

There are other serious problems with the principle of verifiability (Misak 1995, pp. 70-96). For example, on a strict interpretation of the principle, propositions about the past and the future would have to be judged unverifiable and thus meaningless. But they are obviously *not* meaningless. Some verificationists dealt with this problem by suggesting that it would suffice for meaningfulness if someone in principle could *in the future* verify a statement made now about the future, and if future historians could in principle discover evidence for a statement about the past. But what about a statement like "There will be no observers in existence after the year 3000"? It is obviously meaningful, but could not be verified, since if true there will be no one around to verify it. Or what about statements about past events for which all evidence has been destroyed? These too are meaningful despite being unverifiable.

Statements about the mental states of other people would also have to be judged unverifiable and thus meaningless on a strict interpretation of the principle. This is not only absurd on its face, but would undermine most scientific knowledge, since most of even what most scientists know about science depends on the sensory experiences and reasoning processes (and thus mental states) of other scientists. The positivists proposed rendering statements about the mental states of others meaningful by interpreting them as statements about behavior, but this behaviorist program itself faced insuperable problems and is now about as dead as philosophical theories get. (One problem is that accurately characterizing a piece of behavior itself requires reference to the mental state of which the behavior is an expression. Hence there is no way entirely to replace talk about mental states with talk about behavior. Another problem is that behaviorist analyses leave out the subjective or first-person character of mental states like bodily sensations. A third problem is that some mental states have a determinate conceptual content, whereas behavior is indeterminate with respect to the conceptual content one might attribute to it. Cf. Feser 2013a for discussion of the latter issue and Feser 2006, pp. 60-63 for discussion of the others.)

There are further problems with reconciling the verifiability principle with science. For example, the universal laws discovered by science would be rendered meaningless on a strict interpretation of the principle of verifiability, since no finite number of observations could verify a universal statement. Nor is it clear what the principle should *count* as an observation in the first place. On one view, observation statements ought to be interpreted as reports about the observer's private sense data rather than about publicly accessible physical objects, since whether there really are such physical objects is open to skeptical doubt. But this makes the meaning of such statements incommunicable, since an observer can have access only to his own sense data. On another view, observation statements ought to be interpreted as reports about publicly accessible physical objects. But then, precisely because such statements are open to skeptical doubt, they would not be incorrigible in the way that statements about sense data were taken by the positivists to be. We would need to ask what verifies *them*, whereas they were supposed to be the *touchstone* of verification. The question of what exactly is it that is observed is further complicated by the theory-ladenness of observation, which entails a blurring of the distinction between what is observed and what is inferred from observation.

Then there is the issue of scientific realism. Scientific theories sometimes postulate unobservable entities. Because they are unobservable, a strict interpretation of the principle of verifiability would seem to render statements about such entities meaningless if interpreted in a realist way. Hence the logical positivists interpreted statements about such entities in an antirealist way instead. But there are powerful arguments for realism about unobservable entities (to be examined below), which constitutes a further reason to reject the principle of verifiability.

In light of the problems facing the principle, some verificationists liberalized it, so that it required of a meaningful statement only that observation be "*relevant* to the determination of its truth or falsehood" (Ayer 1952, p. 38, emphasis added). The problem with this, however, is that the verifiability principle thus interpreted now lets back in many metaphysical statements it had originally excluded. For example, cosmological arguments for God's existence would have to be regarded as consistent with verificationism, since they begin with empirical premises about the existence of the physical world, the existence of causal series, and so forth (Ewing 1937, p. 351). More to the present point, the entire apparatus of Aristotelian philosophy of nature would also have to be regarded as consistent with verificationism, since, as we have seen, observation is certainly "relevant" to determining its truth or falsity.

In a similar line of criticism, Karl Popper objected to verificationism on the grounds that verification was too easily achievable plausibly to demarcate science from non-science (1968, chapter 1). For even pseudo-scientific theories (astrology, Marxism, and Freudianism being among his stock examples) can easily find "verifying" evidence if their predictions are made vague enough. It is for Popper *falsifiability* rather than verifiability that is the mark of a good scientific theory. In particular, a scientific theory ought to make precise predictions which can be tested against experience. It is a theory's survival of attempts empirically to refute its predictions, rather than the gathering of positive evidence in its favor, that justifies our acceptance of it (and then only tentatively, pending future falsification).

Popper did not present falsificationism as a criterion of meaningfulness, nor did he condemn metaphysical and other non-scientific forms of inquiry as long as they were kept clearly distinguished from science. Despite having more modest ambitions than verificationism, however, falsificationism too faces a number of serious problems (Ladyman 2002, pp. 81-89). For example, there is Pierre Duhem's famous point (1991, chapter 6) that a scientific theory is always tested in conjunction with various auxiliary hypotheses and assumptions about background conditions (for example, that one's experimental equipment is working properly). Hence, if some prediction fails, one could conclude that the theory is false, but one could also conclude instead that one of the auxiliary hypotheses or assumptions about background conditions was mistaken.

In other ways too, falsificationism does not fit the actual practice of science. For one thing, scientists sometimes hold onto a theory even when it makes false predictions, on the grounds that the theory is successful in other respects and no better theory is available. (A stock example is physicists' adherence to Newtonian mechanics despite its inconsistency with the observed orbit of Mercury. Only when Einstein provided an alternative theory that accounted for Mercury's orbit was Newton abandoned.) For another thing, there are a number of key scientific claims that appear to be unfalsifiable. Examples would be the principle of the conservation of energy and the second law of thermodynamics. These are so general and so fundamental to the picture of the world presented by modern science that any apparent counterexample would be taken to be evidence, not that these principles are false, but rather that revision is called for somewhere else in the body of scientific theory.

Verificationism and falsificationism are no longer as influential within philosophy as they once were. A related idea which *is* influential is the thesis that only a "naturalized metaphysics" – that is to say, an approach to the subject that confines itself to articulating the metaphysical assumptions implicit in natural science – is worthy of consideration. The only alternative, on this view, would be to ground metaphysics in "conceptual analysis." But the trouble with that approach, it is claimed, is that we have no guarantee that the "intuitions" or "folk notions" the conceptual analyst appeals to really track reality. Indeed, we have, according to this view, good reason to think they do not track reality, insofar as science often presents us with descriptions of the world radically different from what common sense supposes it to be like. (Cf. Ladyman, Ross, Spurrett, and Collier 2007, chapter 1; Ross, Ladyman, and Kincaid 2013)

But the supposition that the only alternative to natural science is "conceptual analysis" is essentially a variation on Hume's thesis that the only knowable propositions concern either "matters of fact" or "relations of ideas," and the logical positivists' thesis that the only meaningful statements are either synthetic or analytic. And it faces the same basic problem. For the thesis that the only alternatives are natural science and conceptual analysis is not *itself* either a proposition of natural science or a conceptual truth. It presupposes a third cognitive perspective of precisely the sort it purports to rule out – namely the kind of perspective represented by traditional metaphysics and philosophy of nature (whether developed in an Aristotelian way, a rationalist way, or whatever).

The proponent of "naturalized metaphysics" may protest that neuroscience or cognitive science supports his position over that of the traditional metaphysician. Naturally, for such a response to succeed, it would be necessary to spell out exactly *how* neuroscience or cognitive science does this. But a deeper point is that whether neuroscience, cognitive science, or any other science really captures *everything* there is to capture in the natural world is precisely what is at issue between "naturalized metaphysics" on the one hand, and more traditional philosophical approaches such as Aristotelianism on the other. For the Aristotelian, of course, a *complete* account of the nature of physical reality would have to include reference to the distinction between actuality and potentiality, substantial form and prime matter, efficient and final causes, and so forth – not to mention specifically *biological* and *psychological* Aristotelian notions (such as the Aristotelian accounts of the nature of life and of concept formation). Hence since modern neuroscience, cognitive science, and other natural sciences make no use of such notions, they cannot be regarded as complete descriptions of the phenomena with which they deal. The "naturalized metaphysician" will disagree with this, of course, but the point is that for him merely to appeal to neuroscience, cognitive science, or some other science in defense of his position simply begs the question (Feser 2014b, pp. 25-30).

Harold Kincaid (2013) attempts to defend the "naturalized metaphysics" project against such objections. One complaint lodged against naturalism, Kincaid notes, is that beginning at least with Quine (who was himself responding to deficiencies in logical positivism of the sort we've considered), naturalists have had to broaden their conception of evidence so far that it is hard to see how metaphysics of the more traditional sort is ruled out by it (Chakravartty 2013). (Recall what was said above about Ayer's weak requirement that observation be "relevant" to the evaluation of a statement.) In response, Kincaid expresses skepticism about the Quinean thesis (Quine 1980) that the *whole* of our network of beliefs is implicated in the testing of any particular statement. In fact, Kincaid suggests, revisions of our system of beliefs in light of experience typically need involve only beliefs that are fairly close to those being tested. They need not involve large-scale revisions and thus would not justify grand metaphysical assertions. Hence, "if there is metaphysics in science, it will have to be quite local" (Kincaid 2013, p. 8).

It is hard to know what to make of this response. No doubt Kincaid is correct to say that relatively minor and local changes of belief, rather than major general metaphysical claims of a traditional sort, are *often* all that can plausibly be justified by scientific evidence. But is he also saying that major general metaphysical claims of a traditional sort can *never* be so justified? That is what he would need to show in order to respond to his critic, but he does nothing to establish such a sweeping claim. Quine thought that nothing less than mathematical Platonism was rationally justified by the general utility mathematics has in empirical science. How exactly do Kincaid's remarks show that such a bold metaphysical position is *always and in principle* incompatible with a "naturalized" metaphysics?

In any event, the traditional metaphysician (Aristotelian, rationalist, or whatever) need not abide even by Quine's milder naturalist scruples, let alone Kincaid's more austere naturalist scruples. Kincaid's response essentially takes the "naturalized metaphysics" project for granted, and addresses only the question of whether that project ought to be as liberal in its evidential standards as Quineans are. The deeper question, though, is whether we ought to work within the naturalist straightjacket in the first place. We have in this book examined many reasons for concluding that we ought not to, and since Kincaid does not address them, his response simply begs the question against the traditional metaphysician.

The second objection Kincaid considers offers a further reason for refusing that straightjacket, and that is that naturalism confines itself to purely *descriptive* judgments and thus leaves no room for the *normative* epistemic judgments that must enter into the evaluation of scientific claims. Kincaid responds by pointing out that scientists do in fact make normative epistemic judgments, and indeed "the reliability of methods, new and old, is a key scientific question" (p. 8). But this simply misses the point. No one denies that scientists *do* in fact make normative judgments and theorize about methodological and epistemic standards. The question is how they could *consistently* do so *if* they confined themselves to metaphysical assumptions the naturalist regards as respectable.

Now, as Kincaid himself notes, proponents of "naturalized metaphysics" like James Ladyman and Don Ross (Ladyman, Ross, Spurrett, and Collier 2007) insist that "the special sciences cannot contradict or override the results of fundamental physics" (Kincaid 2013, p. 14). Ladyman and Ross also take quantum physics to show that there are no *things* in existence, but only structures. (More on this subject in a later chapter.) Hence, since the special sciences, no less than common sense, speak as if there *are* things in existence, Ladyman and Ross conclude that they are speaking falsely. Similarly, for naturalist Alex Rosenberg (2011), since physics makes no reference to notions like teleology or intentionality, we ultimately have to eliminate such notions entirely from the special sciences. The point Kincaid's opponent is making, then, is that if common sense and the special sciences alike must radically revise themselves in order to conform to the austere ontology of physics, it is difficult to see how any normative notions could survive such a revision. (For example, it is hard to see how the notion of an epistemic state can be made sense of in the absence of intentionality, or how the "reliability" of a method or epistemic faculty can be made sense of in the absence of teleology.)

A third objection Kincaid considers is that naturalism is "question-begging" insofar as it attempts to justify itself by way of epistemic standards that presuppose the truth of naturalism, and that it is "subject-changing" insofar as it simply dismisses, without answering, traditional epistemic concerns about skepticism (Kincaid 2013, pp. 8-9). Amazingly, Kincaid responds to these charges with assertions that are themselves manifestly question-begging and subject-changing. For example, he says that the "sophisticated naturalist" maintains that "disciplined inquiry into the ways of scientific practice" shows that it is "untenable" to suppose that the methods the naturalist would approve of stand in need of justification, and that "the filtering processes of institutionalized peer review" and the like suffice to deal with the legitimate questions that do arise (p. 9). Furthermore, this "sophisticated naturalist" also "denies that there is a particular philosophical method and standpoint from which to make a priori judgments about the requirements for knowledge," and holds that traditional questions about skepticism are "unmotivated by real science and real knowledge production" (p. 9).

The problems with all this should be obvious. First, simply applying honorific labels like "sophisticated," "disciplined," and "real" to the naturalist's preferred methods and suppositions proves exactly nothing. For what is in question is whether these methods and suppositions actually *deserve* these labels and, even if they do deserve them, whether there are also *other*, *non*-naturalist methods and suppositions that are useful or necessary to the investigation of reality. Second, the critic of naturalism is well aware that the naturalist "denies" that there is any epistemic standpoint other than those recognized by naturalism. What is in question is whether this denial is correct or well-founded, and simply reiterating the denial doesn't show that it is. Third, no doubt it is true that scientists generally don't find it necessary to address skeptical and other general epistemological issues when investigating the questions that interest them qua physicists, chemists, biologists, etc. So what? The claim that the only questions worth investigating are the ones scientists themselves investigate is precisely what is at issue in the debate between naturalists and their critics, so that to appeal to what scientists themselves do is, in this context, merely to beg the question.

Fourth, whether processes like "institutionalized peer review" suffice to guarantee that every issue worth addressing will be addressed is also part of what is at issue. For one thing, if what Kincaid has in mind is peer review *among scientists*, specifically, then this will hardly guarantee that all the important distinctively *philosophical* issues will be addressed.

And if Kincaid were to respond that the only philosophical issues *worth* addressing *would* be raised in peer review among scientists, then he would once again be begging the question. For another thing, if what Kincaid has in mind is peer review that includes *philosophers* as well as scientists, then we have to ask who Kincaid would include among this body of philosophers doing the reviewing. Would *non*-naturalist philosophers (Aristotelians, Platonists, rationalists, et al.) be included? In that case, all the important philosophical questions would no doubt be addressed, but that would include questions and answers that naturalists like Kincaid want to rule out of bounds. Or would Kincaid exclude these philosophers and include only naturalist philosophers? But in that case, how could Kincaid justify such an exclusion in a way that doesn't yet again beg the question?

Finally, Kincaid claims that "it is not clear what third activity metaphysics might be if it is not conceptual analysis or scientifically inspired metaphysics" (2013, p. 3). But there are two problems with this claim. First, the fact that Hume's Fork, the logical positivists' analytic-synthetic dichotomy, and the naturalist's bifurcation between conceptual analysis and science all founder on the self-refutation problem suffices to show that there *is* in fact such a "third activity" and that these thinkers are all engaged in it themselves in the very act of denying that it exists. It is certainly reasonable to ask for an elucidation of the *nature* of this third activity, but that it *exists* cannot reasonably be denied.

Second, some of the ideas and arguments put forward in earlier chapters of this book (such as the defense of the reality of change, the theory of actuality and potentiality, and the defense of the principle of sufficient reason) provide examples of this third kind of activity and illustrate how it differs from the other two. On the one hand, such arguments purport to establish necessary truths that cannot be overthrown by the findings of science (such as the propositions that *experience entails change*, that *change entails the actualization of potential*, and that *everything is intelligible*). Hence they do not fall on the "matters of fact" side of Hume's Fork or within the boundaries of science as the naturalist understands it. On the other hand, the arguments do not rest on mere "intuitions," revisable "folk notions," appeals to ordinary language, or the like. The retorsion arguments defended in earlier chapters claim to show that the denial that change occurs and the denial that everything is intelligible each entail a *contradiction*. They are *reductio ad absurdum* arguments purporting to establish truths about *objective reality*, not mere appeals to how we contingently happen to carve up the world conceptually. Nor are the arguments of this "third activity" always *a priori*. The defense of the reality of change, and the theory of actuality and potentiality that is its sequel, are grounded in experience (albeit such *general* features of experience that no *particular* experience or set of experiences could overthrow the conclusions drawn from them). Hence, arguments and ideas like the ones in question do not fall on the "relations of ideas" side of Hume's Fork or within the naturalist's category of "conceptual analysis."

The naturalist will no doubt object to such arguments, but the point is that if he is going to refute them he will have to engage them directly and evaluate them on their own merits. He cannot, without begging the question, sidestep such engagement by simply stomping his foot and insisting dogmatically that there can be no "third activity," so that such arguments simply *must* really be either veiled appeals to intuition or revisable scientific hypotheses.

3.2 Epistemic structural realism

3.2.1 Scientific realism

An upshot of the previous section is that the truth about the natural world outstrips what is empirically verifiable or falsifiable, certainly on a strict construal of verification and falsification. As we have seen, the successors of Hume and of the logical positivists acknowledged this and, accordingly, loosened their conception of empirical testing. As our treatment of "naturalized metaphysics" indicates, though, the truth about the natural world outstrips even this liberalized riff on Hume and positivism. There is more to the natural world than is captured even by the feet of the "arch of knowledge" and its apex taken together.

A closer inspection of the nature of the empiriological theories that constitute that apex – and in particular of the empiriometric theories that are taken to be the gold standard of empiriological inquiry – will reveal that in fact they tell us *relatively little* about the natural world. Now, one way such a claim might be developed is along the lines of instrumentalism or some other form of anti-realism. On this sort of view, scientific theories are merely useful fictions, tools for predicting and controlling the natural world but not accurate descriptions of that world. Taking such a position would in one respect make the work of the Aristotelian philosopher of nature much easier. The naturalist claims (falsely, but still he *claims*) that the results of modern science undermine the traditional Aristotelian picture. If the Aristotelian could maintain that those results tell us *nothing* about nature as it really is, he would thereby defuse this objection in one fell swoop. He could say that science could undermine Aristotelian philosophy of nature only if it told us something about objective reality that is incompatible with the latter, but in fact it tells us nothing about objective reality in the first place.

I do not take this tack myself, however, for two reasons. First, like other contemporary Aristotelians (e.g. Wallace 1983), my inclinations are in fact realist. I simply don't believe that instrumentalist and other anti-realist accounts of scientific theories are correct, at least not in general. Second, it is always preferable to take on a stronger rather than weaker version of an opponent's position. If anti-Aristotelian arguments fail even given a *realist* construal of scientific theories, then it is that much clearer that they fail, full stop.

To a first approximation, scientific realism may be characterized as the view that the statements a theory makes about unobservable entities are *true*, or that the entities in question are *real* and not mere useful fictions. The standard argument for this position is called the "no miracles" argument, and it was given an influential formulation by Hilary Putnam:

> The positive argument for realism is that it is the only philosophy that doesn't make the success of science a miracle. That terms in mature scientific theories typically refer..., that the theories accepted in a mature science are typically approximately true, that the same term can refer to the same thing even when it occurs in different theories – these statements are viewed by the scientific realist... as part of the only scientific explanation of the success of science, and hence as part of any adequate scientific description of science and its relations to its objects. (1979, p. 73)

J. J. C. Smart gave expression to a similar idea when he wrote that if the theoretical entities posited by successful scientific theories do not exist, then "we must believe in a *cosmic coincidence*" (1963, p. 39). (See also Maxwell 1962 and Boyd 1989. For discussion of some differences between these various statements of the "no miracles" argument, see Psillos 1999, pp. 72-77.)

The basic point is that theories exhibiting impressive predictive and technological success behave *as if* they were true, *as if* the theoretical entities they posit were real. Such successes are just what we would expect if those entities actually existed. If the entities don't really exist, we are faced with a coincidence that defies plausible explanation. It would be like a murder case where we have a suspect who had the means, motive, and opportunity to commit the crime, faces a mountain of incriminating circumstantial evidence and has confessed to the murder, and where we have no other plausible suspects – and yet where the suspect is still somehow innocent. It is possible but not plausible. By the same token, it is not plausible that the entities posited by successful theories are not real. Their reality, and the truth of the theory that posits them, is the best explanation of the theory's success.

All of this requires qualification, however, in light of some common objections to realism. Hence, consider the objection from the "underdetermination" of theory by evidence. The idea is that for any body of empirical evidence, there are always alternative incompatible theories that are equally consistent with that evidence. The evidence is not by itself sufficient to determine which, if any, of these rival theories is correct. Hence success in accounting for the evidence cannot suffice to show that the entities a theory posits are real. Scientific realists thus often emphasize that it isn't mere consistency with the known evidence that undergirds the "no miracles" argument. Good scientific theories are often also successful in making *novel predictions* and thus accounting for new evidence that was not available and perhaps not foreseen when the theory was first formulated (Ladyman, Ross, Spurrett and Collier 2007, pp. 76-79; Psillos 1999, pp. 104-8). To return to the murder case analogy, suppose our theory about the suspect in question generates a prediction about where the murder weapon is likely to be found, and then it is indeed found there. It is even less plausible in that case that he is innocent. Similarly, when a scientific theory not only accounts for previously known evidence but generates novel predictions on the assumption that the entities it posits are real, and those predictions are confirmed, it is even less plausible to doubt the existence of those entities.

There is also the objection from theory change. Sometimes scientific theories which exhibit predictive and technological success are nevertheless eventually abandoned in favor of some even more successful theory. Hence success does not suffice to establish realism. In response, some philosophers emphasize that the point of scientific realism is not to claim that successful scientific theories are always in fact true. The point is rather that they do at least *aim* to be true (Godfrey-Smith 2003, pp. 174-79). That is to say, they are not (contrary to what some anti-realists maintain) aiming merely for consistency with the empirical evidence and instrumental success, but are also at least claiming that the theoretical entities they posit are real. Whether they succeed in establishing this in any particular instance is another and more complicated question, and to answer it we would have to evaluate each theory on a case by case basis. But that does not undermine the point that the *aim* of science is a realist one. It is also important to emphasize, in this connection, that one can without inconsistency be *selective* in one's realism. It is possible to hold of one particular theory that it succeeds in establishing the reality of the entities it posits, while at the same time holding of some other particular theory that it fails to do so and is at the end of the day more plausibly interpreted in a non-realist way. The dispute between realism and anti-realism is not an "all or nothing" matter.

Then there is the charge that the "no miracles" argument commits a "base-rate fallacy" (Magnus and Callender 2004). The idea here is that we would need information about how common or rare true and false theories are in the population of theories as a whole (that is to say, we would need to know the "base rates" of such theories) before we could judge on the basis of a "no miracles" argument that some particular theory is true. But the "no miracles" argument has no such independent information to appeal to. Now, this objection takes the "no miracles" argument to be an inference distinct from the inference made to the existence of some theoretical entity. For example, there is the inference from the observed facts about water to the conclusion that water is composed of hydrogen and oxygen, and then (on this interpretation of the "no miracles" argument) there is an additional inference to the conclusion that a realist interpretation of hydrogen and oxygen is the best explanation of the success of the theory that posits them. However, as some defenders of scientific realism have pointed out, this is not the best way to interpret scientific realism (Bird 1998, pp. 141-43; Lewens 2016, pp. 102-3). There are not two inferences in a case like the one in question, first an inference to the existence of oxygen and hydrogen and then a second inference to realism. Rather, there is just the first inference. Realism is simply the thesis that the default interpretation of a theory that posits some entity is a realist one, so that the first inference all by itself gives us defeasible grounds to believe that oxygen and hydrogen are real. There is no second inference that might be accused of committing a "base-rate fallacy" or any other fallacy.

On the understanding of realism that I would endorse, then, there is a *presumption* in favor of believing in the existence of a theoretical entity posited by a scientific theory that is successful with respect to novel predictions or technological applications. That presumption can be overridden. For example, we might judge that the evidence in favor of a certain theory is not strong enough to warrant belief in the entities it posits, or we might judge that a realist interpretation would be incompatible with better established scientific or metaphysical knowledge. But the burden of proof is not on the realist to establish that the entities posited by a successful theory are real. Rather, the burden is on the critic to show that we should *not* believe that they are real. This is especially true of a completely *general* anti-realism. It is not the scientific realist who has the burden of giving an argument for realism (whether a "no miracles" argument or some other kind) but rather the anti-realist who has the burden of arguing against realism.

Now, the arguments against scientific realism in general (as distinct from arguments for a non-realist interpretation of some particular theory) are not compelling. Consider Bas van Fraassen's "constructive empiricism," according to which science aims only for theories that are "empirically adequate" (1980, p. 12) and not at establishing the reality of unobservable entities. Van Fraassen holds that the "no miracles" argument fails insofar as his own anti-realist position can adequately account for the success of science, in a Darwinian manner. It does so by simply noting that scientific theories compete with one another, and the ones that survive are those having instrumental value (1980, p. 40). Whether the entities they posit are real drops out of consideration as irrelevant.

But this simply raises the further question *why* they have this instrumental value, and the realist argues that the existence of the posited entities is what explains that. As Ladyman and Ross put it, what van Fraassen's Darwinian account explains is, in effect, how a certain *phenotype* survives, whereas the realist explains the *genotype* underlying this phenotype (Ladyman, Ross, Spurrett and Collier 2007, pp. 73-74; the same analogy is deployed in Psillos 1999, pp. 96-97). Now, as they go on to note, van Fraassen would deny that we need to pursue explanation this far. As an empiricist, he refrains from asking the sorts of questions the traditional metaphysician would. Ladyman and Ross, who at least to some extent share van Fraassen's misgivings about traditional metaphysics, accordingly do not regard this particular objection to van Fraassen as decisive. But the traditional metaphysician need not make such a concession – and indeed *ought not* to make it, given the principle of sufficient reason, which I defended in chapter 2. For in light of that principle, the point that van Fraassen's position does not explain everything that can be explained *is* a decisive one.

As realists often point out, there is no sharp dividing line between observable entities and unobservable ones (Maxwell 1962). Van Fraassen acknowledges this, but holds that the distinction between observable and unobservable entities is still real and important given that we have paradigm examples of each (1980, pp. 13-19). Now, van Fraassen does not claim that talk about unobservable entities is meaningless or that we can know that such entities do not exist. He emphasizes instead only that we do not have grounds to conclude that they *do* exist, and that acceptance of a scientific theory should not require such a conclusion, but only the judgement that the theory is empirically adequate. But this poses another problem for his position. For if the boundary between observable and unobservable is by his own admission vague, why is van Fraassen so insistent that we take such sharply divergent epistemic *attitudes* toward them? Why not instead acknowledge that science does give us reasons to believe in unobservable entities, but reasons that are less conclusive than the ones we have to believe in observable entities, and reasons about which we should be more tentative the further we get from what is clearly observable? (Cf. Godfrey-Smith 2003, pp. 185-86)

Then there is the problem that the judgment that a theory is "empirically adequate" appears on closer inspection to be no less underdetermined by the evidence than the anti-realist says all scientific theories are. For any claim to the effect that a particular theory is consistent with all the evidence can be countered with the claim that the theory is consistent only with the evidence we happen to be aware of but not with the evidence we have not yet examined, or that it is consistent with all the evidence that exists so far but not with evidence that will come to light a year from now (Ladyman 2002, p. 193). Why should we accept the one claim rather than the other? We cannot appeal to the evidence to help us decide, because that evidence is consistent with both claims. So, if underdetermination is a problem, how is it any less of a problem for constructive empiricism than for realism?

In general, the trouble with underdetermination arguments is that they appear to be arbitrarily selective in their skepticism. The underdetermination of theory by evidence is presented as if it were a special problem for scientific realism. But one could give parallel arguments against belief in the external world, or belief in other minds, or the belief that the world is older than five minutes. For philosophers have, of course, historically examined arguments to the effect that all the evidence available to us could be just as it is even if there were no external world, or no minds other than one's own, or if the world in fact came into being five minutes ago complete with false memories of earlier events and objects that appeared to be older than they are. Yet most opponents of scientific realism do not try to defend these more extreme kinds of skepticism. So why be skeptical about the unobservable entities posited by scientific theories? If other brands of skepticism can be treated as mere intellectual puzzles that shouldn't undermine our confidence in the reality of the external world, other minds, or the past, then to be consistent we should not regard the unobservable entities posited by successful scientific theories as open to skeptical doubt. (Cf. Lewens 2016, pp. 94-95)

However, the argument from theory change is perhaps the main objection to scientific realism. As developed by philosophers like Larry Laudan (1984), the idea is that, from the fact that so many past successful scientific theories have eventually been abandoned, we can by a "pessimistic induction" draw the inference that current theories are likely eventually to be abandoned as well. The success of a theory thus does not give us a good reason to believe that the theory is true or that the entities it posits are real.

But here too there are serious problems. For one thing, it is debatable whether all of the past theories Laudan has in mind really were successful in the first place (Psillos 1999, pp. 104-8). If many of them were not, then the number of examples he can appeal to is arguably too small to ground an inductive inference of the sort he is making. For another thing, even in the case of an abandoned theory that was successful, we need to distinguish the specific *parts* of the theory that were abandoned from other parts that were retained and incorporated into replacement theories. And it is not clear that the problematic abandoned parts had anything to do with the older theory's successes (Kitcher 1993, pp. 140-9; Psillos 1999, pp. 108-14). As that distinction indicates, the changes that have taken place in the history of science are not always as radical as Laudan needs them to be in order for his argument to go through, for abandoned theories are sometimes at least approximations of newer ones and are not necessarily thrown out wholesale but rather reinterpreted.

In particular, what a later theory often preserves, even when it abandons commitment to the specific entities posited by an earlier theory, are the mathematical equations describing the *structural relations* holding between whatever the actually existing entities happen to be. Ernan McMullin illustrates the point as follows:

> Ethers and fluids are a... category... which Laudan stresses. I would argue that these were often, though not always, interpretive additions, that is, attempts to specify what "underlay" the equations of the scientist in a way which the equations (as we now see) did not really sanction. The optical ether, for example, in whose existence Maxwell had such confidence, was no more than a carrier for variations in the electromagnetic potentials. It seemed obvious that a vehicle of some sort *was* necessary; undulations cannot occur (as it was often pointed out) unless there is something to undulate! Yet nothing could be inferred about the carrier itself; it was an "l-know-not-what," precisely the sort of unknowable "underlying reality" that the antirealist so rightly distrusts...
>
> [But] in many parts of natural science there has been, over the last two centuries, a progressive discovery of *structure*. (1984, pp. 17-18, 26)

Hence, even if Laudan's pessimistic induction were taken to justify skepticism about the entities posited by scientific theories (including the successors to Maxwell's ether no less than the ether itself), that would not suffice to establish anti-realism tout court. For if the structural features of the natural world captured in equations carry over from theory to theory, then to that extent at least the history of theory change would be consistent with realism. In particular, it would be consistent with *structural realism*, a version of scientific realism according to which the aspects of reality that our best scientific theories capture are, specifically, its structural aspects (Worrall 1996; Zahar 2007).

3.2.2 Structure

Now, structural realism is precisely the brand of realism I would defend (though only with qualifications to be noted below), particularly where empiriometric sciences like physics are concerned. It is a position that is at least implicit even in the work of two major philosophers of science often classified as *instrumentalists* – namely, Henri Poincaré and Pierre Duhem. Like Laudan, Poincaré notes "the ephemeral nature of scientific theories," and cites as an illustration the transition from Fresnel's account of the nature of light to Maxwell's (1952, p. 160). But Poincaré immediately goes on to note that Maxwell preserves Fresnel's differential equations, and writes:

> It cannot be said that this is reducing physical theories to simple practical recipes; these equations express relations, and if the equations remain true, it is because the relations preserve their reality. They teach us now, as they did then, that there is such and such a relation between this thing and that; only, the something which we then called *motion*, we now call *electric current*. But these are merely names of the images we substituted for the real objects which Nature will hide forever from our eyes. The true relations between these real objects are the only reality we can attain, and the sole condition is that the same relations shall exist between these objects as between the images we are forced to put in their place. If the relations are known to us, what does it matter if we think it convenient to replace one image by another? (1952, p. 161)

So, for Poincaré, while a physical theory does not give us reason to believe in the *entities* it posits (whether motion or electrical current, in this case), it does give us reason to think that the *relations* captured in its equations do indeed reflect "reality." Similarly, Duhem writes:

> [P]hysical theory never gives us the explanation of experimental laws; it never reveals realities hiding under the sensible appearances; but the more complete it becomes, the more we apprehend that the logical order in which theory orders experimental laws is the reflection of an ontological order, the more we suspect that the relations it establishes among the data of observation correspond to real relations among things, and the more we feel that theory tends to be a natural classification. (1991, pp. 26-27)

Like Poincaré, then, Duhem takes the *relations* in question to be "real" and a "reflection of an ontological order," despite echoing Poincaré's denial that physics tells us anything about the non-sensible relata themselves.

So, just as the objections against scientific realism have moved realists to qualify their position, so too have some thinkers who seem at first glance to be anti-realists been moved, by the preservation of structure through theory change, to qualify their own position. Structural realism is essentially the middle ground position that results from the convergence of these tendencies. John Worrall (1996) argues that it therefore gives us the "best of both worlds." It is realist insofar as it acknowledges the force of the "no miracles" argument, but at the same time it acknowledges the force of the "pessimistic induction" by confining its realism to structure. (It should be noted, however, that the proponent of a "best of both worlds" argument *need* not insist that structure is the most that is ever preserved. The point is rather that as long as *at least* structure is preserved, the pessimistic induction cannot justify a thoroughgoing anti-realism.)

As Worrall notes, the claim that structure is preserved needs to be qualified. Sometimes the carryover is fairly straightforward, as in the case of Fresnel and Maxwell. Fresnel thinks of light in terms of an elastic solid ether and Maxwell in terms of a disembodied electromagnetic field, but despite this stark difference in the entities they recognize, the equations they make use of are the same. But in other cases the transition is less smooth, as in the shift from Newton's understanding of gravitation as action-at-a-distance to Einstein's notion of space-time curvature. Here it is not just the entities posited by the theories that are different, but also the equations. However, the older equations can nevertheless be taken to be *limiting cases* of the newer ones, so that the structure captured in the old theory at least *approximates* that of the new.

Worrall's argument for structural realism focuses on the *results* of scientific inquiry, and in particular on the way different theories end up positing identical or similar structural features despite reaching different conclusions in other respects. But other structural realists emphasize the *methods* of scientific inquiry, and in particular of physics. On their view, it's not just that structural relations *happen* to be what survive theory change, but that such relations are all that physics could have revealed to us in the first place (though as we will see, this claim ends up being qualified). This is a theme emphasized by mid-twentieth-century structural realists like Bertrand Russell and Arthur Eddington, and it has to do with the scientist's procedure of abstracting from concrete reality – a procedure carried through most thoroughly in physics. Russell writes:

> Physics started historically, and still starts in the education of the young, with matters that seem thoroughly concrete. Levers and pulleys, falling bodies, collisions of billiard balls, etc., are all familiar in everyday life, and it is a pleasure to the scientifically-minded youth to find them amenable to mathematical treatment. But in proportion as physics increases the scope and power of its methods, in that same proportion it robs its subject-matter of concreteness. The extent to which this is the case is not always realized, at any rate in unprofessional moments, even by the physicist himself... (1927, p. 130)

Eddington illustrates the way such mathematical abstraction from concrete reality proceeds in a vivid passage that is worth quoting at length:

> Let us then examine the kind of knowledge which is handled by exact science. If we search the examination papers in physics and natural philosophy for the more intelligible questions we may come across one beginning something like this: "An elephant slides down a grassy hillside... ." The experienced candidate knows that he need not pay much attention to this; it is only put in to give an impression of realism. He reads on: "The mass of the elephant is two tons." Now we are getting down to business; the elephant fades out of the problem and a mass of two tons takes its place. What exactly is this two tons, the real subject matter of the problem? It refers to some property or condition which we vaguely describe as "ponderosity" occurring in a particular region of the external world. But we shall not get much further that way; the nature of the external world is inscrutable, and we shall only plunge into a quagmire of indescribables. Never mind what the two tons *refers* to; what *is* it? How has it actually entered in so definite a way into our experience? Two tons *is* the reading of the pointer when the elephant was placed on a weighing-machine. Let us pass on. "The slope of the hill is 60°." Now the hillside fades out of the problem and an angle of 60° takes its place. What is 60°? There is no need to struggle with mystical conceptions of direction; 60° *is* the reading of a plumb-line against the divisions of a protractor. Similarly for the other data of the problem. The softly yielding turf on which the elephant slid is replaced by a coefficient of friction, which though perhaps not directly a pointer reading is of kindred nature. No doubt there are more roundabout ways used in practice for determining the weights of elephants and the slopes of hills, but these are justified because it is known that they give the same results as direct pointer readings.
>
> And so we see that the poetry fades out of the problem, and by the time the serious application of exact science begins we are left with only pointer readings. If then only pointer readings or their equivalents are put into the machine of scientific calculation, how can we grind out anything but pointer readings? But that is just what we do grind out. The question presumably was

> to find the time of descent of the elephant, and the answer is a pointer reading on the seconds' dial of our watch.
>
> The triumph of exact science in the foregoing problem consisted in establishing a numerical connection between the pointer reading of the weighing machine in one experiment on the elephant and the pointer reading of the watch in another experiment. And when we examine critically other problems of physics we find that this is typical. The whole subject matter of exact science consists of pointer readings and similar indications. (1958, pp. 251-2)

The result of this method is a description of nature that has stripped away so many of the concrete features of the world as we experience it that it tells us relatively little about nature. Quoting again from Russell:

> It is not always realised how exceedingly abstract is the information that theoretical physics has to give. It lays down certain fundamental equations which enable it to deal with the logical structure of events, while leaving it completely unknown what is the intrinsic character of the events that have the structure. We only know the intrinsic character of events when they happen to us. Nothing whatever in theoretical physics enables us to say anything about the intrinsic character of events elsewhere. They may be just like the events that happen to us, or they may be totally different in strictly unimaginable ways. All that physics gives us is certain equations giving abstract properties of their changes. But as to what it is that changes, and what it changes from and to – as to this, physics is silent. (1985, p.13)

The "events that happen to us" to which Russell refers are *mental* events – for example, our perceptual experiences of the pointer readings spoken of by Eddington. Russell's point is that whereas we are acquainted with the intrinsic nature of the perceptual events that occur in the course of observation, experiment, and the like, when we infer from these perceptions to the theoretical description of the external world afforded by physical theory, what we get is an account only of the mathematical structure of that world and not its intrinsic nature.

In Russell's view, this theoretical description *seems* to give us more than mathematical structure only because we are led by ordinary language and by our imaginations to attribute meanings to the statements of physical theory that they don't actually have. For example, if we are not careful, we find ourselves mistakenly thinking of the entities referred to by particle physics as comparable to everyday observable objects, only smaller (Russell 1927, p. 135; Russell 1948, p. 327). Recourse to austere formal languages is essential to avoiding such misconceptions:

> Ordinary language is totally unsuited for expressing what physics really asserts, since the words of everyday life are not sufficiently abstract. Only mathematics and mathematical logic can say as little as the physicist means to say. As soon as he translates his symbols into words, he inevitably says something much too concrete, and gives his readers a cheerful impression of something imaginable and intelligible, which is much more pleasant and everyday than what he is trying to convey. (Russell 1931, p. 82)

Now, insofar as Eddington speaks of the pointer readings themselves, and not our perceptions of them, as the beginning and end points of the process of theory construction, his account seems to make the scientific picture of the world even more abstract than it is on Russell's. For while Russell certainly puts heavy emphasis on the abstract mathematical character of physical theory, he also came to emphasize (for reasons to be explained presently) that there is nevertheless *some* concrete content to physics, broadly construed, that derives from the perceptual experiences that are its evidential basis:

> When physics is brought to [a high] degree of abstraction it becomes a branch of pure mathematics, which can be pursued without reference to the actual world, and which requires no vocabulary beyond that of pure mathematics. The mathematics, however, are such as no pure mathematician would have thought of for himself. The equations, for instance, contain Planck's constant *h*, of which the magnitude is about 6.55×10^{-27} erg secs. No one would have thought of introducing just this quantity if there had not been experimental reasons for doing so, and as soon as we introduce experimental reasons the whole picture is changed... Physics as verifiable, therefore, uses various empirical concepts in addition to those purely abstract concepts that are needed in "pure" physics...

> [I]f the evidence for physical laws is held to be part of physics, then any minimum vocabulary for physics must be such as to enable us to mention the experiences upon which our physical beliefs are based. We shall need such words as "hot," "red," "hard," not only to describe what physics asserts to be the condition of bodies that give us these sensations but also to describe the sensations themselves. (1948, pp. 247-49)

To summarize, then, for Russell and Eddington, the abstractive method of physics, when applied in a thoroughgoing way, cannot but yield a description of nature which captures only mathematical structure. Now, we have grounds for believing that this structure corresponds to reality only because of the evidence provided by the perceptual experiences we have in the course of carrying out observations and conducting experiments. And at least for Russell, statements which describe these perceptions – that is, which describe the subjective perceptual experiences *themselves* rather than the external physical objects which cause them – *do* capture more than structure. If we include these statements within physics broadly construed, then to *that* extent physics gives us more than structure. But if we are speaking only of the theoretical description of the natural world *beyond* perceptual experience, it is only the mathematical structure of that world that is captured by physical theory. Hence if we interpret physical theory in a realist way, then for Russell and Eddington we must do so in a structural realist way, specifically.

The thesis that physics conveys only abstract structural relations and not the intrinsic nature of the relata was not uncommon among philosophers and scientists in the first half or so of the twentieth century. One finds variations on the theme not only in Poincaré, Duhem, Russell, and Eddington, but in thinkers like Moritz Schlick (1985), Rudolf Carnap (1967), Ernst Cassirer (1956), Hermann Weyl (1934), F. A. Hayek (1952), and Grover Maxwell (1970a, 1970b, 1972). It can hardly be denied that there is considerable truth in the thesis. Abstraction by its very nature involves ignoring the individualizing features of a thing and focusing instead on patterns it has in common with other things. When, in geometry, we consider *triangularity* in the abstract, we ignore the specific color, physical constitution, etc. of any particular triangle. Naturally, then, geometry tells us only about the structure possessed by triangles in general, and nothing about *this* specific green triangle drawn in ink, *that* specific white triangle drawn in chalk, and so forth. When an aircraft engineer determines the average weight of airline passengers, he abstracts from the sex, ethnicity, food and entertainment preferences, etc. of individual passengers. Needless to say, his conclusions about how many passengers could be carried in an airplane of a certain type tells us nothing about whether a certain individual passenger is female, whether that passenger prefers vegetarian meals or foreign movies, and so on. By the same token, when (as in Eddington's example) the physicist considers only the mass of an elephant, the angle of a hill, etc., his results are not going to tell us anything about the intrinsic natures of elephants, hills, or the like. And the more thoroughly the physicist's theoretical description of nature abstracts from such individual features, the less it is going to tell us about concrete reality. Again, the very nature of abstraction suffices to guarantee that.

However, as we have seen, Russell himself qualified the thesis insofar as he allowed that, since the perceptual experiences from which physics takes it evidence are *themselves* concrete realities, statements describing such experiences convey more than mere abstract structure. And that the thesis needs further qualification should be evident even just from the examples Worrall cites. Fresnel's theory made reference to an elastic solid ether, Maxwell's to a disembodied electromagnetic field, Newton's to action-at-a-distance, and so forth, and it is part of Worrall's point that these features *do* go beyond mathematical structure. Worrall's claim is not that the positing of these entities is not part of physics, but rather that it is a part of physics which (in light of the pessimistic induction) we need not interpret in a *realist* manner. Now, perhaps Russell and Eddington would agree with this. In the passages quoted above, Russell says that "*in proportion* as physics increases the scope and power of its methods, *in that same proportion* it robs its subject-matter of concreteness" and "*when physics is brought to [a high] degree of abstraction* it becomes a branch of pure mathematics." Eddington refers to "the *more intelligible* questions" dealt with by "*exact* science." These remarks indicate that some of what falls within physical theory is *not* after all confined to propositions about structure. Presumably, what Russell and Eddington meant to emphasize was simply that the parts of physics that have had the greatest predictive and technological success, and about which we can accordingly be most certain, are those which abstract from everything but mathematical structure. Some of what goes under the name of physics may go beyond structure, but it is also for that reason less "exact" or "intelligible" and lacks the "scope and power" of physics at its most impressive.

But yet *further* qualification of the structural realist thesis is necessary in light of an objection raised against Russell by the mathematician

M. H. A. Newman (Newman 1928; Cf. Demopoulos and Friedman 1989). Newman argued that Russell's position implies that the knowledge physics gives us is completely *trivial* – in particular, that physics, understood as conveying *only* structure and *nothing* about the relata, imposes nothing more than a cardinality constraint on the natural world, viz. a mere requirement that there be a certain minimal number of entities related by the structure described by physics. But this is absurd; surely physics, even by Russell's reckoning, tells us more than *that*.

Now, Russell's concession that statements about perceptual experience, which do go beyond structure, are part of physics broadly construed, was at least in part motivated precisely by the need to deal with the problem raised by Newman. Developing Russell's concession, more recent structural realists like Worrall (2007) and Elie Zahar (2001, pp. 236-45; 2007, pp. 177-91) have argued in response to Newman that there is *observational* as well as *theoretical* content to physics, and that this puts stricter constraints on what the natural world must be like than a mere cardinality constraint.

Whatever one thinks of such moves, Newman's objection arises in the first place only because Russell took the knowledge that physics gives us to be even *more* abstract than what has been said so far lets on. On Russell's version of structural realism, the content of a physical theory is captured by what is called the *Ramsey sentence* of the theory (after the philosopher Frank P. Ramsey). The Ramsey sentence of a theory is arrived at by replacing its theoretical terms with variables and then existentially quantifying over the variables. It turns out that it is really only the *properties* of the relations holding between physical entities, as captured in the language of formal logic (that is, the language in which the Ramsey sentence is formulated), that we can know from physics. Van Fraassen nicely sums up just how *extremely* abstract is the picture that results:

> [T]his structure is exactly, no more and no less, what can be described in terms of mathematical logic. The logic in question is strong, and today we would see it as higher order logic or set theory. But still, how little this is! Science is now interpreted as saying that the entities stand in relations which have such properties as transitivity, reflexivity, etc. but as giving no further clue as to what those relations are. (2008, p. 219)

It is this extreme abstraction that opens Russell up to the charge that he makes the knowledge that physics gives us entirely trivial.

As a result, Russell in the end "capitulated" to Newman (as Van Fraassen puts it at p. 220), giving up the idea that physics gives us knowledge only of the properties of relations, and not merely because we have knowledge of our perceptual experiences. In correspondence with Newman, Russell also allowed that we can have knowledge of an *isomorphism* between the structure of a perceptual experience, on the one hand, and the structure of the physical event in the external world which causes it, on the other (Landini 2011, pp. 331-3). Specifically, Russell held that in perceptual experience we have *knowledge by acquaintance* of spatio-temporal relations, and that we can have *knowledge by description* of spatio-temporal relations between the events that cause the experiences. (Cf. Van Fraassen's discussion of the history of Russell's views on this topic in his 2008, chapter 9.)

More recent writers have suggested that rebutting Newman's objection requires moving away from Russell's extreme interpretation of the abstractness of physical theory in yet other ways – for example, by allowing that physics captures natural kinds (Psillos 1999, pp. 67-69) or modalities (French 2014, pp. 116-24). Such proposals take Russell's brand of structural realism further away from the austere empiricism which originally shaped his formulation of it. Needless to say, the Aristotelian philosopher of nature would endorse neither that empiricism nor the aspects of Russell's structural realism that reflect it. That includes Russell's own way of understanding the thesis that perceptual experience gives us knowledge of something more than abstract structure. Russell endorses what, in chapter 2, I called a "representationalist" account of knowledge, including perceptual knowledge. When you have the experience of seeing a tree (for example), what you have non-structural knowledge of, in Russell's view, is not the tree itself, but only your perceptual experience of the tree. For the reasons given in chapter 2, I reject this account of perception, and thus I reject his application of it to the formulation of structural realism.

The brand of structural realism I endorse, then, is highly qualified and by no means committed to everything that falls under the "structural realist" label in the contemporary literature. I do not claim that physics, much less other sciences, gives us *nothing* but knowledge of structure. I certainly do not claim that we have no knowledge of physical reality at all other than knowledge of its mathematical structure – metaphysics and philosophy of nature would give us further knowledge even if physics and the other sciences did not. I do not claim that the content of the propositions about structure that physics does give us can be entirely captured via the Ramsey sentence method. I do not claim that the non-structural knowledge we do have is confined to subjective perceptual experiences, or to whatever else Russell and later thinkers influenced by him would concede.

What I do claim is merely that, *to the extent* that physics or any other science confines itself to an abstract mathematical description of nature, it cannot give us complete knowledge of concrete physical reality, and that the more abstract the description, the less complete it is. This follows from the nature of abstraction. Hence, even if what is captured in the abstract language of mathematics is the most *certain* part of physics, it is also the *least informative* part of it. It genuinely captures physical reality, but far from all of physical reality. That is the deep insight that Russell grasped and that too many proponents of scientism miss. One implication of this insight is that if some feature is absent from the mathematical description of nature afforded us by physics, it simply does not follow that it is absent from nature itself. We will have reason to return to and elaborate upon this point several times in the chapters to follow.

I say, again, that physics and other sciences are incomplete *to the extent that* they confine themselves to mathematical and other abstractions. It doesn't follow that physics or any other science always does so confine itself, much less that it ought to do so. To be sure, from Galileo to the present day, there has been a constant temptation to think that, to the extent that a science fails to achieve the precision and predictive power of mathematical physics, it is not "real" science. As Galileo famously remarked in *The Assayer*:

> Philosophy is written in this grand book, the universe, which stands continually open to our gaze. But the book cannot be understood unless one first learns to comprehend the language and read the letters in which it is composed. It is written in the language of mathematics, and its characters are triangles, circles, and other geometric figures without which it is humanly impossible to understand a single word of it; without these, one wanders about in a dark labyrinth. (Drake 1957, pp. 237-38)

That Galileo's attitude has deeply informed the thinking of his successors is evident from remarks from some prominent twentieth-century scientists. We have already seen it evinced in Eddington. Then there is Sir James Jeans, who wrote:

> The essential fact is simply that all the pictures which science now draws of nature, and which alone seem capable of according with observational fact, are mathematical pictures...
>
> [T]he final truth about a phenomenon resides in the mathematical description of it; so long as there is no imperfection in this our knowledge of the phenomenon is complete. We go beyond the mathematical formula at our own risk; we may find a model or picture which helps us to understand it, but we have no right to expect this, and our failure to find such a model or picture need not indicate that either our reasoning or our knowledge is at fault. The making of models or pictures to explain mathematical formulae and the phenomena they describe, is not a step towards, but a step away from, reality; it is like making graven images of a spirit. (1931, pp. 111, 129-30)

Similarly, in his lectures on *The Character of Physical Law*, Richard Feynman says:

> Every one of our laws is a purely mathematical statement in rather complex and abstruse mathematics. Newton's statement of the law of gravitation is relatively simple mathematics. It gets more and more abstruse and more and more difficult as we go on... [I]t is impossible to explain honestly the beauties of the laws of nature in a way that people can feel, without their having some deep understanding of mathematics...
>
> [M]athematics is a deep way of expressing nature, and any attempt to express nature in philosophical principles, or in seat-of-the-pants mechanical feelings, is not an efficient way...
>
> Physicists cannot make a conversion to any other language. If you want to learn about nature, to appreciate nature, it is necessary to understand the language that she speaks in. She offers her information only in one form. (1994, pp. 33-34, 51-52)

Galileo, Jeans, and Feynman were speaking most immediately about physics, but they nevertheless characterize "the universe" and "nature" in general, and not merely those aspects of it that are the special interest of the physicist, as understandable only in the language of mathematics. To use Maritain's terminology, the temptation is to judge an "empirioschematic" science to be a mere placeholder until a fully "empiriometric" one can replace it, or to be otherwise second-rate.

But as we will see in chapters to come, not all scientists and philosophers of science take such a view. In any event, if a scientist or philosopher insists that genuine science is confined to what can be captured by empiriometric methods, yet still allows that there is knowledge of nature to be had by way of other methods (such as the philosophy of nature), then the issue would seem to be essentially semantic. "Science" in that case simply becomes the label for knowledge of those specific aspects of nature which can be captured empiriometrically. On the other hand, if a scientist or philosopher says that there is no aspect of nature, or at least no knowable aspect, beyond what can be captured in empiriometric terms, then he is making a substantive rather than merely semantic claim – but a claim which, as the arguments developed in this book show, is question-begging at best and demonstrably false at worst.

The kind of structural realism I am affirming is perhaps best understood as simply an application, to the interpretation of the mathematical models put forward in physics and other sciences, of general Aristotelian realism vis-à-vis universals, mathematical entities, and other abstractions. For the Aristotelian realist, *triangularity* (to take a simple example) does not exist in mind-independent reality *qua* universal, after the fashion of a Platonic Form. Rather, it exists there only in individual triangles, and thus only together with particularizing features such as a certain specific color, size, etc. Qua universal, *triangularity* exists only as *abstracted from* particular triangles by an intellect, and thus as an idea within an intellect. However, contra nominalism and conceptualism, this idea is not the free creation of the intellect. *Triangularity* is really there *in* particular triangles, and in forming its universal idea the intellect *pulls it out*, as it were, so as to consider it in isolation from such particularizing features as color, size, and the like. What the intellect grasps is something that is *really there* in mind-independent reality rather than something the intellect invents, but it does not exist *qua universal idea* until the intellect abstracts it. An abstract object, on this view, is by virtue of being abstract something *abstracted from* concrete reality *by an intellect*. (For discussion and defense of this view, see Oderberg 2007, pp. 81-85, and Feser 2017, chapter 3.)

I am proposing that, in a similar way, the abstract mathematical structure of nature described by physics and other sciences does not exist *qua* abstract mathematical structure in mind-independent reality. Rather, it exists there only *in* a concrete natural order which has various features that go beyond the ones that can be captured in the mathematical description (just as concrete particular triangles have features – a certain color, a specific size, etc. – which go beyond the ones captured by the concept *triangularity*). The mathematical structure is not the free creation of the intellect (contra instrumentalism and other forms of anti-realism), but it exists in mind-independent reality not as pure abstract structure but only together with various concrete non-mathematical features. Qua pure mathematical structure, the world as the physicist describes it exists only as *abstracted from* or *pulled out of* the concrete natural order by the intellect, which considers it apart from the world's concrete features (just as geometry considers *triangularity* apart from the concrete features of particular individual triangles).

So, the structural realism I am advocating is indeed a kind of realism rather than a kind of anti-realism (whether instrumentalist, constructive empiricist, relativist, or whatever) in just the way that Aristotelian realism about universals is a kind of realism rather than a kind of nominalism or conceptualism. But just as Aristotelian realism about universals is not *Platonic* realism, neither does the structural realism I advocate hold that structure exists or can exist on its own, apart from concrete individual things. That brings us to a further qualification.

3.2.3 Epistemic not ontic

In recent discussions of structural realism, a distinction is often drawn between *epistemic* structural realism and *ontic* structural realism. Epistemic structural realism holds that even if physics or some other science gives us *knowledge* only of the structure of the natural world, there is nevertheless more to the natural world itself than just structure. This is the kind of structural realism Russell defended, and it is the kind I defend. Ontic structural realism makes a much bolder claim. It holds that there is *nothing more to* the natural world than the structure revealed by physics or other sciences. Structure is all that actually *exists*, not merely all that science tells us about (French 2014; Ladyman, Ross, Spurrett, and Collier 2007; Tegmark 2014).

Ontic structural realism cannot be correct, in part for reasons implicit in earlier chapters. I have argued that the existence of the experiences and cognitive processes of the conscious embodied subject entails the existence of change and thus of the actualization of potentiality, of final and efficient causality, and of substantial form and prime matter. Since all of that goes beyond what is captured in a description of the mathematical structure of nature, what has been said in earlier chapters suffices to show that there *is* more to the world than such structure. I have also argued against a Cartesian interpretation of the relationship between the conscious subject and the natural world, so that it will not do for an ontic structural realist even to hold that there is no more to the *external, physical* world than structure, even if there is more to the conscious subject than structure.

But there are fatal problems with ontic structural realism even apart from all that. A standard objection is that knowledge of the mathematical structure of the world is essentially knowledge of *relations*, and that there cannot be relations without *relata*. Hence it cannot be that mathematical structure exhausts reality. There must be relata which are related by the mathematical relations described by physics. I think this is a good objection, though Anjan Chakravartty (who has put forward the objection himself) suggests that ontic structural realists might regard it as question-begging, on the grounds that their point is precisely that we need to revise our concept of a relation in such a way that it can exist without relata (Chakravartty 2007, p. 77). But by itself this is hardly a powerful response. If I assert that there could be round squares and offer nothing more in response to the charge that this is incoherent than the bare suggestion that we need to revise our concepts of roundness and squareness, I have hardy made my assertion more plausible. I need to give some positive account of exactly *how* such a revision might be accomplished.

Ladyman (2014) has suggested that there are at least two ways in which we might be able to make sense of the idea of relations without relata. To understand the first, consider a universal like the relation *being larger than*. We could grasp this universal even if we were to deny that there are any things that instantiate it, any two things such that one is larger than the other (just as we grasp the universal *unicorn* even though we know that it is not instantiated). Hence, we have a sense in which we might conceive of a relation without relata. An obvious problem with this suggestion, though, is that when talking about the natural world it is precisely instantiations and *not* universals we are interested in. Even if we regarded the natural world as a single four-dimensional object, we could distinguish between the world itself as a concrete particular and the universal that it instantiates – a universal which, unlike the natural world itself and qua universal, is abstract rather than concrete, in principle multiply instantiable, causally inert, and so on. And the problem is that there is no way to make sense of concrete *instances* of relations without relata.

Nor will it do for the ontic structural realist to try to dodge this problem by suggesting that the natural world really just *is* a kind of universal or Platonic object. Among other problems, this would make it utterly mysterious how physics is or could be an *empirical* science any more than mathematics is and would thereby threaten to undermine the very evidential basis of physics.

The second way Ladyman suggests we might be able to make sense of relations without relata is by thinking of the purported relata as *themselves* analyzable in terms of further relations, and those in terms of yet further relations, all the way down as it were. But it is hard to see how this solves the problem. Either relations require relata or they do not. If, as ontic structuralists claim, relations do not require relata, then what is the point of positing an infinite regress of relations to serve as relata? Why not just stop with the top level set of relations and be done with it? But if relations do require relata, then how does positing an infinite regress of relations serve to *identify* those relata (as opposed to endlessly *deferring* an identification)? What non-question-begging reason could the ontic structural realist give for claiming that an infinite regress of relations is any less problematic than relations without relata?

Ontic structural realism essentially blurs the distinction between the concrete and the abstract, between physics and mathematics. But even scientists who emphasize the role of mathematics in physics acknowledge the difference. For example, Feynman writes:

> Mathematicians are only dealing with the structure of reasoning, and they do not really care what they are talking about. They do not even need to *know* what they are talking about...
>
> But the physicist has meaning to all his phrases. That is a very important thing that a lot of people who come to physics by way of mathematics do not appreciate. Physics is not mathematics, and mathematics is not physics. One helps the other. But in physics you have to have an understanding of the connection of words with the real world. It is necessary at the end to translate what you have figured out into English, into the world, into the blocks of copper and glass that you are going to do the experiments with. Only in that way can you find out whether the consequences are true. This is a problem which is not a problem of mathematics at all. (1994, pp. 49-50)

And as Stephen Hawking famously wrote:

> Even if there is only one possible unified theory, it is just a set of rules and equations. What is it that breathes fire into the equations and makes a universe for them to describe? The usual approach of science of constructing a mathematical model cannot answer the questions of why there should be a universe for the model to describe. (1988, p. 174)

The very asking of Hawking's question presupposes a distinction between the abstract patterns represented by mathematical equations, and the concrete physical world which instantiates the patterns.

Now, Ladyman and Ross frankly acknowledge that ontic structural realism, which they take to follow from the "naturalized metaphysics" they endorse (and which we considered earlier in this chapter), blurs the distinction between the concrete and the abstract. But they propose dealing with the problem by simply dispensing with the distinction (Ladyman, Ross, Spurrett, and Collier 2007, pp. 159-60). They justify this bold proposal by suggesting that the distinction is incompatible with physics as they understand it. Concrete objects are said to be causally efficacious and to be located in space and time, whereas abstract objects are neither. But "these categories seem crude and inappropriate for modern physics" given that causality is commonly held to be problematic in light of quantum mechanics, and that "the structure of spacetime itself is an object of physical investigation" but "can hardly be *in* spacetime" (p. 160). As to the difference between mathematics and physics, they write:

> What makes the structure physical and not mathematical? That is a question that we refuse to answer. In our view, there is nothing more to be said about this that doesn't amount to empty words and venture beyond what the PNC [Principle of Naturalistic Closure] allows. (p. 158)

The "Principle of Naturalistic Closure" is Ladyman and Ross's thesis that a metaphysical claim should be taken seriously only when it allows us to combine some hypothesis drawn from fundamental physics with some other scientific hypothesis to yield an explanation we wouldn't otherwise have – where what counts as a serious scientific hypothesis is in turn determined by "institutionally *bona fide* scientific activity," which is "fundable by a *bona fide* scientific research funding body," and so forth (pp. 37-38). In other words, the only respectable metaphysics is the "naturalized" kind, the kind that can be teased out of science as "science" is defined by people of the sort Ladyman and Ross would consider mainstream. Hence if the traditional metaphysical distinction between concrete and abstract (or for that matter, it seems, the distinction between physics and mathematics) ends up being at odds with the deliverances of this kind of metaphysics, so much the worse for the distinction.

But this reply is muddleheaded and dogmatic. It is dogmatic insofar as, as we saw when considering the "naturalized metaphysics" project earlier in this chapter, there appears to be no way to defend that project without begging the question. Certainly Ladyman and Ross do that here. It is no good to respond to the objection about blurring the abstract and the concrete by appealing to naturalized metaphysics, because the very coherence of naturalized metaphysics, at least in the ontic structural realist way Ladyman and Ross understand it, is precisely part of what is in question. The reply is muddleheaded insofar as it amounts to responding to a *reductio ad absurdum* by embracing the *absurdum*. And in a strict *reductio*, that is not an option, philosophically speaking. One might, as a matter of *psychological* fact, embrace the self-contradiction to which one's position leads. But *logically* speaking it remains a self-contradiction, and therefore falsifies the proposition that entails it.

That the conclusion to which Ladyman and Ross are led *is* self-contradictory (and not merely odd or surprising) follows from the nature of abstraction. They cite the lack of causal efficacy and of spatiotemporal location as the marks of being abstract, but these are secondary characteristics. The *essence* of being abstract is simply *not being concrete*. The north side of the Great Pyramid is one concrete individual thing, and a billiard rack is another. *Triangularity* is abstract insofar as it is not a further concrete individual thing alongside these, but rather the pattern that remains when the stone of the pyramid, the wood of the rack, the colors of the stone and wood, etc. are stripped away in thought. Because having spatiotemporal location and causal efficacy entails being a concrete individual thing, *triangularity* in the abstract lacks these features, but this lack is a *consequence* of its being abstract rather than constituting its abstractness. Its being abstract is, again, essentially just its being non-concrete. To suggest that there is no distinction between being concrete and being abstract is, then, implicitly to say that there is no distinction between being concrete and not being concrete – which is a contradiction.

Even if it were not self-contradictory, it is still hard to see what it could even *mean* to say that there is no distinction between (for example) *triangularity* in the abstract and concrete triangles like the north side of the Great Pyramid and a particular billiard ball rack. Yet that is the sort of thing it seems we would have to say if we reject the distinction between the abstract and the concrete. Ladyman and Ross thus owe us some account of what we are to put in place of all the applications we ordinarily make of the distinction. Even apart from its begging the question, an appeal to the implications of naturalized metaphysics cannot solve the problem. For the problem is not merely that of rationally *justifying* the claim that there is no distinction between abstract and concrete, but that of simply *giving some coherent content* to that claim.

Nor will it do to dismiss the demand for clarification as presupposing a suspect method of "conceptual analysis," outmoded "folk" notions, appeals to "intuition," etc. For one thing, what I am talking about here is not some positive thesis being put forward by the critic of ontic structural realism (on the basis of intuitions, conceptual analysis, or whatever), but rather simply a request by the critic that the ontic structural realist explain exactly *what he means* when he says that there is no distinction between the abstract and the concrete. After all, we cannot be expected to affirm or even consider a thesis until we know what the thesis is. For another thing, the claim that *all* so-called "folk" assumptions, conceptual truths, and the like are revisable is part of what is at issue in the dispute between naturalized metaphysics and its critics, so that to dismiss the demand for clarification on the basis of that claim would once again be to beg the question.

There is a further problem with ontic structural realism, though pursuing it in any depth would take us too far afield from the concerns of this book. Abstractions presuppose a mind which does the abstracting, or so Aristotelians would argue. (See Feser 2017, chapter 3.) But you needn't be an Aristotelian to think that abstract objects exist only in minds. Even Platonism, which is commonly interpreted as locating abstract objects in a third realm distinct from both minds and material things, gave way historically to the *Neo*-Platonist thesis that the Forms exist in a divine intellect. But if abstracta exist only in minds, and the natural world is (as ontic structural realism claims) nothing more than an abstract structure, then it would follow that the natural world is essentially mind-dependent. Ontic structural realism thus appears to be implicitly idealist. (Indeed, Eddington and Jeans took their emphasis on the mathematical nature of the physical world in an idealist direction. Cf. the critical discussion of their views in Stebbing 1958.) Hence if idealism is problematic, so too is ontic structural realism. Needless to say, that idealism *is* problematic is a claim that would require argumentative support, and such argumentation would go beyond the scope of this book. The point for present purposes is merely that since philosophical naturalists like Ladyman and Ross would presumably have no truck with idealism, the potentially idealistic implications of their ontic structural realism should give them pause.

3.3 How the laws of nature lie (or at least engage in mental reservation)

There is one further qualification to be made to the brand of scientific realism I would endorse. As noted in chapter 1, the founders of the mechanical world picture replaced the Aristotelian mode of explanation in terms of essences and teleology with explanation in terms of laws of nature. Now, I am defending an Aristotelian philosophy of nature, but have also endorsed a structural realist interpretation of the results of modern science. So does this realism extend to laws of nature? Yes and no. It depends on how they are interpreted.

The standard view of laws of nature regards them as universal regularities, ordered in something like a pyramidal structure, and where at least the laws at the apex of the pyramid are ontologically fundamental in the sense that they don't presuppose anything else (except God, for proponents of the standard view who are theists). They are universal in the sense that they hold everywhere and always. They form a pyramid insofar as the laws of one science are taken to be subsumed under those of another, which are in turn subsumed under those of another, and so on until we reach some level of laws under which all other laws are subsumed. For example, if there are laws of psychology at the base of the pyramid, then above them in the pyramid would be laws of biology from which the laws of psychology follow as a special case. These laws of biology would in turn fall under laws of chemistry, which in turn fall under laws of physics, with some most general set of physical laws at the top of the pyramid. (I borrow the pyramid analogy from Cartwright 1999, p. 7.) The laws at the top are ontologically fundamental in that everything else that exists is taken to be explained by reference to these laws, with the laws themselves not being explicable in terms of something more basic. When we reach the laws at the top of the pyramid, we have (if you'll pardon the mixed metaphor) reached metaphysical bedrock. (For the atheist, anyway. Again, the theist who is committed to this picture of laws would say that God is the cause of the laws. Even for such theists, though, there is nothing *in the natural world* that is more basic than the laws.)

There is another way to understand laws of nature, however, which is most famously associated with Nancy Cartwright and first set out in the essays collected in her influential book *How the Laws of Physics Lie* (1983). On Cartwright's view, each of the tenets of the standard view is false. First, laws are not universal regularities. Or to be more precise, if interpreted as universal regularities, laws turn out not to be strictly true; whereas if they are interpreted in a way that makes them come out true, they are no longer strictly universal. For example, the law of universal gravitation will not correctly describe the behavior of bodies that are charged or subject to air friction. Newton's law of inertia holds only in circumstances where no forces are acting on a body – circumstances which never in fact obtain. Kepler's first law tells us that planets move in ellipses, but this is only approximately true insofar as planets are always acted upon by the gravitational pull of other bodies. And so on. Laws are true only *ceteris paribus*, only when certain conditions obtain. In that case, though, they correctly describe the behavior of the entities they govern only under those particular conditions, and are not true of the entities universally.

Those who claim that laws are universal would respond that the *ceteris paribus* qualifications reflect only our ignorance rather than reality itself. If we knew enough about all the complex details that affect how a thing will behave, then (so the argument goes) we would be able to formulate laws that are strictly universal. But there is in Cartwright's view nothing in the actual empirical evidence for laws that supports this claim over her own interpretation of the situation. It reflects a background philosophical commitment to a certain view about the nature of laws, rather than the actual findings of science.

A second way Cartwright departs from the standard view is by denying that laws are ontologically fundamental. What are fundamental to the entities studied by physics and the other sciences are rather their *natures* and *capacities* (Cartwright 1999, pp. 59-73, 78-90). By virtue of these natures and capacities, entities "try" or "tend" to behave in certain distinctive ways (1999, pp. 28-29), and the tendencies of one entity can combine with those of another to produce a joint effect. A relatively stable arrangement of entities jointly exercising their capacities can give rise to relatively stable patterns of behavior within the resulting system. For example, a star, planets, moons, etc. constitute a relatively stable arrangement of entities, jointly exercising their characteristic capacities and manifesting their distinctive tendencies in a way that gives rise to the patterns of behavior typical of a solar system.

Such an arrangement constitutes what Cartwright calls a "nomological machine" (1999, chapter 3). Laws are essentially descriptions of the regularities characteristic of a certain kind of nomological machine. For example, Kepler's laws of planetary motion are a description of the regularities characteristic of a solar system. In that case, though, laws have a derivative rather than fundamental ontological status. They presuppose the existence of nomological machines, which in turn presuppose the existence of entities possessing distinctive natures, capacities, and tendencies. Moreover, the laws correctly describe a system only given a certain model of the system, and models are constructed precisely so that they will conform to the laws. A model presents an idealized version of a concrete situation, which leaves out complicating details and might introduce features that are there in order to facilitate use of the model rather than because they correspond to reality (Cartwright 1983, chapters 7 and 8).

The third way Cartwright's position differs from the standard view is that she takes laws to form a "patchwork" rather than a pyramid (1999, chapter 1). There are the laws describing the behavior of this nomological machine and the laws describing the behavior of that one, but we have no reason to believe that anything unites them all. In particular, we have no reason to believe that laws are arranged in a hierarchy or that there is some one most basic law or set of laws from which all the others follow. The natural world is "dappled" rather than uniform; or at any rate, those who claim otherwise are, once again, motivated not by the empirical evidence but by philosophical commitments.

Now, there are three general considerations which, I maintain, together provide a powerful argument for Cartwright's account of laws, or something like it, over the standard view. First, and as already noted, there is nothing in the actual findings of modern science that favors the standard view over hers. *Empirically* speaking, the rival views are evenly matched at best, with the choice between them essentially philosophical rather than scientific.

But second, there are serious philosophical problems with the standard view. For one thing, the reductionism implicit in the pyramid model faces well-known difficulties (summarized in Cartwright 2016, pp. 30-32). Many higher-level natural kinds are neither reducible to lower-level kinds nor plausibly eliminable. For example, biological kinds are defined in functional terms, which are notoriously difficult to analyze in terms of the behavior of micro-level parts. Even inorganic phenomena are more resistant to reductionist analysis than is often realized. For instance, thermodynamics is often thought reducible to statistical mechanics, but the latter leaves out a crucial thermodynamic property, viz. the direction of entropy. In general, macro-level phenomena often have properties that are not predictable from lower-level features nor entirely understandable apart from yet larger systems of which they are themselves parts. These facts block the reduction of the laws of the various special sciences to fundamental laws of physics, à la the pyramid model. (We will examine these issues in detail in later chapters.)

For another thing, the very notion of a law of nature becomes problematic when removed from the theological context in which (as we saw in chapter 1) it had its original home (Cartwright 2005). If a law is not a divine decree, then what is it? One answer is the regularity theory of laws of nature. To state this view in its simplest form, a law of nature is simply a regular pattern that we happen to find in nature. It's not that God or anything else *causes* this regularity to exist in nature. It is just there in nature, and that's that. An object at rest stays at rest, and an object in motion stays in motion at a uniform speed in the same direction unless acted upon. Planets have elliptical orbits. Radium has a half-life of 1600 years. And so on. That's just how the world is. In the philosophy of science, this view is often traced back to David Hume, and it seems to be the view taken by at least many contemporary scientists. For example, Feynman seems to be committed to something like it when he gives physical laws the following characterization:

> There is... a rhythm and a pattern between the phenomena of nature which is not apparent to the eye, but only to the eye of analysis; and it is these rhythms and patterns which we call Physical Laws. (1994, p. 3)

The basic idea of the regularity theory is very simple and many scientists seem to think it obvious and unproblematic. But philosophers of science who defend it have had to qualify it significantly, because on closer inspection the regularity theory is subject to several serious objections. (Cf. Bird 1998, chapter 1)

The first problem is that a pattern's being regular is in fact not *sufficient*, not by itself enough, to make it a law of nature. To take a stock example, consider the following two regularities: (1) *Every lump of gold is smaller than one cubic mile in size*, and (2) *Every lump of uranium-235 is smaller than one cubic mile in size*. Both of these statements are true, but there is a crucial difference between them. Though there is in fact no lump of gold as large as a cubic mile, such a lump is at least theoretically possible. But a lump of uranium-235 that large is *not* theoretically possible, because a chain reaction would occur before the lump could get that big. So, though the regularity concerning uranium-235 plausibly counts as a law of nature, the regularity concerning gold does not. So there must be something more to a law of nature than merely being a regularity.

Or consider Nelson Goodman's famous "grue" example (1983). Suppose it were a law that *all emeralds are green* and also a law that *all sapphires are blue*. (This is not quite correct, but for the sake of simplicity suppose it were. Or substitute a different example if you wish.) Now consider the attribute of being *grue*, which something has if it is observed before December 31, 2050 and is green, or observed after December 31, 2050 and is blue. And consider further the attribute of being an emerire, which something is if it is observed before December 31, 2050 and is an emerald or is observed after December 31, 2050 and is a sapphire. Then it will be true that *all emerires are grue*. But it seems implausible to regard this regularity as a law of nature. Of course, it might be objected that attributes like *being grue* or *being an emerire* seem silly and are obviously "made up" rather than capturing some objective feature of nature. But that's precisely the point. Since, precisely because of its artificiality, a regularity like "All emerires are grue" does not plausibly count as a law of nature, there must be more to a law of nature than simply being a regularity.

The actual existence of a regularity also does not appear to be *necessary* for something to be a law of nature. For example, consider a law to the effect that particles of a certain kind have a fifty percent probability of decaying within a certain period of time *t*. It might seem that there is a regularity that makes this a law, namely that among any collection of particles of the type in question, a certain proportion will in fact have decayed by time *t*. But suppose there happened to be only one such particle. It is perfectly possible that that particle will not in fact decay by time *t*. In that case we would not have a certain proportion of particles decaying by time *t*, and thus would not have any actual regularity for the law to describe. But there nevertheless would still be a physical law to the effect that any particle of that type has a fifty percent probability of decaying by time *t*.

Consider also that there are chemical elements that do not exist in nature but would have to be produced artificially, in the lab or by nuclear explosions, if they are to exist at all. Fermium would be an example. As with other elements, there are physical laws that describe the properties and behavior of fermium. But suppose fermium had never in fact been produced. Then the laws of nature describing fermium would still be true, even though they corresponded to no actual regularities found anywhere in nature. For it still would have been true, even under those circumstances, that *if* fermium were to exist, it would behave in such-and-such a way.

It might seem that some of these problems could be dealt with if we added *counterfactual conditionals* to our statement of a law. A counterfactual conditional is a statement about what *would have* happened if a certain situation that did not in fact exist *had* existed. Hence, as I just indicated, even in a world without any actual fermium we could state laws governing fermium by saying that *if* fermium had existed, then it *would have* behaved in such-and-such a way. Or we could say that *if* we had tried to produce a lump of uranium-235 as big as a cubic mile, it *would have* caused a chain reaction before it could form. Since no such counterfactual conditional would be true of a lump of gold the size of a cubic mile, it might seem that we could use counterfactuals to capture the fact that the regularity concerning uranium-235 is a genuine law while the regularity concerning gold is not.

However, this will not work, because it gets the relationship between laws and counterfactual conditionals the wrong way around. Counterfactual conditionals will be true only *given* certain background assumptions, including assumptions about what the laws of nature happen to be. Hence, consider the counterfactual conditional statement to the effect that if a certain object had been set in motion, then it *would have* continued in motion at a uniform speed. This counterfactual will be true only on the assumptions that Newton's first law is in fact true, and that the object in question was not being acted upon by an outside force. So, we cannot analyze laws of nature in terms of counterfactual conditionals, because we have to analyze counterfactual conditionals in terms of laws of nature.

A way to try to deal with some of these problems that was developed by philosopher David Lewis (1973, pp. 72-77) is to suggest that a physical law is not just any old regularity, but is, specifically, a regularity that covers a *broad range* of phenomena and yet can be captured in a relatively *simple* description. One problem with this approach is that it still doesn't rule out all regularities that are not plausibly thought of as physical laws. For example, the regularity captured by the statement that *all emeralds are green* is no more simple or broad in scope than the regularity captured by the statement that *all emeralds are grue.* But while the former is a plausible candidate for a law of nature, the latter is not. Nor is it clear exactly how we are to evaluate the criterion of simplicity. Recall what it means for something to be grue. By one standard of simplicity, the statement that *all emeralds are grue* is simpler than the statement that *all emeralds are green if they are observed before December 31, 2050 and blue if observed after December 31, 2050.* But since these statements amount to the same thing, the greater simplicity of the first formulation hardly makes it a more plausible candidate for a law of nature than the second and more verbose formulation. A further problem is that *simplicity* in the statement of a law and the *breadth* of the phenomena covered by the law may come into conflict. If we add details to the statement of a law it may cover a wider range of phenomena, but at the same time be less simple in its formulation.

But there is an even deeper and more serious problem with the regularity theory of laws of nature, however many qualifications we add to it. The problem is that if a physical law is a mere regularity, then it doesn't really *explain* anything. All it does is *re-describe* things. Suppose you say: "Planets always move in elliptical orbits. I wonder what explains that?" Suppose I answer: "Kepler's first law explains that." You then ask: "Oh, how interesting. What is Kepler's first law?" And I respond by telling you that Kepler's first law states that planets always move in elliptical orbits. Obviously, we've gone around in a circle. I haven't really explained the regularity in question at all, but merely slapped the label "law" on it.

If laws are mere regularities, then slapping a new label on a phenomenon is all I *could* be doing. Again, the regularity theory tells us that a law simply describes a regular pattern we find in nature. To say that it is a *law* that "All As are Bs" is just a fancy way of saying that as a matter of fact all the As that exist in the world happen to be Bs. If I tell you that all the chairs in this room are beige, that would obviously be no *explanation* of the fact that the chair to my left is beige or of the fact that the chair to my right is beige. By the same token, if I say that all planets move in elliptical orbits, that does not provide an explanation of the fact that Mars moves in an elliptical orbit or that Venus movies in an elliptical orbit. It merely summarizes the facts to be explained, rather than actually explaining them.

Now, one might be tempted to say that the appeal to Kepler's laws really is a genuine explanation of the motion of the planets, because Kepler's laws can be interpreted as a special case of Newton's laws, and Newton's laws make reference to concepts like force, mass, and acceleration that can illuminate why the planets move. But if the regularity theory of laws were true, this would be an illusion, because Newton's laws too would really merely describe regularities rather than explain them, even if the description is a more general one. Go back to my example of the chairs. Suppose you ask me why the chair to my left is beige, and I answer "Because all the chairs in this room are beige." Suppose you object that this does not really explain the color of the chair at all, and I reply: "But the fact that all the chairs in the room are beige is actually a special case of the more general fact that all the *furniture* in the room is beige, and to point this out brings in a new concept – the concept of furniture – which illuminates the fact that all the chairs are beige." Obviously, this doesn't really illuminate anything. And by the same token, even if you can derive Kepler's laws from Newton's, and then take Newton's in turn to be an approximation of Einstein's laws, you still will not really have explained anything *if* physical laws are mere regularities. All you will be doing is *describing* the phenomena to be explained using more general concepts, rather than actually explaining the phenomena.

The bottom line is that if the regularity theory is true, then the fundamental laws of nature cannot be the ultimate explanation of things, as the standard view that Cartwright is criticizing requires. (Nor will it do to suggest that there just is no ultimate explanation of things, given PSR, which I defended in chapter 2.)

An alternative to the regularity view of laws of nature would a *Platonist* view. The easiest way to explain this view is as follows. Suppose we think of the key properties referred to in a scientific theory as something like Platonic Forms. For example, suppose we think of mass, force, and acceleration as Platonic Forms. There is the Form or abstract pattern of having mass, the Form or abstract pattern of having force, and the Form or abstract pattern of having acceleration. All the particular physical objects that there are participate in these Forms. Then laws of nature, on this view, can be thought of as necessary connections holding between these Forms. For example, Newton's second law of motion, $F = ma$, would be understood as describing a necessary connection holding between the

Form of mass, the Form of force, and the Form of acceleration. The law is like a higher-order Form in which these Forms participate. So, since all physical objects participate in the Platonic Forms of force, mass, and acceleration, they also participate in the higher-order Form that we call Newton's second law.

The problem with this Platonic account is that it doesn't really explain *why* the natural world behaves in accordance with physical laws. Consider that, if Platonism is true, then there are Forms corresponding to all sorts of things that don't in fact exist. For example, there is a Form of being a unicorn and a Form of being a Tyrannosaurus Rex, just like there is a Form of being a lion. The difference, of course, is that there actually are things that participate in the Form of being a lion, but there are no longer things that participate in the Form of being a Tyrannosaurus Rex, and there never was anything that participated in the Form of being a unicorn. Now, by the same token, not only the laws that actually govern the world, but also alternative possible physical laws that don't in fact govern it, all presumably exist together in the Platonic realm of abstract objects. So, what explains why the world participates in just the specific physical laws it does, rather than one of the alternative sets of laws, or no laws at all? Abstract objects like Forms are causally inert. By themselves they don't do anything, and so if we think of laws of nature as Forms or abstract objects, we still need to appeal to something in addition to the laws in order to explain why the world actually participates in these Forms.

Suppose we say that we can explain this in terms of higher-level laws of some sort, again understood on the model of Platonic Forms. The trouble with this, of course, is that we would now need to explain why the world operates according to these higher-order laws, which raises the same problem all over again and threatens an infinite regress. Suppose instead that we say that God causes the world to operate according to the laws, using the Forms as a blueprint for creation, as Plato suggested in the *Timaeus*. Then we're essentially back to the theological conception of laws of nature, to which the Platonic view was supposed to be an alternative. Or suppose we say that it is just an inexplicable regularity that the world operates according to physical laws conceived of in Platonic terms. Then we're essentially back to the regularity view of laws, and the Platonic view was also supposed to be an alternative to that. Or suppose instead that we interpret the Forms that the laws relate to one another in an Aristotelian fashion, as immanent to the natural world rather than located in a transcendent "third realm." (Cf. Armstrong 1983.) Then we would essentially be interpreting laws as a reflection of the natures or capacities of material objects – which is precisely to interpret them *Cartwright's* way rather than according to the standard view that she is criticizing.

Of course, one could in principle return to the theological view of laws of nature. A theological problem with this move, however, is that it makes the principles by which things operate entirely *extrinsic* to them. It isn't really anything in stones, trees, or dogs themselves that makes them behave the way they do, but rather God as a lawgiver who imposes on them from without a certain way of acting. It is hard to see how this picture can avoid collapsing into the occasionalist view that only God ever really causes anything. (Cf. Feser 2013b. For a critique of occasionalism, see Feser 2017, pp. 232-38.) In any event, philosophical naturalists, whose version of the mechanical world picture is the main target of this book, can hardly salvage their position by appealing to the theological conception of laws of nature.

So, again, a second consideration in favor of Cartwright's position is that the rival accounts of laws of nature all face grave philosophical difficulties.

The third consideration that favors Cartwright's account of laws of nature over the standard view is that there are powerful positive and independently motivated arguments for the conception of nature that it embodies. As Stephen Mumford has noted (2004, p. xiv; 2009, p. 267), whereas Cartwright's theory of natures and capacities was motivated by considerations in philosophy of science, other contemporary philosophers such as C. B. Martin (2008, Chapter 2) and George Molnar (2003) developed similar ideas on the basis of considerations drawn from general metaphysics rather than science.

Now, the confluence of the work of these and other analytic metaphysicians and philosophers of science has in recent philosophy led to a revival of interest in Aristotelian essentialism and causal powers. (Cf. Tahko 2012, Groff and Greco 2013, and Feser 2013c for representative samples of the literature.) Naturally, as a proponent of Aristotelian philosophy of nature, I am bound to sympathize with this work. (I expound and defend some of it in Feser 2014b.) Equally naturally, a proponent of the standard view of laws of nature is bound to be unsympathetic to it – though, as I argued in chapter 1, the usual modern objections to the Aristotelian conception of nature are without force. In any event, the point for present purposes is simply that the arguments for natures and capacities developed by contemporary analytic metaphysicians are *independent* arguments. That is to say, even *apart from* the considerations from science adduced by Cartwright, we have philosophical grounds for thinking that something like her account of nature is correct.

To summarize, then: First, the actual findings of science fit Cartwright's account of laws of nature better than they do the standard account, or at least are neutral between the accounts, making the dispute an ultimately philosophical rather than scientific one. Second, there are serious philosophical difficulties facing the standard account. Third, there are positive independent philosophical grounds for preferring something like Cartwright's account to the standard one. Again, these considerations taken together provide a powerful argument for Cartwright's position. But even if they did not, they would suffice to defuse any glib suggestion, on the part of proponents of the mechanical world picture, to the effect that laws of nature are sufficient to make the natural world intelligible, making the Aristotelian philosophy of nature otiose. As Cartwright's arguments show, the notion of a law of nature is itself a contested one, and can be interpreted in a neo-Aristotelian way rather than a mechanistic way. *Merely* to suggest that explanation in terms of laws of nature suffices to make an Aristotelian conception of nature unnecessary would therefore beg the question.

But might Cartwright's position not face difficulties of its own? Carl Hoefer suggests that while her account "offers us a picture of science and its possibilities that is very faithful to the current state of theory and practice," that is actually a "weakness" rather than a strength (2008, p. 320). The reason is that what matters, in Hoefer's view, is not what *current* scientific theory says, but rather what a *completed* and more thoroughly mathematicized scientific picture of the world will say. Such a picture is, in his view, *not* likely to support Cartwright's description of the world in terms of natures, capacities, and related causal notions. Hoefer also thinks that a proponent of the standard view need not be committed to the reductionism of the pyramidal model of laws.

Now, one problem with Hoefer's position is that it will not do for a proponent of the standard account of laws *merely* to reject reductionism. He will have to show how he can reject reductionism *without at the same time* implicitly returning to the Aristotelian hylemorphist view that there are fundamental discontinuities in nature (between the rational or human form of animal life and non-rational forms, between sentient and non-sentient forms of life, and so forth). (Again, more on this in later chapters.) The main problem, though, is that while Hoefer is correct to hold that a thoroughly mathematicized conception of nature would leave no place for the causal notions Cartwright endorses, that does not, for all Hoefer has shown, show that such notions have no application. It would do so only if a thoroughly mathematicized description of nature were an *exhaustive* description. But as I have been arguing in this chapter, that is precisely what it is not. A mathematicized description of nature leaves out causal notions precisely because it *abstracts from* concrete physical reality. It no more captures all there is to physical reality than an aircraft engineer's description of the height and weight of the average airline passenger captures all there is to airline passengers. Cartwright's point is precisely that the closer we stick to concrete physical reality, the less accurate mathematical laws become and the more we have to bring causal notions back in to our description.

Cartwright's critique of the standard view of laws of nature raises the question of whether the notion of a law of nature is worth preserving. To be sure, one could argue that laws of nature are real, but that what they boil down to are descriptions of how the capacities that a thing has, given its nature, will manifest. As David Oderberg puts it, "laws of nature are truths about how objects must behave" and "to say the laws are *of nature* is to say that they are *of the natures* of things" (2007, p. 144). But one could also argue, as Mumford does (2004), that in that case we can say everything that needs to be said in terms of natures, capacities, and the like, so that the notion of *laws* of nature becomes otiose and might as well be abandoned.

For present purposes it doesn't matter which position we take. Suffice it to note that the former approach shows that realism (of a sort) about laws of nature is compatible with an Aristotelian philosophy of nature – though on either Oderberg's or Mumford's view, laws certainly become less interesting and important than is usually supposed.

One further issue regarding laws of nature merits at least a brief comment. As indicated earlier, one of the difficulties facing the non-theological version of the standard view of laws is explaining where the fundamental laws come from. Several physicists have proposed that the fundamental laws might be explicable in terms of *evolution.* Lee Smolin is one such physicist, and he attributes a similar view to Paul Dirac, John Wheeler, and Richard Feynman (Smolin 2013, pp. xxv-xxvi). The idea is that the laws that now govern the universe may have arisen from previous, different laws, and those in turn from yet other laws. Smolin proposes that a kind of "cosmological natural selection" guides this process (2013, p. 123).

However, there are serious problems with this view. First, we need to ask if this proposed evolutionary process is *itself* law-governed. If it is not, then it seems that this process has no explanation but is just a brute fact (unless we appeal to God, or to the natures and capacities of the concrete objects whose behavior the laws describe – options that are not open to the non-theological version of the standard view of laws). But this would violate the principle of sufficient reason, which I defended in an earlier chapter (and to which, as it happens, Smolin himself is committed). So, we have to say that the evolutionary process in question *is* law-governed. But now we have another problem, which is that the laws that govern the evolutionary process now *themselves* stand in need of explanation. If we say that *they* have no explanation, then we not only would once again violate the principle of sufficient reason, but we will have rendered pointless the initial appeal to evolution. For if we are going to allow that the laws that govern *the evolutionary process* have no explanation, then we might as well say that the laws of nature that now govern *the universe*, which we were proposing to explain in terms of evolution, have no explanation. But if instead we say that the laws that govern the evolutionary process do have an explanation, and posit some further, higher-order evolutionary process to explain those laws, then it seems we are led into a vicious regress.

Smolin recognizes that his positon faces this "meta-laws dilemma," as he labels it (2013, pp. 243-44). He proposes two possible solutions. The first would be to posit what he calls a "principle of the universality of meta-law." The idea here is that it might turn out that all the possible meta-laws that could govern the proposed evolutionary process are equivalent to one another insofar as they would generate the same results. But it is hard to see how this solves the problem. For one thing, no reason is given for believing that there is any such principle. It appears to have no motivation other than the *ad hoc* one of solving the meta-laws problem.

For another thing, the principle wouldn't solve that problem even if it were true. The most it would show is that, *if* there is an evolutionary process governed by a meta-law, *then* any meta-law will be as good as any other. But that doesn't explain what makes it the case that there *is in fact* such a process. If you see me eating vanilla ice cream and ask me why I am eating it, I would not be giving a complete explanation if I told you that the only ice cream available was vanilla. That would explain why I am eating *vanilla* ice cream, specifically, but not why I am eating any ice cream at all. Similarly, the most that Smolin's proposed principle could explain would be why the evolutionary process is governed by such-and-such a meta-law, specifically. What remains to be explained is why there is any evolutionary process in the first place. And if Smolin appealed to a *meta*-meta-law in order to answer that question, that would simply land him in a higher-order version of the same problem.

Smolin's other proposed solution to the meta-law dilemma is to propose "a marriage of law and configuration" (2013, p. 244). The idea here is that there are not two things, a concrete evolutionary process and a distinct meta-law that governs it. Rather, there is just the one concrete reality, with the meta-law being immanent to it. Now, Smolin's proposal here appears to be roughly in the spirit of the Aristotelian approach to laws of nature advocated by Oderberg. The meta-law would simply be a description of the way that, given its nature, the concrete physical universe manifests a capacity to give rise to new laws. But this does not solve the meta-law problem, because once again we still face the question of what makes it the case that there is such a concrete physical universe (whether governed by laws immanent to it or not) in the first place. And again, positing meta-meta-laws to deal with that problem will only raise the same problem over again at a higher level.

The bottom line is that all such proposals will inevitably face problems like these as long as they confine themselves to a terminus of explanation that is *contingent*. If they opt for a terminus that exists of absolute *necessity*, the problem will be solved. However, that will commit them to theism – at least given the traditional thesis of natural theology that what exists of absolutely necessity must have the divine attributes, a thesis I would defend (Feser 2017). Smolin, it seems (2013, p. 265), would not welcome this result, and neither, of course, would defenders of the non-theological version of the standard view of laws of nature. An evolutionary account of laws thus cannot afford a third option to the naturalist seeking to avoid both the theological account of laws and Cartwright's neo-Aristotelian position.

3.4 The hollow universe

A blueprint can tell you a lot about a building, and has considerable practical utility insofar as it enables you to predict which rooms you will see as you enter the building and when you make your way up the stairs, how big the rooms will be, etc. All the same, the blueprint hardly tells you everything there is to know about the building, such as the color of the walls or the temperature inside the rooms. Furthermore, some aspects of the blueprint do not reflect what is really there in the building at all, but are rather mere artifacts of the blueprint's mode of representation. For example, there are no literal blue lines along the edges of the walls or floors of the actual building, there are no curved lines (of the kind that in a blueprint indicate the presence of a door) to be found on the floors of the actual rooms, and so on.

What I have argued in this chapter is that the representation of the physical universe afforded us by empiriometric science, and by mathematical physics in particular, is like a blueprint. It tells us a great deal about physical reality and thereby allows us to predict and control nature to a considerable extent. But the representation physics gives us does not tell us everything there is to know about physical reality, and some of what it tells us reflects merely its mathematical mode of representation rather than objective physical reality itself. Like the aircraft engineer of my earlier example, the physicist abstracts from concrete physical reality those aspects susceptible of mathematical representation and ignores the rest. The resulting representation no more tells us everything about nature than the engineer's calculations tell us everything about airline passengers. Or to take another analogy presented earlier, the physicist's description of nature no more tells you everything about nature than the geometry of triangles tells you everything about pyramids, dinner bells, and billiard ball racks. Moreover, just as the engineer's calculations may make reference to the average weight of passengers even though there may be no actual passenger with that exact weight, and just as a triangle as described in geometry has perfectly straight sides even though actual physical triangles do not, so too might the mathematical abstractions of physics make reference to properties that do not correspond to objective physical reality.

Unsurprisingly, these points were emphasized by twentieth-century Aristotelian-Thomistic philosophers of nature. For example, Charles De Koninck (1964) aptly characterized the world as represented by mathematical physics as a "hollow universe," the rich concrete qualitative detail having been squeezed out of it like juice is squeezed out of an orange, leaving only a desiccated husk. (Cf. also Mullahy 1946.) But many thinkers outside the Aristotelian-Thomistic orbit have made similar points. We have already seen how much emphasis Russell put on the theme. Eddington wrote that:

> [T]he exploration of the external world by the methods of physical science leads not to a concrete reality but to a shadow world of symbols, beneath which those methods are un-adapted for penetrating. (1929, p. 73)

Susan Stebbing noted that the physicist's "theme" is "Nature... in certain of its aspects, namely, those aspects that are susceptible of mathematical treatment" (1958, p. 115). Developing the point, she wrote:

> '[T]he world of physics', i.e. the constructive descriptions of the physicists, is necessarily restricted in it reference to experience. This restriction is a logical consequence of the nature of the methods appropriate to the study called 'physics'; it is not an unfortunate result of the inscrutable nature of the world. On his own admission, the physicist starts from the familiar world of tables, stars, and eclipses, aims at constructing a complex of metrical symbols which shall symbolize the recurrences in this familiar world, and has found it necessary, in order to fulfill this aim, to introduce symbols that have no exact counterpart in sensible experience and thus cannot be translated into the language of common sense. The methods of physical science are not adequate, and are not intended to be adequate, to the description of all that is experienced...
>
> [P]hysics is, and has always of necessity been, abstract... [insofar as] the physicist deals with a selection of the properties of what there is in world, and... his success in his investigations depends upon his isolating those properties and considering them on their own account. He has never been concerned with *chairs*, and it lies beyond his competence to inform us that the chairs we sit upon are abstract. (1958, pp. 116, 278)

More recently, Paul Feyerabend (1999) has emphasized that there is a strong tendency in modern science to substitute abstractions for what he called the "richness of being" or the "abundance" that actually exists in nature. C. B. Martin notes that the "mathematicizations" of the physicist involve only a "partial consideration" of what is actually there in nature (2008, p. 75). Smolin acknowledges that in physics:

> [W]e artificially mark off and isolate a phenomenon from the continual whirl of the universe. We seek insight into universals of physics through restricting our attention to the simplest of phenomena... I call it *doing physics in a box*...
>
> [T]o apply mathematics to a physical system, we first have to isolate it and, in our thinking, separate it out from the complexity of motions that is the real universe...
>
> This kind of approximation, in which we restrict our attention to a few variables or a few objects or particles, is characteristic of doing physics in a box. The key step is the selection, from the entire universe, of a subsystem to study. The key point is that this is always an approximation to a richer reality. (2013, pp. 38-39)

As such citations indicate, the abstract and incomplete character of the description of nature afforded by physics is by no means unacknowledged within mainstream philosophy and science. All the same, other mainstream philosophers and scientists often say things that imply that physics *does* give us an exhaustive description of the world. The radical eliminativist scientism of Alex Rosenberg (2011) – who advocates jettisoning intentionality, consciousness, the self, ethics, and much else, on the grounds that physics makes no reference to them – is an extreme example. But even more moderate thinkers evince a similar attitude when they suggest, for example, that relativity theory shows that time and change are illusory, or that quantum mechanics shows that causality is illusory. For all such arguments take the absence of the phenomena in question from the *description* of nature afforded by physical theory to imply that the phenomena are absent from nature *itself* – an inference that makes sense only on the supposition that physics captures everything that is really there in nature. Inferences of this sort commit what Alfred Korzybski (1933) famously characterized as the error of mistaking a map for the territory mapped (even if the reference to Korzybski is ironic given his hostility to Aristotelianism). In particular, they commit what Alfred North Whitehead called "The Fallacy of Misplaced Concreteness" – the error of confusing an abstract mathematical representation of reality with the concrete reality represented (1967, pp. 51 and 58).

In the previous chapter, we saw that the methods of empiriometric science cannot capture all there is to the nature of the scientist qua conscious and thinking subject, and that doing so in fact requires deployment of the basic concepts of Aristotelian philosophy of nature – actuality and potentiality, form and matter, efficient and final causality, and so forth. In the current chapter we have seen that the methods of empiriometric science also cannot capture all that there is to the nature of the physical world *external to* the conscious and thinking subject. Since it cannot do so, the absence from empiriometric science of any reference to the basic Aristotelian concepts in question *by itself* gives no reason whatsoever to conclude that those concepts don't in fact have application to the external physical world. Indeed, since we know that they *do* have application to the conscious, thinking subject, and that that subject is part of the physical world, we have reason to expect that the concepts do have application also to the wider physical world.

To know exactly *how far* they apply to that wider world requires consideration of the various specific ways that questions about change, time, substance, teleology, and the like arise in physics, chemistry, biology, and the other natural sciences. The chapters to follow will address these issues. But what has been said so far suffices to show that an Aristotelian position on these issues cannot be ruled out in advance by glib general assertions about what scientific method permits or what the results of modern science have shown.

4. Space, Time, and Motion

4.1 Space

4.1.1 Does physics capture all there is to space?

It is in the study of space, time, and motion that modern physics, in the work of Galileo and Newton, had its most impressive early successes – and, in the work of Einstein, one of its two most profound recent successes. Could there be anything more to space, time, and motion than what physics reveals? Could the Aristotelian still have anything to say on these subjects? Yes, and yes. Let's begin with space.

In chapter 3, we examined the view of Bertrand Russell that our knowledge of the physical world external to the mind is knowledge only of abstract structural relations rather than of the intrinsic nature of the concrete entities that bear those relations. We also saw that Russell and those influenced by him ended up having to qualify this position considerably, in light of an objection raised by M. H. A. Newman. *If* we think that the content of physics is exhausted by what it captures in the language of mathematics, then we cannot avoid the conclusion that there must be more to the natural world than is described by physics.

Galen Strawson (2008, pp. 29-33) takes the nature of *space* to be a specific example of something that outstrips what can be captured in the purely structural description Russell and other structural realists think physics affords us. If you hold up your hands and take note of the spatial extension between them, says Strawson, what you grasp is something more than what he calls the mere "abstract dimensionality" or "abstract metric" conveyed by a mathematical representation. For such a representation "won't distinguish *space* from any other possible three-dimensional 'space', e.g. the emotional state-space of a species that have just three emotions, love, anger, and despair" (2008, p. 31). Or, to borrow an example given by Craig Bourne in a different context:

> [J]ust because something is represented spatially, we cannot draw the conclusion that it is a spatial dimension or that it is in anyway [sic] analogous to a spatial dimension. For consider... a three-dimensional colour space which illustrates the possible

> ways in which things can match in colour... [I]t would be misconceived to draw the conclusion that brightness, hue, and saturation were each spatial dimensions, just because they were represented spatially. (2006, p. 158)

Neither "emotional state-space" nor "three-dimensional color space" is literally *space*, even if at some very abstract level of description their dimensions are analogous to those of space. Since only what is common to both literal space and these other "spaces" is captured by a purely mathematical representation, there is more to the nature of space than what such a representation conveys.

Can we say exactly what that is? Beyond noting that *extension* is at the core of our concept of space, Strawson says little more than that "in the present context I am inclined just to hold up my hands again" (2008, p. 31). Now, we can in fact say more than that, and will do so presently. But it is worth emphasizing that even if we could not, that wouldn't cast doubt on Strawson's point. For one thing, the difference between literal space and the "spaces" just referred to suffices to show that there must be more to space than abstract dimensionality, *whether or not* we can say what it is. For another, recall from chapter 2 that some of what we know is of a tacit and embodied nature. We "know how" to move through space, to reach into space, to avoid things coming toward us through space, etc. whether or not we "know that" space has such-and-such a nature. There must be facts about space in virtue of which it is possible for us to have such capacities or "know how," whether or not we can explicitly represent those facts.

To be sure (and as Strawson notes) even the notion of extension illustrated by the space between one's hands is abstract insofar as it outstrips any *particular* experience. That is why we can acquire the same one concept of space through either sight or touch. The point is that the concept is nevertheless not *as* abstract as that of mere dimensionality. Yet might we not take an eliminativist line and simply jettison this less abstract, commonsense concept of space in favor of the more abstract one expressible in purely mathematical language?

No, we may not, because we will in that case be left with a *mere* abstraction rather than a concrete reality of any sort. We will thereby have eliminated *the natural world itself*, and not merely some feature that common sense attributes to it. In that case, we will no longer be doing empirical science, but instead some *a priori* ersatz. (To be sure, we saw in chapter 3 that Ladyman and Ross advocate an interpretation of science that blurs the distinction between the abstract and the concrete, but we also saw that this position is incoherent.)

In particular, we will not have captured concrete physical space any more than the "emotional state-space" of Strawson's example captures emotion, or the "three-dimensional color space" of Bourne's example captures color – or, to borrow some examples from the end of the previous chapter, any more than the abstract concept *triangularity* captures the nature of a pyramid, or the aircraft engineer's calculations capture the nature of airline passengers. As John Campbell writes:

> We can distinguish between a pure geometry, which is a purely formal exercise in mathematical computation, and an applied geometry, which is a body of doctrine about the world in which we live. What turns one into the other is the assignation of some physical meaning to the spatial concepts, for example, the identification of a straight line as the path of a light ray *in vacuo*... It is only its figuring in an intuitive physics of one's environment, through regularities connecting spatial properties with other physical properties, that makes [spatial reasoning] reasoning that is not purely mathematical but rather about the space in which one lives. (1994, p. 25)

As Campbell goes on to argue, the "other physical properties" to which spatial properties are crucially connected include the *causal* properties we attribute to physical objects (pp. 26-29). The notion of a world of physical objects entails a distinction between causal factors internal to such objects and causal factors external to them. For both factors play a role in determining what happens within such a world. Why did a certain glass break when struck? Part of the answer concerns the brittleness of the glass, and part of it concerns the solidity and force of the object that struck it. Had the glass been less brittle, then it would not have broken even if the solidity and force of the other object remained the same. If the solidity or force of the other object were much less great, the glass would not have shattered even if it were just as brittle. Now, these causal relationships, Campbell says, "give physical meaning to spatial connectedness" and "physical significance to the metric for the space within our intuitive physics" (p. 28).

Suppose the notion of a world of physical objects and the notion of spatial extension do indeed go hand in hand. (Cf. P. F. Strawson 1959,

Chapter 1.) Might we nevertheless be able to make sense of a world that has neither physical objects nor spatial extension yet is still objective or mind-independent? A famous thought experiment from P. F. Strawson (not to be confused with his son Galen Strawson) might seem to show that this is possible. Strawson proposed that there could, in principle, be a world of objective but non-spatiotemporal particulars all of which were sounds (1959, chapter 2). The idea is that there might be criteria which, by reference to pitch alone, would allow for the re-identification of a sound when it is not heard and for distinguishing between sounds which are numerically identical and not merely qualitatively identical.

Now, whether this is in fact possible even in principle is far from clear. Richard Gale identifies a number of problems with Strawson's proposal (1991, pp. 329-40), such as the fact that on a closer analysis, *causal* criteria seem necessary in order to identify and distinguish sounds. For example, consider a musician who begins to play a note on one instrument just before another musician playing the same sort of instrument stops playing the same note. To someone who can only hear and not see the two musicians, it might seem that it is one continuous note being played, whereas someone who can see what is going on will know that it is really two different notes that are overlapping. Only our knowledge of the physical causes of the notes allows us to determine which scenario is correct. Similarly, arguably we are able to distinguish the notes we hear in a chord only because we know that they have different causes.

Even if Strawson is right, however, that would not help our imagined eliminativist. For one thing, no one would claim that *our* world is a purely auditory world of the sort described by Strawson. It is instead a world of rocks, trees, animals, stars, molecules, atoms, and other physical objects, which, unlike the sounds described by Strawson, would have to exist in spatial relationships to one another. For another thing, even Strawson's world of sounds is a world of *particular* things, and thus would contain more than the abstract dimensionality or metric our eliminativist would want to make do with.

4.1.2 Abstract not absolute

The results of the argument so far are summed up by Andrew van Melsen as follows:

> Scientific language... schematizes both object and place. The place of a body in science is usually indicated as the place of a

> point in a system of coordinates. Science has, of course, a very good reason for using such a concept of place, but it is clear that what science is describing is not the real place, but only a schematized one. The real place has extension since the object of which it is the place is extended, too. (1954, p. 164)

Galen Strawson, as we've seen, also takes extension to be an essential aspect of space that the merely schematic and metrical notion of physics fails to capture. Now, common sense and Aristotelian philosophy of nature alike would add that the concept of space is that of extension considered specifically as a kind of *receptacle* for physical objects, insofar as such objects are *contained and move about within* space (Bittle 1941, pp. 155-56; McWilliams 1950, p. 101).

This container or receptacle is not to be confused with a *void* or strict nothingness. To be sure, some Aristotelians think a void is possible at least in theory (Phillips 1950, pp. 93-96), though others doubt this (McInerny 2001, pp. 192-3). Certainly the notion is highly problematic. If there were a void between two purportedly spatially separated material objects, why would they be *separated*? A void is nothingness, so that there would in this case be nothing between them and thus nothing to separate them. Hence they should be in contact. Phillips responds that what would keep the two objects apart is not any sort of matter between them, but just the fact that they are at different positions in space. But this seems to miss the point. For if space were strictly a void or nothingness, then there wouldn't *be* anything there in the first place in which there are different positions to occupy. There are further problems. Barry Dainton (2010, pp. 147-48) argues that if space were a void, there would be nothing to restrict us to movement in only three dimensions. But we are so restricted. Furthermore, Dainton argues, if there were a void between two material objects, there would be literally *nothing* to connect them, in which case they would each constitute a self-contained world that could have no influence on the other (pp. 149-51). Whether or not a void is ultimately possible in theory, though, the notion of space as a receptacle does not identify the space of the actual world with a void.

The idea of space as a kind of receptacle or container can be elucidated by noting what it rules out, such as the views of Descartes and Leibniz. (Cf. Bittle 1941, p. 152.) If space is what *contains* extended physical substance, then (contra Descartes) it cannot be identified with extended physical substance *itself*. Space qua container can either be filled or empty in a way a physical substance itself cannot be. Neither, in that case (and contra Leibniz), can space be identified with relations between physical objects. For if space qua container can be empty of physical objects, then it can exist in the absence of such relations (since in empty space there would be nothing to serve as relata, and – as I argued in chapter 3 – there can be no relations without relata). Nor can a relationalist view plausibly account for the *boundaries* of space. E. J. Lowe asks us to consider a universe in which the only physical objects are three particles arranged in the form of an equilateral triangle (2002, p. 265). On a relationalist view there would presumably be space within the boundaries of the triangle. But would space extend beyond those boundaries? If not, why not? And if so, how far would it extend, and why that far exactly? There do not seem to be any facts about the particles themselves that could justify any particular answers to these questions.

Lowe suggests that a relationalist may respond by holding that the boundaries of space in such a case are determined by where it is possible for a particle to be relative to the center of the triangle, where what is possible is in turn determined by the laws of nature that govern the universe (2002, p. 266). (In a somewhat different way, Nick Huggett (2010, pp. 98-100) also appeals to laws of nature to defend relationalism.) But without further elaboration, this seems to be an attempt to solve the problem by sheer stipulation. What does the claim that the laws of nature determine this amount to? Should we interpret it in terms of a theological view of laws, viz. as the claim that God has simply laid it down that the boundaries of space will extend thus far? The trouble with this suggestion is that even God cannot make just any old thing a law of nature. For example, even God could not establish a law of nature according to which squares are round. What laws are possible is constrained by what is possible for the things governed by the laws, given their nature. So, God could establish a law determining that space will have such-and-such boundaries only if the nature of space allows for such a limitation. But whether space, as the relationalist conceives of it, does in fact allow for such a limitation is precisely what the relationalist was supposed to be explaining. So an appeal to laws of nature construed as divine decrees wouldn't solve the problem at all.

Should we appeal instead to an Aristotelian view of laws, according to which a law of nature is a summary of how a thing can behave given its nature or essence? On this proposal, the idea would be that it is simply in the nature of the particles in our imagined universe that it is possible for them to be in some places relative to the center of the triangle, and not in others. But the problem with this should be obvious. The critic of relationalism, remember, objected that there seemed to be nothing about the natures of physical objects that could determine the boundaries of space. That is why the relationalist was appealing to laws of nature in the first place, as a *further* factor, *additional to* the natures of the objects, that could account for the boundaries. If laws of nature themselves are now to be explained in terms of the natures of physical objects, then we are back where we started. We will be implicitly appealing after all to the natures of physical objects in order to account for the boundaries of space, when the whole point was supposed to be to avoid having to do so given that there seems to be nothing in the nature of physical objects that *could* account for this.

But doesn't Einstein's general theory of relativity favor the relationalist position? As Lowe points out, that is not necessarily the case (2002, pp. 266-67). While Newton's absolutism was indeed Einstein's target, Lowe suggests that it is really Newton's conception of material objects, rather than the absolute space in which he located them, that is incompatible with general relativity. For Einstein's theory can be interpreted as regarding material objects as local deformations in the fabric of space-time, thereby making of matter "a purely geometrical feature of space itself" (p. 267). On this interpretation, space is as absolute as it is on Newton's account, and it is only Newton's supposition that material objects are *distinct from* space that is rejected by general relativity.

Be that as it may (and we will come back to Einstein before long), the Aristotelian is by no means committed to a Newtonian absolutist conception of space. To be sure, Aristotelian philosophy of nature regards space as *real*. It is not, in other words, merely *ideal* (in Kant's sense of the term), viz. a sheer artifact of the human mind with no foundation in objective reality. So to treat it would be incompatible with the realism about physical objects defended in chapter 2 and the scientific realism defended in chapter 3. If physical entities of the ordinary and/or theoretical sort are real and causally related to one another, and the existence of these entities and their causal relations presupposes the existence of space, then space must be real.

It would be an error, however, to suppose that if one rejects both the relationalist and idealist conceptions of space, then the only remaining option is something like a Newtonian absolutist conception. The Aristotelian position purports to be a further alternative, and it is best understood on analogy with the Aristotelian realist approach to the problem of universals. Take, for example, *humanness*. There is the humanness of

Socrates, the humanness of Plato, the humanness of Donald Trump, and so on. The mind abstracts *humanness* qua universal out from these particular instances, disentangling it, as it were, from the individualizing features with which it is mixed in Socrates, Plato, Trump, et al. Now, contrary to nominalism, this is not sheer invention. There really is something in mind-independent reality that the mind gets hold of when it abstracts *humanness* out from the individuals. But contrary to Platonic realism, *humanness* qua universal does not exist outside the mind, in some "third realm." There is the humanness of Socrates, the humanness of Plato, the humanness of Trump, et al., and there is the universal *humanness* that the mind entertains when it notes what is common to these individuals, but there is no further thing over and above these. Abstract entities like humanness exist only as *abstracted by* the mind *from* concrete particulars, not in some independent way.

Space qua extended receptacle or container is like that. In concrete reality outside the mind, the receptacle is always filled with material objects, just as humanness exists in concrete mind-independent reality only in actual particular human beings. Space qua extended receptacle is nevertheless real and distinct from material objects and their relations, just as humanness qua universal is real and distinct from particular human beings like Socrates, Plato, and Trump. But all the same, space qua extended receptacle does not exist in some third kind of way, entirely apart from either material objects or the mind, just as humanness qua universal does not exist in some third, Platonic kind of way independent of both concrete particulars and the abstracting mind. From the Aristotelian point of view, the absolutist view of space Platonizes space in something like the way Platonic realism Platonizes universals.

Now, this absolutist or "Platonizing" view of space cannot be right. (Cf. Bittle 1941, pp. 150-52; Koren 1962, pp. 108-9; Phillips 1950, pp. 82-85.) If space qua extended receptacle existed in the way absolutism supposes, then it would be either a substance or an attribute, and it cannot be either. For suppose it is a substance. In that case it is either a material substance, or an immaterial one. If it is immaterial, then it cannot have any extended parts. But of course, any region of space does have extended parts, namely the smaller regions of space within it. So it cannot be an immaterial substance. If it is a material substance, though, then precisely because it is a material substance it will require an extended receptacle to contain it. And yet space was *itself* supposed to be the extended receptacle that contained material substances! So to think of it as a material substance is simply a category mistake. Moreover, if we do say that it is a material substance, we will be led into a vicious regress. For in that case it will, again, require an extended receptacle of its own to contain it. But that extended receptacle or high-order space will for the same reason require a further extended receptacle to contain *it*, and so on ad infinitum. So, space cannot be a material substance any more than it can be an immaterial one. But neither can it be an attribute. For it would be an *extended* attribute and thus presuppose an extended *substance*, which would land us again in the difficulties just described.

It is telling that even Newton's conception of space is less absolutist than it might at first appear, insofar as he identified space with God's sensorium. This essentially makes of it a divine attribute. Theologically this is problematic insofar as it is hard to see how it can avoid collapsing into pantheism, and pantheism cannot be correct, for several reasons. For example, God is absolutely simple or without parts, so that all his attributes are ultimately one and the same thing looked at from different points of view. But space, and the world generally, has parts. Hence God cannot be identified with the world in general or space in particular. (Cf. Feser 2017.) The point for present purposes, though, is just to note how difficult it is consistently to treat space as absolute.

As Adrian Bardon notes (2013, p. 53), Newton's absolutist view of space (and of time and motion) was motivated by his model of scientific explanation as a matter of identifying *universal* laws of nature. Newton's laws account for changes of motion in terms of force, and mere relative motions do not require force for their explanation. Hence the need in his system for absolute motion, which entailed a need for absolute space and time. (Cf. Weatherall 2016, pp. 32-33) Now, as we saw in chapter 3, for neo-Aristotelians like Cartwright, laws of nature are not universal, and they are not ontologically fundamental. What are fundamental are the natures and capacities of things, and laws describe how a thing will behave given its nature and capacities, not necessarily universally but under certain conditions. Natures and capacities (and thus laws, rightly understood) are *immanent to* things rather than standing above and apart from them like Platonic forms or divine decrees. And for the Aristotelian, space too, though real rather than ideal, is *immanent to* the world of material things rather than being either a quasi-Platonic entity or a divine attribute. (It is worth adding that Newton's view of laws, like his view of space, tends toward pantheism, or at least occasionalism. For if a law is essentially a divine decree and physical objects have no immanent capacities or powers, then it seems that it is really *God* rather than physical objects who is doing everything in the natural world.)

4.1.3 The continuum

Common sense takes some of the objects that occupy space to be made of discrete parts and others to be continuous or uninterrupted. For example, a stack of wooden blocks is made of discrete parts, whereas a single block is continuous. Common sense also takes objects of the continuous sort to be the more fundamental kind, insofar as non-continuous objects have continuous ones as their parts. However, upon reflection it might seem that even continuous objects must really be made up of parts. After all, a single wooden block can be divided in half, those halves can be divided themselves, and so on. Hence, it might be concluded, though some objects *appear* to be continuous, this is an illusion.

But a number of traditional arguments show that this cannot be right, and that common sense is correct. (Cf. Hugon 2013, pp. 240-47; Phillips 1950, Chapter V.) For the parts of which a continuous object is purportedly composed would be either extended or unextended, and either supposition leads to absurdity. Suppose first that the parts are unextended. These unextended parts are either at a distance from each other or they are not. If they are at a distance from one another, then they would not form a *continuum*, but would rather be a series of discrete things (like the dots in a dotted line, only without even the minute extension such dots have). Suppose then that they are not at a distance from one another, but are instead in contact. Then, since they have no extension at all and thus lack any extremes or middle parts, they will exactly coincide with one another (like a single dot, only once again without even the minute extension of such a dot). All these parts together, no matter how many of them there are, will be as unextended as an individual part. In that case too, then, they will not form a true continuum.

So, if a continuous object is made up of parts, they will have to be extended parts. Now these purported extended parts would either be finite in number or infinite. They cannot be finite, however, because anything extended, no matter how small, can always be divided at least in principle into yet smaller extended parts, and those parts into yet smaller extended parts *ad infinitum*. So if a continuous object is made up of extended parts, they will have to be infinite in number. But the more extended parts a thing has, however minute those parts, the larger it is. Hence if a continuous object is made up of an infinite number of extended parts, it will be of infinite size. This will be so of every continuous object, however small it might seem. For example, it would follow that the single wooden block of our example is infinite in size. But this is absurd. Hence a continuous object can no more be made up of extended parts than it can be made up of unextended parts.

Arguments of this sort trace back to Zeno's *paradox of parts*, which we briefly considered in chapter 1. Some of Zeno's other paradoxes reinforce the point. Hence consider the *dichotomy paradox*, in which a runner tries to get from point *A* to point *B*. To get to *B*, the runner first has to get to the midpoint between *A* and *B*. But to get to that point, he first has to get to the quarter point between *A* and the midpoint, and so on *ad infinitum*. Now, since the continuum between *A* and *B* is infinitely divisible, if that continuum is made up of extended parts then there will be an actually infinite number of distances the runner will have to traverse in order to get from *A* to *B*. Indeed, there will be an actually infinite number of distances he will have to traverse even to get his foot off the ground. Hence he will be unable to get to *B*, and indeed unable even to get started.

Now, Zeno, of course, drew skeptical conclusions from his paradoxes. The paradox of parts was intended to show that there are no distinct objects, and the dichotomy paradox to show that there is no such thing as motion. But the Aristotelian draws a different conclusion. Applying the theory of actuality and potentiality, he argues that what the paradoxes really show is that the parts of a continuum are in it only *potentially* rather than actually. That is not to say that they are not there at all. A potentiality is not nothing, but rather a kind of reality. That is why a wooden block (for example) is *divisible* despite being continuous or uninterrupted in a way of stack of blocks is not. But until it is actually *divided*, the parts are not actual. If they were, the Aristotelian argues, then we'd be left with Zeno's bizarre consequences. Extended objects would all be infinitely big, runners would be unable to move, and so on. Affirming that reality includes both potentialities as well as actualities allows us to acknowledge the reality of the parts of a continuum while at the same time avoiding paradox.

The Aristotelian point is sometimes misunderstood. For example, David Foster Wallace claims that when Aristotle held that the subintervals between *A* and *B* are potentially infinite, what he had in mind was comparable to the way a measurement can always potentially be made more precise (2010, pp. 66-67). For instance, for practical purposes, we might say that a certain child's height is 38.5 inches, but if we wanted to we could state it more precisely as 38.53 inches, or even more precisely still by taking the measurement to further decimal places *ad infinitum*.

Nevertheless, any *actual* measurement will only be taken to some finite number of decimal places. Similarly, Aristotle's point (so Wallace seems to think) is that though we could potentially identify ever smaller units within an interval *ad infinitum*, we will in actuality only ever identify some finite number of units.

But that is not the Aristotelian's point, or not the main point anyway. The claim is not fundamentally about what *we* might do by way of identifying parts of a continuum, but rather about what is there in the continuum *itself*. Wallace's way of stating the idea appears to leave it open that a continuum might have an actually infinite number of parts, even if we will in practice never be able to pick them all out. But in fact the Aristotelian's claim is that the parts are not actually there to pick out in the first place, but only potentially there.

However, even critics who correctly understand the Aristotelian position sometimes suggest that the appeal to the theory of actuality and potentiality is not necessary in order to resolve Zeno's paradoxes. We can, it is supposed, simply apply modern mathematics instead (Russell 1963, pp. 63-64). For example, calculus tells us that a convergent infinite series can sum to 1. Hence, the argument goes, there is no difficulty in understanding how the runner in the dichotomy paradox can traverse the finite distance between *A* and *B* despite there being an infinite number of ever smaller distances between them.

But as even many philosophers with no Aristotelian ax to grind have noted, there are several problems with such claims. First, while a convergent infinite series like the one at issue here has a *limiting value* of 1, it does not follow that it has a *sum* of 1 in the ordinary sense of the word "sum" (Black 2001, p. 70; Ray 1991, p. 13). As Barry Dainton notes:

> In defining the sum of an infinite series in terms of its limit, mathematicians are really introducing a new *stipulation* as to how the term "sum" can be used in a new context: that of infinite – as opposed to finite – series of numbers. (2010, p. 275, emphasis added)

Now, if the question raised by Zeno's dichotomy paradox is how the traversal of an infinite number of ever smaller distances between *A* and *B* can amount to the traversal of the distance between *A* and *B* itself, it is no answer simply to *stipulate* that we will *count* the former as amounting to the latter (Bardon 2013, p. 16; Salmon 2001, p. 29). But that is what the appeal to calculus essentially does. It merely tries to define the problem away. As Wallace wryly observes, this is a "deeply trivial" solution, "along the lines of 'Because it's illegal' as an answer to 'Why is it wrong to kill?'" (2010, pp. 51-52).

Second, even if we do count the limiting value as a sum in the ordinary sense, that still does not solve Zeno's problem. For this still only gives us a mathematical abstraction, and what we need to know is why we should take that abstraction to tell us anything about concrete physical reality (Mazur 2008, pp. 7-8; Ray 1991, pp. 13-14; Salmon 2001, pp. 33-34). Aristotle's solution, by contrast, is precisely an attempt to tell us what it is about concrete reality itself – namely, the fact that concrete reality includes potentialities as well as actualities, and that the parts of a continuum are present only as potentialities – that blocks Zeno's conclusion.

Third, calculus essentially gives us a solution to a *practical* problem, whereas what paradoxes like Zeno's raise is a *metaphysical* problem. *Given that* an object really does get from *A* to *B*, calculus gives us a way to describe this motion mathematically, and we can go on to deploy this description for practical and scientific purposes. But Zeno is not concerned with the practical needs of everyday life and with science. What he wants to know is what it is about objective reality that could make it the case that an object gets from *A* to *B*. Merely noting that an object *does* in fact get to *B* and then going from there does not answer this question. As J. B. Kennedy writes, "Newton's calculus *simply assumes* that the series does add to one exactly... The calculus did not solve, but rather suppressed, Zeno's paradoxes" (2003, p. 125, emphasis added). In short, the appeal to calculus simply changes the subject.

Other potential objections to the Aristotelian positon grounded in modern mathematics similarly miss the point. For example, as Dainton points out, it is now the "standard view" that space is composed of dimensionless points, and this view exhibits a "consilience with the mathematical methods deployed with such spectacular success by physicists and engineers over the past few centuries" (2010, p. 301; Cf. Grünbaum 2001, p. 166). This success might be thought to tell against the Aristotelian view, defended above, that the continuum cannot be made up of unextended parts. As with the appeal to calculus, however, the *practical* utility of a mathematical model simply does not by itself resolve *metaphysical* problems of the sort that motivate the Aristotelian position. Indeed, as Dainton goes on to point out, many philosophers and scientists – he cites Poincaré, Gödel, Weyl, Brentano, and Peirce as examples – reject the claim that the standard view resolves the metaphysical questions, precisely because it raises problems like the ones cited by the Aristotelian. Dainton writes:

> Mathematicians may have worked out fruitful ways of assigning magnitudes to infinite collections of *abstract entities*, but this is of limited relevance to the real problem posed by Zeno. For it remains very difficult to comprehend how a *concrete material thing* could be entirely composed of dimensionless parts. (2010, p. 303. Emphasis added.)

Now, the Aristotelian position is concerned precisely to address what must be true of concrete material reality in order for Zeno's paradoxes to be rebutted. No mere mathematical result can *substitute* for such an account, even if it might supplement it.

4.2 Motion

4.2.1 How many kinds of motion are there?

As explained in chapter 1, Aristotelian philosophy of nature traditionally draws a distinction between four kinds of motion or change. There is *local motion* or change with respect to location or place, as when you throw a banana across the room. There is *qualitative change*, as when a green banana turns yellow as it ripens. There is *quantitative change*, as when an old banana shrinks in size as it dries out. And there is *substantial change*, as when the banana is thrown into a fireplace and reduced to ash.

The tendency in modern philosophy and science has been to try to reduce all change to the first kind, and here the moderns take their inspiration from ancient atomism. For the atomist, there are only atoms and the void, and all change is analyzable in terms of the local motion of atoms. Color, heat, cold, and other qualities aren't really in atoms or collections of atoms, but only in our conscious awareness of them. Hence when the banana ripens, it is not that it loses its greenness and acquires yellowness. All that is really going in the banana itself is a change in the positions of the banana's constituent atoms with respect to one another. The greenness and yellowness we see are only in our minds, not in the banana itself, and thus there is no need to posit any genuine qualitative change in the banana. When the banana shrinks in size, that is just a matter of its losing certain of its atoms, which also reduces to their changing their spatial location relative to the remaining atoms. Hence quantitative change is analyzable in terms of local motion. When the banana is burned and reduced to ash, that does not involve prime matter losing substantial form, but rather just the further distancing from one another of the atoms of which the banana had been constituted. Hence substantial change too can be analyzed in terms of local motion. (Cf. McGinn 2011, Chapter 4)

Modern science has, of course, abandoned many of the details of the ancient atomist picture of the world, but the general idea that all change is reducible to local motion has remained largely intact. But the idea does not withstand careful scrutiny. (Cf. Feser 2014b, pp. 177-84; Madden 2013, pp. 229-35.) Note first that the atomist account of substantial change does not really *analyze* it so much as *eliminate* it and replace it with what the Aristotelian would call accidental change. Dogs, trees, stones, and other natural objects turn out, on the atomist analysis, to be mere aggregates rather than true substances. Only the atoms themselves are genuine substances, and a dog-like, tree-like, or stone-like arrangement is merely an accidental form that collections of these substances can acquire or lose. But for all the atomist has shown, the atoms *themselves*, qua genuine substances, would still be composites of prime matter and substantial form, and as such would be capable in principle of a kind of change (namely the prime matter's loss of one substantial form and acquisition of another) that is not analyzable in terms of the local motion of more fundamental particles.

To be sure, the atomist *claims* that atoms cannot change in such a way, but it is one thing to assert this and quite another to make it plausible. An atom, as the ancient atomists understood it, is supposed to be indivisible. That was, of course, the original meaning of the term "atom." However, an atom is also supposed to be extended rather than a dimensionless point. But if it is *extended*, then it is divisible into smaller extended parts, at least in principle. Now, for reasons set out in the discussion of Zeno, it cannot be that the parts into which a so-called atom can be divided exist *actually* in the atom. If they did, then these parts would be either unextended or extended. If they were unextended, then they could never constitute an extended thing like an atom. If they were extended, then they too would be divisible into yet further parts *ad infinitum*, which would entail, absurdly, that an atom has an infinite number of parts and is therefore of infinite size. So, the parts into which a so-called atom is divisible can exist only *potentially* rather than actually in an atom. But once we acknowledge potentiality as well as actuality in the structure of a so-called atom, we are implicitly acknowledging that it is in fact susceptible of substantial change as the Aristotelian understands it. Again, atomism doesn't really analyze or eliminate the Aristotelian notion of substantial change at all, but simply relocates it to the micro-level. The same thing is true, *mutatis mutandis*, for anything modern physics would put in place of the atomists' notion of an atom.

Quantitative change too, on closer inspection, is not analyzable in terms of local motion. Consider the development of a living thing, which is where, on the traditional Aristotelian view, quantitative change in the strictest sense is to be found (O'Neill 1923, pp. 245-46). When a tree grows a branch, the atomist claims, all that is happening is that certain particles which were not originally in close proximity to the tree come into close proximity to it. But when we hang a child's swing from a tree, it might equally well be said that certain particles which initially were not in proximity to the tree come into close proximity to it. Yet when the tree grows a branch, it thereby increases in size, whereas when we hang the swing from the tree, it does *not* thereby increase in size. Hence there is more to quantitative change of this sort than a mere change of location of particles. This additional ingredient is a kind of *immanent teleology*. The processes that result in the growth of the tree arise from within the tree and are directed toward its flourishing and completion qua tree, whereas the processes that result in the hanging of the swing neither arise from within the tree nor are aimed at the tree's completion or flourishing.

Then there is qualitative change. The atomist maintains that when the banana goes from being green to being yellow, the only change that occurs in the banana itself is a change in the arrangement of atoms and their impact on the sense organs. Neither the greenness nor the yellowness we see is really there in the banana in the first place, but only in the conscious experience of the perceiver. Something similar can be said, in the atomist view, of apparent changes in the sounds a thing makes, its taste or odor, its temperature, and so on. But in fact this neither reduces qualitative change to local motion nor eliminates it, but merely *relocates* it. For example, the qualitative change from green to yellow is now, in effect, located in the conscious perceiver himself rather than in the banana. It is a transition from the perceiver's experiencing greenish qualia to his experiencing yellowish qualia.

The atomist might respond that this change too can be analyzed in terms of local motion, insofar as perceptual experiences can be identified with motions in the brain. But there are three serious problems with such a response. First, part of the *point* of relocating colors, sounds, heat, cold, etc. to the mind of the observer was presumably to *avoid* having to analyze them in terms of local motion. If the greenness and yellowness we see are not really in the banana in the first place, then, the atomist was telling us, we needn't worry about how to analyze the banana's change from green to yellow in terms of atoms changing position. There is in the banana itself just the change in position and nothing more. If we are now told that a yellowish or greenish experience can be identified with motions of atoms in the *brain*, what was the point of moving colors out of the banana? Why not just identify them with motions of the atoms in the banana itself?

But of course, there is a good reason why they could not be so identified. The atoms are said to be intrinsically *colorless*, as well as intrinsically soundless, odorless, tasteless, etc. It is precisely because it is, to say the least, difficult to see how color could ever arise out of what is intrinsically colorless that the atomist was moved to assert that color is not really in objects after all, but only in the mind. But the atoms that make up the brain are no less colorless, soundless, odorless, tasteless, etc. than those that make up a banana. So if it is difficult to see how color could arise out of the intrinsically colorless atoms of the banana, how is it any less difficult to see how color qualia could arise out of the intrinsically colorless atoms of the brain?

Now, everything said here about ancient atomism once again applies *mutatis mutandis* to whatever modern physics would put in place of it. Today's successors to the atomists would identify color with the surface reflectance properties of a physical object. But it is not yellow or green *as common sense* understands those features that is really in the banana, even on this account. It is only yellow or green as redefined for purposes of physics that is there in the banana. Yellow and green *as common sense* understands them exist only in the mind of the observer, for the modern successors of the atomists no less than for the atomists themselves. That leaves us with the same basic problem facing the atomist. If matter outside the brain is devoid of color *as common sense* understands it, and the matter that makes up the brain is no less devoid of color in that sense, then how can the matter that makes up the brain give rise to color qualia?

That brings us to the third problem, which is that modern attempts to provide a materialist analysis of qualia face notorious problems – the zombie argument, the knowledge argument, the absent qualia argument, the inverted spectrum argument, and so on (Chalmers 1996; Feser 2006, Chapters 4 and 5; Levine 2001). This shouldn't be surprising. If you define matter so that it is essentially devoid of color, sound, odor, heat, cold, etc. as common sense understands these features – as modern materialists do no less than their ancient atomist forebears – then (as noted in earlier chapters) you are bound to find it difficult to analyze the qualia associated with experiences of color, sound, odor, heat, cold, etc. in materialist terms. In that case, though, the qualitative *change* that occurs in conscious experience is bound to resist analysis in terms of the local motion of material elements inside the brain.

Nor is the qualia problem the only obstacle to analyzing all qualitative change in terms of local motion. A conscious subject's transition from one *thought* to another is another sort of qualitative change, and since materialist accounts of the propositional attitudes are no less problematic than materialist accounts of qualia, this sort of change is no more plausibly analyzable either in atomist terms or in whatever terms the modern successors of the atomists would put in place of a crude atomist account.

Needless to say, fully to defend the claims I've been making in this section would require a more detailed treatment of the hylemorphist account of substance, the metaphysics of life, the metaphysics of color, the metaphysics of qualia and of the propositional attitudes, and so on. We will return to most of these issues in later chapters. The point for the moment is that the traditional Aristotelian classification of motion or change into four basic sorts is no less defensible today than in previous eras. The results of modern science at most affect *how* we apply this distinction, not *whether* we need to apply it.

4.2.2 Absolute and relative motion

Having said that, in the remainder of my treatment of motion in this chapter I will focus on some special problems that arise in connection with local motion. Recall that the Aristotelian position on the nature of space is neither relationalist, nor idealist, nor absolutist. Space is real rather than ideal, it is not reducible to relations between objects, but neither does it exist independently of all objects. Aristotelian realism about space differs from Newtonian absolutism in something like the way Aristotelian realism about universals differs from Platonic realism. Now, since local motion is change with respect to location in space, it might be expected that an Aristotelian positon on the nature of local motion would be analogous to the Aristotelian position on space, and I would argue that that is indeed the case. Local motion is real rather than ideal, and it is not reducible to relations between objects. However, this does not entail Newtonian absolutism about motion, any more than realism about space entails Newtonian absolute space. Local motion does not have a quasi-Platonic reality any more than space does (Gardeil 1958, p. 88). It exists only *in* a system of physical objects, just as space does, even if, like space, it cannot be reduced to relations between those objects.

Though he is not an Aristotelian, Colin McGinn (2011, Chapter 3) develops an aporia concerning local motion which, I propose, supports this Aristotelian position. On the one hand, McGinn argues, there are serious difficulties with the thesis that all local motion is relative, common though this view is. For one thing, the main argument for the thesis is no good. The argument is that since we cannot *empirically detect* the motion of any object except relative to other objects, we should reject the idea that anything actually moves except relative to other objects. The premise of this argument is true, but as McGinn notes, the conclusion would follow from it only given verificationism, and (as we saw in the previous chapter of this book) verificationism is false. That relative motion alone is empirically detectable may be a good reason for *physics*, given its narrow methods and interests, to confine itself to local motion. But that does not suffice to show that the *philosophy of nature* should recognize only relative motion, especially given that (as I have argued in earlier chapters) the methods of physics simply cannot in the first place capture all there is to physical reality.

For another thing, McGinn argues, there are difficulties with the thesis *itself*, never mind the argument for it. First of all, on analysis it appears to be incoherent. Consider a universe with just two objects, *A* and *B*. Suppose that from *A*'s frame of reference, *A* is stationary and *B* is moving toward *A*, whereas from *B*'s frame of reference, *B* is stationary and *A* is moving toward *B*. According to the relationalist, there is no fact of the matter about which is really moving. Relative to *A*, *B* is moving and *A* is not, and relative to *B*, *A* is moving and *B* is not, and that is all that can be said. But remember that local motion is change with respect to place or location. For *B* to move, then, is for it to be at location *L1* at one moment and at a different location *L2* at the next. Now, since *B* is indeed moving from *A*'s frame of reference, the locations *L1* and *L2* that *B* is at at each moment must be different locations. But since *B* is not moving from *B*'s frame of reference, the locations *L1* and *L2* that *B* is at at each moment must *not* be different locations. So *L1* and *L2* are both identical and not identical. But that is absurd.

Second, McGinn argues that the relativity of motion becomes implausible once we factor in considerations other than motion. If we are considering only their motion, we could say either that the sun is at rest and that the earth is moving relative to the sun, or that the earth is at rest and the sun is moving relative to the earth. However, when we factor in the different masses of the sun and the earth, this is no longer the case. For given its far greater mass, the sun exerts a gravitational pull on the earth that is much greater than the pull that the earth exerts on the sun. Hence it is the sun that is causing the earth to move relative to itself, rather than the other way around. The motions considered in the abstract may be symmetrical, but the causal factors are not, so that there is a fact of the matter about which is really moving relative to which.

Nor should the fact that absolute motion is not empirically detectable trouble us, McGinn suggests, because this is just what we should expect given the nature of space. Space *itself*, with all objects subtracted from it, is *featureless*. There is nothing in it we could point to by reference to which motion could be discerned, since anything we *could* point to would be just another physical object which could be subtracted from space. (As I put it earlier, space is essentially an extended receptacle for physical objects rather than a physical object itself, though this is not McGinn's way of making the point.) McGinn suggests that our epistemic situation here is analogous to the one we are in with respect to the external world and other minds. All the evidence of the senses is compatible with the external world being illusory and with other human beings being zombies (in the philosopher's sense of that term) rather than conscious, just as it is compatible with all real motion being relative rather than absolute. But few would suggest that this is a good reason seriously to doubt that the external world or other minds are real, and neither should it lead us to doubt that absolute motion is real.

So, McGinn judges that such considerations support the conclusion that not all real motion (as opposed to apparent motion) can be relative. However, he also argues that there is a powerful consideration that points in the opposite direction. For on closer inspection it is not clear that we really can make sense of absolute motion. Consider again a universe in which there are only two objects, *A* and *B*, moving relative to one another in absolute Newtonian space. If *A* really is moving in this scenario, then it might seem coherent to suppose that it continues moving if we subtract *B*. However, the reason this seems coherent, McGinn suggests, may be because we are, without realizing it, smuggling into our conception of the situation elements that a consistently absolutist conception of motion would have to leave out. For example, the absolutist might be thinking about space as if it had a *boundary*, and of A's motion as real insofar as it gets closer to or farther away from that boundary. But that would be *relative* rather than absolute motion, so to be consistent the absolutist would have to subtract any such boundary from his conception. Or the absolutist might be inadvertently thinking of *A* as moving relative to some position that *he*, the *absolutist*, occupies in space. For example, he might have a mental image of *A* receding into the distance in otherwise empty space, without realizing that he is thereby putting *himself* into space as an observer of A's motion, and *relative* to whom *A* is moving. So, to be consistent, the absolutist would also have to subtract from his conception any point of view of an observer within space from whom *A* might seem to be receding. But when all such subtractions are made, it is hard to see what is left in the situation that would make it the case that *A* really is moving.

So, McGinn judges, there appear to be strong considerations both in favor of the conclusion that change of relative position is not sufficient for motion, and in favor of the conclusion that absolute motion is not intelligible. What should we conclude from this? McGinn's conclusion is *epistemological*. He suggests that while local motion is real, we do not know its intrinsic nature. When we try to describe it, we end up with either a relational or absolutist conception of motion, but neither one is ultimately intelligible. What we can know of motion is only its mathematical structure as captured by physics. In short, McGinn's position vis-à-vis motion is an epistemic structural realist position.

But I propose that we can draw *metaphysical* conclusions from McGinn's aporia. We can conclude, first of all, that change of relative position is not *sufficient* for motion (as McGinn's first set of considerations indicates) but that it is nevertheless *necessary* for motion (as his second set of considerations indicates). We can conclude that while local motion is real (contrary to an idealist view of motion), it cannot be reduced to the relations between physical objects (contrary to a relationalist account of motion) but also cannot exist entirely apart from a system of relations between objects (contrary to an absolutist conception of motion). We can, in short, conclude that a fourth, Aristotelian position is the correct one, in the theory of local motion as in the theory of space.

4.2.3 Inertia

4.2.3.1 Aristotle versus Newton?

In Book VII of the *Physics* (1930), Aristotle maintains that "everything that is in motion must be moved by something." Aquinas, in *Summa Theologiae* I.2.3 (1948), asserts a similar thesis, to the effect that "whatever is in motion is put in motion by another." Let us call this the "principle of motion" (Wippel 2000, p. 453). Newton's First Law states that "every body continues in its state of rest or of uniform motion in a straight line, unless it is compelled to change that state by forces impressed upon it." (This is a common rendering of Newton's statement, in Latin, of his First Law in *Philosophiae Naturalis Principia Mathematica*.) Call this the "principle of inertia."

It is widely thought that the principle of motion is in conflict with the principle of inertia, and that modern physics has therefore put paid to this aspect of the Aristotelian philosophy of nature. The assumption is that Aristotle, followed by Aquinas and other Scholastics, held that an object cannot keep moving unless something is continuously moving it, but that Newton showed that it is simply a law of physics that once set in motion an object will remain in motion without any such mover. (Cf. DeWitt 2004, p. 109; McGinn 2011, p. 111.)

Common though this view is, it is not only mistaken, but unfounded. To think otherwise requires reading into each of the principles in question claims they do not make. When we consider what Aristotelian philosophers have actually said about the principle of motion and what modern physicists have actually said about the principle of inertia, we will see that they do not contradict one another. Indeed, when we consider the philosophical issues raised by motion, by the idea of a law of nature, and so forth, we will find that there is a sense in which the principle of inertia *presupposes* the principle of motion.

4.2.3.2 Why the conflict is illusory

There are at least five reasons to think that any appearance of conflict between the two principles is illusory:

1. *No formal contradiction*: Suppose that "motion" is being used in the two principles in the same sense. Even given this assumption, there is no *formal* contradiction between them. Newton's law tells us that a body *will* in

fact continue its uniform rectilinear motion if it is moving at all, as long as external forces do not prevent this. It does not tell us *why* it will do so. In particular, it does not tell us one way or the other whether there is a "mover" of *some* sort which ensures that an object obeys the First Law, and which is in that sense responsible for its motion. As G. H. Joyce writes:

> Newton, indeed, says that a body in motion will continue to move uniformly in a straight line, unless acted upon by external forces. But we need not understand him to deny that the uniform movement itself is due to an agency acting *ab extra*; but merely [to deny] that it is produced by an agency belonging to that category of agents which he denominates "external forces"... forces whose action in each case is of necessity confined to a particular direction and velocity. (1924, p. 100)

Of course, one might ask what sort of "mover" an object obeying the principle of inertia could have if it is not an "external force" of the sort Newton intended to rule out. One might also ask whether such a mover, whatever it might be, really serves any explanatory purpose, and thus whether we ought to bother with it given Ockham's razor. Those are good questions, and we will return to them. But they are beside the present point, which is that the principle of motion and the principle of inertia do not actually contradict one another, *even if* we assume that they are talking about the same thing when they talk about motion.

2. Equivocation: In any event, we shouldn't make that assumption, because they are *not* talking about the same thing, or at least not exactly the same thing. "As usually happens when science appears to contradict philosophy," notes Henry Koren, "there is here an ambiguity of terms" (1962, p. 95). Newton's principle of inertia is concerned solely with *local* motion, change with respect to place or location. But "motion" in the traditional Aristotelian usage meant change of *any* kind. This would include local motion, but also includes change with respect to quantity, change with respect to quality, and change from one substance to another. More to the point, for the Aristotelian all such change involves the actualization of a potential. Hence what the principle of motion is saying is that *any potential that is being actualized is being actualized by something else (and in particular by something that is already actual)*.

So understood, the principle of motion is, so the Aristotelian would say, something we can hardly deny. For a potential, being merely potential, can hardly actualize itself or anything else. In any event, the principle is, we see once again, not in formal contradiction with the principle of inertia because they are not talking about the same thing. When the Newtonian principle states that a body in motion will tend to stay in motion, it isn't saying that a potential which is being actualized will tend to continue being actualized. Even if it were suggested that the principle *entails* this claim, the point is that that isn't what the principle of inertia itself, as understood in modern physics, is *saying*. Indeed, modern physics has defined itself in part in terms of its eschewal, for purposes of physics, of such metaphysical notions as actuality and potentiality, final causality, and the like. So, it is not that modern physics has falsified the principle of motion so much as that it simply makes no use of it.

Now one might ask whether modern physics has not for that very reason made the principle of motion otiose and of nothing more than historical interest. We will return to this question as well, but it is also beside the present point, which is that there is no *necessary* conflict between the principle of motion and the principle of inertia.

3. The "state" of motion: Having said all that, we must immediately emphasize that there is a sense in which the Newtonian principle implicitly *affirms* at least an aspect of the Aristotelian principle it is usually taken to have displaced. To see how, consider first that modern physics characterizes uniform motion as a "state." Now this has the flavor of paradox. Reginald Garrigou-Lagrange objects:

> Motion, being essentially a change, is the opposite of a state, which implies stability. There is no less change in the transition from one position to another in the course of movement, than in the transition from repose to motion itself; if, therefore, this first change demands another cause, the following changes demand it for the same reason. (1939, p. 273)

Yet the modern physicist would respond to this objection precisely by collapsing the distinction between repose and motion. As Lee Smolin writes:

> Being at rest becomes merely a special case of uniform motion – it is just motion at zero speed.
>
> How can it be that there is no distinction between motion and rest? The key is to realize that whether a body is moving or not has no absolute meaning. Motion is defined only with respect to an observer, who can be moving or not. If you are moving past

> me at a steady rate, then the cup of coffee I perceive to be at rest on my table is moving with respect to you.
>
> But can't an observer tell whether he is moving or not? To Aristotle, the answer was obviously yes. Galileo and Newton were forced to reply no. If the earth is moving and we do not feel it, then it must be that observers moving at a constant speed do not feel any effect of their motion. Hence we cannot tell whether we are at rest or not, and motion must be defined purely as a relative quantity. (2007, pp. 21-22)

Now, as the discussion of absolute and relative motion earlier in this chapter indicates, this sort of move raises philosophical problems of its own. As Smolin goes on to note:

> This is a powerful strategy that was repeated in later theories. One way to unify things that appear different is to show that the apparent difference is due to the difference in the perspective of the observers. A distinction that was previously considered absolute becomes relative...
>
> Proposals that two apparently very different things are the same often require a lot of explaining. Only sometimes can you get away with explaining the apparent difference as a consequence of different perspectives. Other times, the two things you choose to unify are just different. The need to then explain how things that seem different are really in some way the same can land a theorist in a lot of trouble. (2007, pp. 22-23)

Indeed, as I have suggested, the attempt to explain away what Aristotelians mean by "motion" by means of such relativizing moves faces limits in principle.

But the point to emphasize for the moment is that, precisely because the principle of inertia treats uniform local motion as a "state," it treats it thereby as the *absence* of change. Moreover, it holds that external forces *are* required to move a thing out of this "state" and thus to bring about a change. One more quote from Smolin:

> There is an important caveat here: We are talking about uniform motion – motion in a straight line... When we change the speed or direction of our motion, we do feel it. Such changes are what

> we call *acceleration*, and acceleration *can* have an absolute meaning. (2007, p. 22)

But then the Newtonian principle of inertia hardly conflicts with the Aristotelian principle that "motion" – that is to say, change – requires something to cause the change. The disagreement is at most over whether a particular phenomenon *counts* as a true change or "motion" in the relevant sense, *not* over whether it would require a mover or changer if it *did* so count.

4. Natural motion: If Newton is closer to the Aristotelians than is often supposed, so too are the Aristotelians (or at least Aristotle and Aquinas) closer to Newton than is often supposed. As James A. Weisheipl (1985) has shown, the idea that Aristotle and Aquinas held that no object can continue its local motion unless some mover is continuously conjoined to it is something of an urban legend. To be sure, this was the view of Averroes and of some Scholastics, but not of Aristotle himself or of St. Thomas. On the contrary, their view was that a body will of itself tend to move toward its natural place by virtue of its form. That which generates the object and thus imparts its form to it can be said thereby to impart motion to it, but neither this generator nor anything else need remain conjoined to the object as a mover after this generation occurs. Aquinas comments:

> [Aristotle] says, therefore, that what has been said is manifested by the fact that natural bodies are not borne upward and downward as though moved by some external agent.
>
> By this is to be understood that he rejects an external mover which would move these bodies *per se* after they obtained their specific form. For light things are indeed moved upward, and heavy bodies downward, by the generator inasmuch as it gives them the form upon which such motion follows... However, some have claimed that after bodies of this kind have received their form, they need to be moved per se by something extrinsic. It is this claim that the Philosopher rejects here. (*Sententia de caelo et mundo* I.175, as translated in Aquinas 1964)

Even Aquinas's understanding of projectile motion is more complicated than modern readers often suppose:

> An instrument is understood to be moved by the principal agent so long as it retains the power communicated to it by the princi-

> pal agent; thus the arrow is moved by the archer as long as it retains the force wherewith it was shot by him. Thus in heavy and light things that which is generated is moved by the generator as long as it retains the form transmitted thereby... And the mover and the thing moved must be together at the commencement of but not throughout the whole movement, as is evident in the case of projectiles. (*Quaestiones disputatae de potentia Dei* 3.11 ad 5, as translated in Aquinas 1952)

To be sure, even though that which initiated a projectile's motion need not remain conjoined to it for the motion to continue, Aquinas still thought projectiles required other, conjoined movers given that a projectile's motion is not motion toward its *natural* place but is rather imposed on it contrary to its natural tendency. But as Thomas McLaughlin points out, the motions of projectiles require such conjoined movers in Aquinas's view

> because of the *kinds* of motions that they are and *not* because of a general conception of the nature of motion itself. In this respect, projectile... motions resemble accelerated motions in Newtonian physics, for accelerated motions require a force to act on a body throughout the time that it is accelerating. (2004, p. 243. Emphasis added.)

And insofar as *natural* motions require no such conjoined mover, the Aristotelian-Thomistic view sounds to that extent quite Newtonian indeed: "Thus, the Law of Inertia in the sense of absence of forces is similar to Aristotle's concept of natural gravitation, which is very remarkable" (Moreno 1974, p. 323).

Obviously, the Aristotelian notion of an object having some specific place toward which it tends naturally to move is obsolete, as is Aquinas's view that projectile motions require a continuously conjoined mover. Though modern writers should not be too quick to ridicule the latter notion. As Benedict Ashley comments:

> Aristotle... suppos[ed] that when the ball is struck some force is communicated to the medium through which it moves, which then keeps it moving after it has left the bat that put it in motion. This seems to us absurd, but we should recall that today science still relies on the notion of "field," that is, a medium, to explain the motion of bodies through that field. (2006, p. 99; Cf. Sachs 1995, p. 230)

There are also questions to be raised about Aquinas's view that the generator of a natural object moves that object instrumentally by virtue of having imparted to it its form. For how can the generator move the object as an instrument if by Aquinas's own admission it is no longer conjoined to it?

We will return to this question. The point for now is just to emphasize yet again that when one examines the principles of motion and inertia more carefully, the assumption that they are *necessarily* in conflict can readily be seen to be unfounded.

5. *Natural science versus philosophy of nature*: That certain key aspects of Aristotelian physics have been falsified is not in dispute. However, as I noted in chapter 1, the moderns have been too quick to throw the Aristotelian metaphysical baby out with the physical bathwater. Though Aristotle and pre-modern Aristotelians did not clearly distinguish the metaphysical aspects of their analysis of nature from the physical ones (in the modern sense of "physical"), these aspects *can* in fact be clearly distinguished. In particular, questions about what the natural world *must* be like in order for any natural science at all to be possible must be distinguished from questions about what, as a matter of *contingent* fact, are the laws that govern that world. The latter questions are the proper study of physics, chemistry, biology, and the like. The former, as I have argued in this book, are the proper study of that branch of metaphysics known as the philosophy of nature. Geocentrism, the ancient theory of the elements, and the notion that objects have specific places to which they naturally move, are examples of Aristotelian ideas in physics that have been decisively superseded. But the theory of actuality and potentiality, the doctrine of the four causes, and the hylemorphic analysis of material objects as composites of form and matter are examples of notions which have (so the contemporary Aristotelian argues) abiding value as elements of a sound philosophy of nature.

Now the principle of motion is, the Aristotelian will insist, another thesis whose import is *metaphysical*, a corollary of the distinction between actuality and potentiality which is the foundation of the Aristotelian philosophy of nature. The principle of inertia, by contrast, is a claim of natural science. Since the domains they are addressing are different, there can be no question of any conflict between them, certainly no direct or obvious conflict.

As I argued in chapter 3, physics, as that discipline is understood in modern times, abstracts from concrete material reality and describes the natural world exclusively in terms of its mathematical structure. Newton's laws of motion reflect this tendency, insofar as they provide a mathematical description of motion suitable for predictive purposes without bothering about the origins of motion or the intrinsic nature of that which moves. Indeed, that is arguably the whole point of the principle of inertia. As Weisheipl writes:

> Rather than proving the principle, the mechanical and mathematical science of nature *assumes* it... [and] the mathematical sciences must assume it, if they are to remain mathematical...
>
> The basis for the principle of inertia lies... in the nature of mathematical abstraction. The mathematician must equate: a single quantity is of no use to him. In order to equate quantities he must assume the basic irrelevance or nullity of other factors, otherwise there can be no certainty in his equation. The factors which the mathematician considers irrelevant are... motion, rest, constancy, and unaltered directivity; it is only the *change* of these factors which has quantitative value. Thus for the physicist it is not motion and its continuation which need to be explained but change and cessation of motion – for only these have equational value...
>
> In the early part of the seventeenth century physicists tried to find a physical cause to explain the movement [of the heavenly bodies]; Newton merely disregarded the question and looked for two quantities which could be equated. In Newtonian physics there is no question of a cause, but only of differential equations which are consistent and useful in describing phenomena...
>
> [T]he nature of mathematical abstraction... must leave out of consideration the qualitative and causal content of nature... [S]ince mathematical physics abstracts from all these factors, it can say nothing about them; it can neither affirm nor deny their reality... (1985, pp. 42 and 47-48; Cf. Wallace 1956, pp. 163-64)

The philosophy of nature, however, and in particular the principle of motion and the other components of the Aristotelian metaphysical apparatus, are concerned precisely to give an account of the intrinsic nature of material phenomena and their causes, of which modern physics gives us only the abstract mathematical structure.

Some related remarks from McGinn, which follow up his reflections on relative and absolute motion cited earlier, are worth quoting at length:

> Physics... deals with operational definitions, and these precisely involve what matter does, not what it is intrinsically. The reason physics is so obsessed with motion is simply that motion is what matter does – and operationalist physics must be about what matter does... [T]o be an experimental science, physics must be operationalist, and operationalism will lead to a focus on motion... What [this] shows is that physics is not a complete science (or study) of matter: to complete it we need to do metaphysics, or whatever we call the subject that takes up where empirical physics leaves off. We must supplement operational definitions with intrinsic or constitutive ones...
>
> Motion is obviously extremely fundamental to the universe we inhabit, but its salience in physics results not merely from that truth but also from methodological requirements. The very methodology of physics skews its picture of the nature of the physical universe.
>
> I think, then, that there is a kind of double narrowing going on in physics, centering on motion, as a consequence of essentially epistemological scruples, which results in a distorted picture of the nature of the physical universe. First, there is the restriction to the notion of relative motion, which stems from verificationist assumptions: nonrelative motion is also a reality, undetectable and incomprehensible as it may be. There is simply more to motion, as it exists objectively, than change of relative position. Second, the focus on motion itself results from the epistemological commitments of operationalism: if operational definitions serve to make the phenomena thus defined measurable and observable, they also deflect attention from the intrinsic nature of what is so defined. But that nature still exists as an ingredient of reality, hard to fathom as it may be. There is more to matter and motion than these epistemological and methodological restrictions permit us to appreciate. In effect, physics gives us the kind of biased view of matter that behaviorism gave us of mind. After all, twentieth-century behaviorism in psychology was partly stimulated by Bridgman's operationalism in physics; and the latter is just as faulty as the former as a guide to what the

> world is really like. Philosophy can help in diagnosing how the methods of physics might bias the picture of the physical world that it presents. (2011, pp. 94-95)

When considering the relationship between the principle of motion and the principle of inertia, then, we need to keep in mind not only that physics is not addressing the same questions that the philosophy of nature is, and that physics is concerned only with the mathematical structure of motion rather than its intrinsic nature, but also that the description of motion that physics affords may distort as much as it reveals.

4.2.3.3 Is inertia real?

Now, on the basis of such considerations, some Aristotelians have gone so far as to insinuate that the principle of inertia really has only an instrumental import, with the Aristotelian philosophy of nature alone providing a description of the reality of motion. Hence Joyce writes that "the mathematician may for practical purposes regard motion as a *state*. Philosophically the concepts of movement and of a state are mutually exclusive" (1924, p. 95). And Garrigou-Lagrange claims: "[T]hat the motion once imparted to a body continues indefinitely, is a *convenient* fiction for *representing* certain mathematical or mechanical relations of the astronomical order" (1939, p. 275, note 24; emphasis in the original).

Certainly a realist construal of inertia is at least open to challenge, not least because the principle is not directly susceptible of experimental test. As William Wallace writes:

> It is never found in ordinary experience that a body in uniform motion continues in such motion indefinitely. All the bodies met with in ordinary experience encounter resistive forces in their travel, and sooner or later come to rest. Nor does refined experimentation and research supply any instances where such resistive forces are absent. (1956, p. 178)

And as N. R. Hanson emphasizes, the problem is not merely that we *have* not observed bodies that are force-free and thus operate in accordance with the principle of inertia, but that we *could* not observe them, given Newton's own Law of Universal Gravitation. The law of inertia thus "refers to entities which are unobservable as a matter of physical principle" (Hanson 1963, p. 112; cf. Hanson 1965a).

To be sure (and as Wallace and Hanson acknowledge) the principle can be argued for by extrapolating from observational data to the limiting case, and Galileo and Newton argued in precisely that way. But no such argument can provide a true demonstration. Wallace's remarks are worth quoting at length:

> The observational data are certainly true, but the only way in which it may be maintained that the limiting case is also true would be by maintaining that what is verified in the approach to a limit is also verified at the limit itself. The latter statement, however, cannot be maintained, because it is not universally true. There are many instances in mathematics where it is known to be violated. One illustration is the approach of polygon to circle as the number of sides is increased indefinitely. All through the approach to the limit, assuming the simple case where all figures are inscribed in the limiting circle, every figure constructed that has a finite number of sides is a polygon. The limiting case is a figure of a different species, it is no longer a polygon, but a circle. It is not true to say that a polygon is a circle; the difference is as basic and irreducible as that between the discrete and the continuous. In this case, what is verified in the approach to the limit (polygon), is not verified at the limit itself (circle).
>
> Now if it is not *always* true that what is verified during the approach is necessarily verified at the limit... then the fact that the observational base for the principle of inertia is true cannot be used to prove, or demonstrate, that the limiting case stated in the principle is also true. (1956, pp. 179-80; Cf. Weisheipl 1985, pp. 36-37)

Nor need one be an Aristotelian to wonder about the epistemic credentials of Newton's principle. Einstein wrote:

> The weakness of the principle of inertia lies in this, that it involves an argument in a circle: a mass moves without acceleration if it is sufficiently far from other bodies; we know that it is sufficiently far from other bodies only by the fact that it moves without acceleration. (1988, p. 58)

Eddington is even more pithy, and sarcastic to boot: "Every body continues in its state of rest or uniform motion in a straight line, except in so far as it doesn't" (1958, p. 124). Isaac Asimov makes the same point and at least insinuates an instrumentalist conclusion:

> The Newtonian principle of inertia... holds exactly only in an imaginary ideal world in which no interfering forces exist: no friction, no air resistance...
>
> It would therefore seem that the principle of inertia depends upon a circular argument. We begin by stating that a body will behave in a certain way unless a force is acting on it. Then, whenever it turns out that a body does not behave in that way, we invent a force to account for it.
>
> Such circular argumentation would be bad indeed if we set about trying to prove Newton's first law, but we do not do this. Newton's laws of motion represent assumptions and definitions and are not subject to proof... The principle of inertia has proved extremely useful in the study of physics for nearly three centuries now and has involved physicists in no contradictions. For this reason (and not out of any considerations of "truth") physicists hold on to the laws of motion and will continue to do so. (1993, pp. 25-26; Cf. Ellis 1965 and Hanson 1965b)

Yet while the difficulty of proving the principle of inertia should certainly give further pause to anyone who claims that modern physics has refuted the Aristotelian principle of motion, that difficulty hardly *forces* a non-realist interpretation on us. Still, it might seem that the Aristotelian's commitment to natural teleology, and in particular to the idea that a potential is always a potential *for* some definite actuality or range of actualities, would require a non-realist construal of inertia. Andrew van Melsen writes:

> If the law of inertia, that a local motion never stops of its own account, is true, then the conclusion seems obvious that a motion does not have an "end" in the Aristotelian sense of this term... [I]t seems that the analysis of motion in terms of potency and act assumes the existence of a definite end of each motion as the natural achievement or perfection of that motion... [But in] such [inertial] motions there seem to be eternal potency but no act. (1954, p. 174)

And as van Melsen indicates, this might lead some Aristotelians to argue that

> such motions as the law of inertia describes do not exist. The law of inertia is not supposed to speak of real motions, for it assumes the absence of physical forces, which, as a matter of fact, are never absent in reality. Since Aristotle's analysis deals with real motions, the difficulty [of reconciling Aristotle with Newton] does not exist. (Ibid.)

But van Melsen immediately goes on to reject such a non-realist interpretation of inertia, as have other Aristotelians. In van Melsen's view, it is an error to assume in the first place that the Aristotelian's commitment to teleology must lead him to conclude that what moves must come to rest:

> Aristotle himself... would have referred to the eternal circular movement of heavenly bodies as an instance of ceaseless motion. So it must be possible to apply analysis in terms of potency and act to motions which are endless...
>
> There may be... no *final* act which gives the motion its unity, but such a final act is not necessary for motion to possess unity. The process of gradual actualization in a definite direction is sufficient. (1954, p. 175)

To be sure, there are other questions that an Aristotelian might raise about the idea of ceaseless motion, as we shall see presently. But in any event, an alternative position is suggested by John Keck, who, while like van Melsen affirming a realist interpretation of inertia, also argues that all natural motion does in fact tend toward a definite state of rest, namely the unity of the thing moving with the larger material world. (2011; cf. Keck 2007). That there is no conflict between these claims can in his view be seen when we recognize that inertia is a *passive* and *incomplete* aspect of an object's motion, which cannot by itself account for the object's actual determinate movement but needs completion by an external agent. (Compare the Aristotelian conception of matter as something which, though a real constituent of things, is essentially passive and incomplete until actualized by form.)

So, an Aristotelian need not deny the reality of inertia, and I think most Aristotelians would not. A mathematical description of nature is not an exhaustive description, but it can capture real features of the world. And that the principle of inertia has been especially fruitful in physics is reason to think that that it does capture them. As Thomas McLaughlin writes:

> Because inertia is common to so many different kinds of bodies, the proper principles of many different natures can be neglected for various purposes and nature can be analyzed at a minimal level. That a given inertial body is a pumpkin is irrelevant for some purposes, and this is not only a consequence of the mathematization of nature. Inertia is undoubtedly a thin treatment of nature, but that is not the same as treating a body as if it had no nature nor need it exclude a fuller treatment of a body's nature. Failure to recognize this point may mislead a thinker into maintaining that the principle of inertia denies inherent principles of nature. (2008, p. 259)

In short, just as acceptance of the Newtonian principle of inertia does not entail rejection of the Aristotelian principle of motion, neither need the Aristotelian take an instrumentalist or otherwise anti-realist approach to the Newtonian principle. They can be regarded as describing nature at different but equally real levels. (For a debate over realism about inertia and related matters conducted from a non-Aristotelian point of view, see Earman and Friedman 1973 and the response in Sklar 1985.)

4.2.3.4 Change and inertia

But what, specifically, does this claim amount to? If the principle of motion and the principle of inertia are not at odds, how exactly are they related?

Whatever else we say in answer to these questions, the Aristotelian will insist that real change of any sort is possible only if the things that change are composites of actuality and potentiality. And since no potential can actualize itself, whatever changes is changed by another. In this way the principle of motion, as a basic thesis of the philosophy of nature, is necessarily more fundamental than the principle of inertia – at least if we allow that the latter principle does indeed apply to a world of real change. (More on this caveat presently.) Determining how the principle of motion and the principle of inertia are related, then, has less to do with how we interpret the former principle than with how we interpret the latter. And here there are two main possibilities:

1. Inertial motion as change: We have noted that writers like Garrigou-Lagrange object to the idea that inertial motion is a kind of "state." Suppose then that we took that to be merely a loose way of speaking and regarded inertial motion as involving real change, the actualization of potential. As van Melsen describes it:

> The moving body goes continuously from one place to another, say from A towards B, from B towards C, etc. If this body is actually in place A, then it is *not* in place B, but is moving towards B. Therefore, there is a definite potency of being at B. The arrival at B means the actualization of that potency... However, the arrival at B includes the potency of going on to C, etc. In other words, each moment of the motion has a definite tendency towards some further actualization, and it is this which gives the motion its unity. (1954, p. 175)

The question, then, is what actualizes these potentials. Now the very point of the principle of inertia is to deny that the continued uniform rectilinear local motion of an object requires a continuously operative external force of the sort that first accelerated the object; so such forces cannot be what actualize the potentials in question. But could we say that the force which first accelerated the object is itself what actualizes these potentials? For example, suppose a thrown baseball were not acted upon by gravitational or other forces and thus continued its uniform rectilinear motion indefinitely, with the actualization of its potential for being at place B followed by the actualization of its potential for being at place C, followed by the actualization of its potential for being at place D, and so on *ad infinitum*. Could we say that the thrower of the baseball is, in effect, himself the actualizer of all of these potentials?

It might seem that Aquinas could sympathize with such a view, since as we have seen, he regarded the motion of an object to its natural place as having been caused by whatever generated the object. The notion of a natural place is obsolete, but if we substitute for it the notion of *inertial* motion as what is natural to an object, then – again, so it might seem – we could simply reformulate Aquinas's basic idea in terms of inertia. That is, we could say that the inertial motion of an object, which involves an infinite series of actualized potentials with respect to location, is caused by whatever force first accelerated the object (or, to preserve a greater parallelism with Aquinas's view, perhaps by whatever generated the object *together with* whatever accelerated it).

But there is a potential problem with this proposal. Natural motions, as Aquinas understood them, are finite; they end when an object reaches its natural place. Inertial motion is not finite. And while there is no essential difficulty in the notion of a finite cause imparting a finite motion to an object, there does seem to be something fishy about the idea of a finite cause (such as the thrower of a baseball) imparting an *infinite* motion to an object (Garrigou-Lagrange 1939, p. 274). Furthermore, as noted above, Aquinas also regarded the motion of an object toward its natural place as being caused *instrumentally* by the generator of the object, even though the generator does not remain conjoined to the object. And this seems problematic even when modified in light of the principle of inertia. For how could the inertial motion of the baseball in our example be regarded as caused *instrumentally* by the thrower of the baseball, especially if the ball's motion continues long after the thrower is dead (Joyce 1924, p. 98)?

So, there are problems with the idea that inertial motion, *if* interpreted as involving real change, could have a *physical* cause. But as we implied above, even if its lacks a physical cause, there is nothing in the principle of inertia that rules out a *metaphysical* cause. Indeed, *if* inertial motion involves real change, then given the principle of motion together with the absence of a physical cause, such a metaphysical cause would be necessary.

Of course, that raises the question of what exactly this metaphysical cause is. One suggestion would be that it is something *internal* to the object – an "impetus" imparted to it by whatever initiated its inertial motion, and which continuously actualizes its potentials with respect to spatial location. But as Joyce notes, there are serious problems with the impetus theory (1924, pp. 98-99). For one thing, a finite object (such as the baseball of our example) can only have finite qualities. And yet an impetus, in order to have local motion *ad infinitum* as its effect, would at least in that respect be an infinite quality. In other respects it would be finite (it would, for example, be limited in its efficacy to the object of which it is a quality) but that leads us to a second problem. For an impetus would continually be bringing about new effects and thus (as a finite cause) itself be undergoing change; and in that case we have only pushed the problem back a stage, for we now need to ask what causes these changes in the impetus itself.

Another possibility, though, would be a metaphysical cause *external* to the moving object. Now, we already have a model for such a cause in the Aristotelian tradition. For the motions of celestial bodies were in that tradition regarded as unending, just as inertial motion is (barring interference from outside forces) unending; and while this view was associated with a mistaken astronomy, a metaphysical kernel might be thought extractable from the obsolete scientific husk. The causes of celestial motion in this earlier Aristotelian tradition were, of course, intelligent or angelic substances. Such substances are regarded as *necessary* beings of a sort, even if their necessity is ultimately derived from God. What makes them necessary is that they have no natural tendency toward corruption the way material things do (even if God could annihilate them if He so willed). Given this necessity, such substances have an unending existence proportioned to the unending character of the celestial motions they were taken to explain. And while it turns out that celestial objects do not as such move in an unending way, *inertial* motion (including that of celestial bodies, but that of all other objects as well) *is* unending. Hence one could argue that the metaphysical cause of inertial motion – again, at least *if* such motion is considered to involve real change and *if* both a physical cause and an internal metaphysical cause are ruled out – is a necessarily existing intelligent substance or substances, of the sort the earlier Aristotelian tradition thought moved celestial objects. (Unless it is simply God Himself causing it *directly* as Unmoved Mover (Cf. Wallace 1956, p. 184). Though it might be objected that to regard God as the immediate cause of inertial motion goes too far in the direction of occasionalism.)

2. Inertial motion as stasis: Needless to say, that would be a pretty exotic metaphysics (and I would not endorse it myself). But alternatively, of course, we could take seriously the idea that inertial motion is a state, involving no real change and thus no actualization of potential. That is, after all, what modern physics appears to say about it. In defense of the thesis that inertial motion is a state rather than a real change, McGinn puts forward two considerations (2011, p. 117). First, when a body continues in rectilinear motion without acceleration, *that* fact about it remains constant despite its change of position, and so to that extent the body can be said to be in stasis rather than changing. Second, change of position is itself a mere *Cambridge* change rather than a real change. The idea of a Cambridge change is illustrated by your going from being taller than your son to being shorter than him as a result of his growing in height. There is no intrinsic change to *you* at all, but only an intrinsic change to your son. The change to your son is real, but the "change" you undergo is a mere Cambridge change. McGinn proposes that an object's changes of position are like that.

Now, if inertial motion is a kind of stasis, then the question of how the principle of motion and the principle of inertia relate to one another does not even arise, for there just *is* no motion in the relevant, Aristotelian sense going on in the first place when all an object is doing is "moving" inertially in the Newtonian sense. To be sure, *acceleration* would in this case involve motion in the Aristotelian sense, but as we have seen, since Newtonian physics itself requires a cause for accelerated motion, there isn't even a prima facie conflict with the Aristotelian principle of motion.

It might seem, however, that the question of how the Aristotelian principle of motion relates to the Newtonian principle of inertia has been rendered moot by further developments in modern physics. If we suppose that all change reduces to local motion and that local motion of an inertial sort amounts to a kind of stasis, then it would follow that real change in the Aristotelian sense was already at least *largely* banished from nature by Newton. As McGinn points out (2011, p. 120), the Newtonian picture thereby takes us partway back to Parmenides. If we then add to this picture the Minkowskian interpretation of relativity as yielding a four-dimensional block universe, it might seem that we will thereby go the rest of the way, and have to deny that there is any change in the Aristotelian sense *at all* in the world – not even the sort Newtonian physics would allow occurs with the acceleration of an object. For on this model, past, present, and future events are all equally actual and there is no potentiality in need of actualization. It is not just inertial motion that turns out to be a kind of stasis. The *entire physical world*, on this picture, is in stasis.

Now, that this is too quick should be obvious already from what has been said about the impossibility of reducing all change to local motion. But more needs to be said. Saying it, however, requires turning to the topic of time.

4.3 Time

4.3.1 What is time?

Recall that on an Aristotelian analysis, a real change (as opposed to a mere Cambridge change) involves the gain or loss of some attribute, but also the persistence of that which gains or loses the attribute. When a banana goes from being green to being yellow, the greenness is lost and the yellowness is gained, but the banana itself persists. Time, on the Aristotelian analysis, is just the measure of change with respect to the succession of such gains or losses. When we say that it took a certain banana four days to go from being green to fully yellow and then another eight days to turn brown, what this temporal description captures is the rate at which the changes in question followed upon one other. Absent such a succession of changes, there would be nothing to measure, and thus no time.

Each of the components of this characterization of time calls for elaboration. First, the claim that time presupposes change is controversial. For example, Sydney Shoemaker (1969) suggests that we can conceive of a world in which the inhabitants of three regions A, B, and C each occasionally observe the other regions go into a "frozen" or unchanging state for a period of a year. Given that each region is observed by the inhabitants of the others to do so according to a regular pattern (every three, four, and five years respectively) they would have reason to conclude that every sixty years the regions must all be "freezing" together, and thus that in their world no change is occurring for a year's time. Hence (it is argued) it is possible for time to exist without change.

But even leaving aside the tendentious assumption that we can deduce what is really possible from what is conceivable, this argument fails for reasons noted by Lowe (2002, pp. 247-49). For one thing, it reasons from the claim that the inhabitants of Shoemaker's imagined world would have *evidence* for time without change to the conclusion that it is *possible* for there to be time without change. But this gets things the wrong way around, for we first have to know that something really is possible before we can know that there might be evidence for it. For another thing, if we suppose that A, B, and C really are all frozen or unchanging, then it becomes utterly mysterious what *causes* Shoemaker's world ever to become unfrozen, or unfrozen after exactly a year's time rather than at some earlier or later time.

Though time presupposes change, it is not identical with change, as is evident from the fact that change can have features that time does not (Bittle 1941, p. 204; Phillips 1950, p. 119). Local motion, for example, can be vibratory or rotational, but it makes no sense to attribute such characteristics to time. A movement can speed up, slow down, cease temporarily and then start up again, with time continuing to pass at the same rate. A change can be reversed without time reversing. For example, when I walk from one side of a room to the other and back again, or when my skin turns red from a sunburn and then returns to its normal color, I don't thereby go from time t_1 to time t_2 and then back to t_1.

Time, again, is the measure of change *with respect to succession.* If I say that a banana both turned brown and began to smell bad, I have numbered the changes it underwent at two. But I have not thereby numbered *time*, as I would be if I said that the banana first turned brown and then began to smell bad after a further two days. Just as time cannot be identified with change, though, neither can it be identified with succession (Phillips 1950, p. 119). Numbers succeed one another, but not temporally. Los Angeles is north of San Diego and south of San Francisco, but this is a matter of spatial rather than temporal succession.

To say that time is a *measure* of change with respect to succession raises the question of who is doing the measuring, and thus of whether time is mind-dependent. The first thing to say is that the Aristotelian rejects the Newtonian absolutist version of realism about time just as he rejects Newtonian absolutism about space, and for analogous reasons. The absolutist view of space, on which space would still exist even if no material objects did, treats space itself as a kind of substance. Similarly, the absolutist view of time, on which time could exist apart from any material objects, treats time as a kind of substance. But like the absolutist view of space, this view of time entails a vicious regress (Bittle 1941, pp. 199-200; Koren 1962, p. 121). For since there is in this putative substance a continual transition from one moment to another, it is changing. But if it is changing, then there must be some *higher-order* time which is the measure of this change. Now, this higher-order time is itself either a substance or it is not. If it is not, then we might as well conclude that first-order time is not really a substance either and abandon absolutism. For what considerations could show that first-order time is a substance but that higher-order time is not? If we do say that it is a substance, though, then since it too will be changing, we will need to post a *third-order* time as well, and so on *ad infinitum.*

Like the Newtonian view of space, the Newtonian view of time is also not as consistently absolutist as it at first appears. For just as Newton regarded space as God's sensorium, so too did he treat time as God's eternity. Now, as with Newton's view of space, this view of time is theologically problematic. For one thing, since it makes of time a divine attribute, it is hard to see how the view can avoid collapsing into pantheism. For another, it is incompatible with the thesis that divine eternity is strict timelessness. (See Feser 2017 for criticism of pantheism and a defense of divine timelessness.) But the point to emphasize for current purposes is that Newton's view illustrates how difficult it is consistently to treat time in an absolutist way. Even Newton ultimately thought of time and space as inhering in something else (namely God) than as substances in their own right.

At the same time, the Aristotelian rejects the idealist view that time is entirely mind-dependent (Bittle 1941, pp. 200-2; Phillips 1950, pp. 124-5). As with space, the reason has to do with the general Aristotelian commitment to realism, and as with space, the realism in question is of a distinctively Aristotelian sort that is usefully understood on analogy with Aristotelian realism about universals. (Cf. Bardon 2013, p. 13.) The view that time is entirely ideal or mind-dependent is like the nominalist view that universals are the free creations of human thought and language. The absolutist view of time is like the Platonic realist view that universals exist entirely apart from concrete particulars and from all minds. The Aristotelian view of time, like the Aristotelian view of universals, is a middle position. Time as the absolutist conceives of it exists only as an abstraction of the mind, just as *humanness* qua universal exists only as abstracted by a mind. But in both cases what the mind abstracts is something that really is there *in* the concrete world and is abstracted *from* that world rather than being a sheer invention. Like space, time is a kind of receptacle for material objects, but like space, it is a receptacle that has no reality apart from the material objects for which it is the receptacle.

Time is also like space in being *continuous*. We speak of instants of time just as we speak of points in space, but time can no more be *made up of* discrete instants than space can be made up of discrete points (Oderberg 2006, p. 106). For just as points are unextended, instants are durationless. Hence, just as space, which is extended, cannot be analyzed as a collection of unextended points, so too time, which has duration, cannot be analyzed as a collection of durationless instants. But neither can time be analyzed even as a collection of discrete intervals having duration, for reasons analogous to the reasons a continuous physical object cannot be analyzed as a collection of discrete extended parts. For since an interval is always divisible into shorter intervals, such an analysis would lead to paradoxes of the kind Zeno raises for the parallel analysis of physical objects. As with the parts of continuous physical objects, shorter intervals within a stretch of time, though real, exist within it *potentially* rather than actually. As David Oderberg suggests, a durationless instant can be thought of as a limit case vis-à-vis the divisibility of any interval of time into shorter units, analogous to prime matter as a kind of limit case vis-à-vis the divisibility of a physical object into the parts that exist in *it* potentially (Oderberg 2006, pp. 107-8).

However, time is, in the Aristotelian view, unlike space in two crucial ways (Koren 1962, pp. 117-18, 129-30, 137). First, every part of space exists all at once, but moments of time exist only successively, one after the other. Second, space can be traversed in all directions, but time can be moved through in one direction only, forward and irreversibly. Of course, these claims are controversial in contemporary philosophy of time, as are some of the other claims made so far. To relate the Aristotelian position to its modern rivals, it will be useful to introduce some terminology owing to J. M. E. McTaggart (1927) which has now become standard.

McTaggart distinguished, first, between what he called *A-determinations* and *B-relations.* A-determinations are captured by predicates like "is past," "is present," and "is future." For example, the event of your birth is in the past, your reading of this sentence is a present event, and your death is an event that will occur in the future. B-relations are expressed in such phrases as "is earlier than" and "is later than." For example, the crucifixion of Jesus occurred later than the death of Socrates but earlier than the birth of Thomas Aquinas.

Next we have the distinction between the *A-series* and the *B-series.* The A-series is the series of events ordered as being in the *distant past*, in the *recent past*, in the *present*, in the *near future*, and in the *far future*. This series is *tensed*, and events are constantly changing their position in the series. For example, your reading of the previous paragraph is an event in the recent past, your reading of this sentence is a present event, and your reading of the next paragraph is an event that will occur in the near future. But once you have read the sentence the event then slips into the near past, a year from now the event of your reading the previous paragraph will have slipped into the distant past, and so forth. The B-series is the series of events ordered as either earlier than or later than other events. This series is *tenseless*, and events retain their position in it in a fixed way. For example, the crucifixion never stops being later than the death of Socrates and earlier than Aquinas's birth.

Then we have the distinction between the *A-theory* (or *tensed theory*) of time and the *B-theory* (or *tenseless theory*) of time. According to the A-theory, the A-series is both real and more fundamental to the nature of time than the B-series. There is, on this view, an objective fact of the matter about whether some event is *now* or present, and an event's place in time is to be understood in terms of its relationship to what is now. The A-theory comes in three main varieties. The classical form taken by the

A-theory is *presentism*, according to which only the present is real, with past events no longer existing and future events not yet existing. A second kind of A-theory is known as the *"growing block" theory*, according to which the present is real and all events that were present but are now past also in some way still exist, but future events do not yet exist. The third main kind of A-theory is the *"moving spotlight" theory*, which holds that past, present, and future events all in some way exist, but only present events fall under the "spotlight" of the *now*, which "moves" along, and successively illuminates, this series of events. The B-theory, meanwhile, takes the B-series to be more fundamental to the nature of time, and the A-series to be either reducible to the B-series or eliminable altogether. The B-theory takes an "eternalist" view according to which all events – whether past, present, or future – are all equally real, as different parts of a single unchanging Parmenidean "block." Temporal passage and the *now* are, on this view, illusory. An event is *now* or *present* only relative to our consciousness of it, and not as a matter of objective fact.

Now, the first thing to say is that at least the *most general* concepts of Aristotelian philosophy of nature could in principle be reconciled to *some* extent with any of these theories. This is true even of the B-theory, though the fit is hardly comfortable. For even if we think of the universe as a single unchanging block, with all events equally actual parts of this block, the universe would still be contingent (since there still might have been no universe at all or a different one) and it would still be a particular physical object (even if a unique and four-dimensional physical object). Hence, while there would be no actualization of potentiality *within* the universe, the universe *itself, considered as a single whole*, would have to be actualized. It could be regarded as one big physical substance composed of substantial form and prime matter. Hylemorphism and the theory of actuality and potentiality would still apply, even if in an eccentric way. Obviously, they would apply even more naturally on any version of the A-theory, since on any of them there would also be the actualization of potentiality *within* the universe.

Having said that, there can be no question, given what has been said so far, that the most natural position for the Aristotelian to take is an A-theory, and presentism in particular. Again, for the Aristotelian, time is the measure of change with respect to succession, and change is the gain or loss of an attribute by a subject which persists through that gain or loss. Change, in turn, entails the actualization of potentiality. Hence, consider once again the yellow banana of our earlier example, which had been green and will be brown. The banana is actual, and its yellowness is actual. Its greenness is no longer actual, and its brownness is not yet actual. Naturally, then, the event of its *becoming* brown is not actual, and neither is its state of being green or the event of its becoming yellow. Thus, neither the *past* nor the *future* of the banana is actual. What is actual is just what is actual *now*, the *present* of the banana.

As Brian Ellis has suggested (2009, p. 120), it is less misleading to speak of temporal *generation* than of temporal passage. Future events are not like ships that are already out there but haven't yet passed yours, nor like visiting guests who exist but haven't yet arrived. They are more like children you have not yet begotten. Similarly, past events are not like visitors who have left your home but continue to exist at their own. They are rather like ancestors who have died. The past and future exist now only in the loose sense that they are, as it were, *causally contained in* what exists now – with the *principle of causality*, the *principle of proportionate causality*, and the *principle of finality* (which were set out in chapter 1) all relevant to understanding the containment in question. Future entities, states, and events are contained within the present as *potentials* which might be actualized. Past entities, states, and events are contained within the present insofar as their *effects* on the present remain. The present *points forward* to a range of things which might yet be caused to exist. The present also *points backward* towards formerly existing things qua causes proportionate to the effects that now exist. But again, what actually exists in the strict sense is what exists now. Past entities, states, and events are those that *did* exist; future entities, states, and events are those that *will* exist; and present entities, states, and events alone are those that *do* in fact exist.

The defense of this position will proceed in two stages. First, I will defend the A-theory, as a general approach to understanding time, against its rival the B-theory. Second, I will defend presentism against rival versions of the A-theory.

4.3.2 The ineliminability of tense

4.3.2.1 Time and language

That Socrates drank hemlock is a historical fact. It is also a fact expressed in terms of tense. Innumerable facts are like that. Now, if there are facts that are tensed, then tense is a real feature of the world, as the A-theory maintains and the B-theory denies. B-theorists have deployed two main strategies to deal with this sort of argument. The first is to argue that any true sentence that is tensed can be translated into a sentence that is not, without any loss of meaning. This is called the "old tenseless theory" (Russell 1915; Smart 1963, Chapter VII). The second is to argue that, even if such translation is not possible, the truth of tensed sentences can be accounted for without having to affirm that there are any tensed facts. This is called the "new tenseless theory" (Mellor 1981).

According to one version of the old tenseless theory, a sentence like "Socrates drank hemlock" can be translated into a sentence like "Socrates is drinking hemlock in 399 B.C." According to another, the sentence should be translated instead into a sentence such as "Socrates is drinking hemlock earlier than the utterance of this sentence." (This second analysis is said to be "token-reflexive" insofar as it gives us a particular instance or token of a general sentence type, and one which refers to itself.) Note that if such translations work, this would show that anything that can be described in the language of the A-series can be described instead in terms of the B-series. But it is now commonly acknowledged that analyses of these sorts do *not* work (Craig 2000a, Chapter 2; Craig 2001, pp. 117-19; Smith 1993, Chapters 2-3).

One problem is that it just isn't the case that the proposed translations capture everything conveyed by the originals. As John Perry (1979) has pointed out, the tenseless sentence "The meeting starts at noon" does not contain all the information that is in the tensed sentence "The meeting starts now," as is evident from the fact that someone who intends to go to the meeting will be moved to action by an utterance of the latter sentence in a way he might not be by the former sentence. Furthermore, the purported translations also contain information that is not in the originals. For example, a person who has lost track of time might know that it is true that "The meeting starts now" without thereby knowing that it is true that "The meeting starts at noon." Another problem is that a token-reflexive sentence implies its own existence in a way that the tensed sentence that it is purportedly a translation of does not. For example, the sentence "Socrates is drinking hemlock earlier than the utterance of this sentence" implies that a certain sentence is being uttered, whereas "Socrates drank hemlock" does not. Hence the two sentences are not logically equivalent.

It is sometimes suggested, however, that the temporal indexical "now" is like the spatial indexical "here" in reflecting merely the subjective point of view of a conscious subject rather than anything in objective reality. However, as William Lane Craig points out (2001, pp. 127-29), this by no means shows that tense itself is merely subjective. For one thing, not all tensed language is indexical or reflects the point of view of a conscious subject. For example, the word "is" in "Socrates is drinking hemlock" or the word "was" in "Socrates was drinking hemlock" are not indexical nor, for all the objection under consideration shows, do they merely reflect the point of view of a subject. For another thing, there is a disanalogy between the two indexicals in that "here" can be analyzed in terms of *where I am now located*, whereas the indexicals "I" and "now" are arguably not in turn similarly susceptible of reductive analysis.

In any event, the new tenseless theory concedes that the old theory fails, but denies that this gives any support to the A-theory. According to the new theory, though the *meaning* of a tensed sentence is not captured by a tenseless sentence, its *truth conditions* are nevertheless captured by the latter. For example, the sentence "Socrates is drinking hemlock earlier than the utterance of this sentence" does not *mean* the same thing as "Socrates drank hemlock." But, claims the new tenseless theory, what makes the latter sentence *true* is captured by the former sentence. Hence, the theory concludes, we needn't affirm the reality of tensed facts in order to account for the truth of tensed sentences.

However, this approach too faces grave problems (Craig 2000a, Chapter 3; Craig 2001, pp. 119-29). One such problem is logical. Suppose Bob and Fred each utter a token or instance of the sentence "Socrates drank hemlock." Let's label Bob's utterance of the sentence *B*, and Fred's utterance of the sentence *F*. According to the new tenseless theory, *B* is logically equivalent to the sentence "Socrates is drinking hemlock earlier than *B*," which gives *B*'s truth conditions. Similarly, *F* is logically equivalent to the sentence "Socrates is drinking hemlock earlier than *F*," which gives *F*'s truth conditions. Now, *B* and *F* are also logically equivalent to each other. In other words, what Bob says when he says "Socrates drank hemlock" is true if and only if what Fred says when he says "Socrates drank hemlock" is also true. So, the sentences "Socrates is drinking hemlock earlier than *B*" and "Socrates is drinking hemlock earlier than *F*," since they are logically equivalent to *B* and *F* respectively, should be logically equivalent to each other as well. However, they are *not* logically equivalent, because it *could* have turned out that Bob uttered his sentence while Fred did not, or vice versa. So, the new tenseless theory's analysis fails.

A second problem is that the theory cannot account for sentences of which there are no tokens or instances. Consider a sentence like "There are now no tokens or instances of any sentences," which would be true at a time when no one happens to be uttering any sentences. The new tenseless theory entails that the truth condition for this sentence would be that it is uttered at a time when there are no tokens or instances of any sentences. But of course, it never *could* be uttered when there are no tokens or instances of any sentences (since for someone to utter it would just be to produce a token or instance of a sentence). The new tenseless theory thus implies that the sentence could never be true. Thus, since the sentence *could* in fact be true, the theory is false.

A third problem is that, as Craig (1996) has emphasized, the new tenseless theory seems to confuse *truth conditions* with *truth makers*. Truth conditions are *semantic*, and describe logical connections between statements. Truth makers are *ontological*, and concern the facts that make it the case that a statement is true. Now, a truth condition for a statement does not necessarily reveal the truth maker for that statement. Consider the statement "Bachelors are unmarried if and only if 2 + 2 =4." Since this biconditional is true, it could be said that it gives truth conditions for the sentence "Bachelors are unmarried." But the fact that 2 + 2 = 4 is hardly what makes it true that bachelors are unmarried.

In light of objections like these, new tenseless theory advocate D. H. Mellor (1998) proposes an alternative account that focuses on truth makers rather than truth conditions, and appears also to focus on sentence types rather than sentence tokens. What makes the sentence type "Socrates drank hemlock" true today is the tenseless fact that at some time before today, Socrates is drinking hemlock. As Craig argues, however (2000a, pp. 91-96; 2001, pp. 124-6), this approach too is problematic. One problem is that a sentence can have multiple truth makers, and Mellor doesn't show that tensed facts are not *also* among the truth makers of tensed sentence types. Another problem is that the account does not really tell us what makes the sentence "Socrates drank hemlock" true. Rather, it tells us what makes it the case that that sentence is true *today*.

Of course, this only scratches the surface of the enormous literature on the subject of language and time, and there are further moves that a B-theorist might make in order to deal with difficulties like the ones canvassed. But it will suffice for present purposes to emphasize that tenseless accounts of temporal language are so convoluted and problematic that it is difficult to see how they could constitute a compelling *standalone* argument against the A-theory. Such accounts seem worth trying to salvage only if one already has *other* strong grounds for favoring the B-theory over the A-theory. Taken in isolation, the linguistic considerations point toward the A-theory rather than away from it.

4.3.2.2 Time and experience

Like language, ordinary conscious experience taken at face value would naturally lead us to affirm the A-theory rather than the B-theory. What we experience we experience *as present* or *now*. What is past we might remember as *having been* now, and what is future we might anticipate as what *will be* now, but what we are consciously aware of is what we take *actually to be* now. To be sure, sometimes the thing an experience is an experience of is *not* in fact now. For example, since light from stars takes a long time to reach us, what we see when we look at the night sky are events that occurred long ago rather than presently. Mellor argues, on this basis, that we don't in fact perceive the presentness of events (1998, p. 16). Even if that were true of *all* events *external to* our experiences, however (which it is not), it would not be true of the event that is the experience *itself*. I may judge that the star I am looking at is no longer present, but my *experience of seeing it*, while I am actually having that experience, is taken by me to be present. The B-theorist will say that even the apparent presentness of an experience is illusory, but the point for the moment is that as a matter of phenomenological fact an experience always *seems* to be present, whether this is an illusion or not.

It is sometimes suggested that the claim that what we experience we experience as present is true, but trivial (Le Poidevin 2007, pp. 77-78; Mellor 1981, p. 54). But this is an odd claim coming from a B-theorist. For one thing, the B-theory denies that anything is *really* present, but only appears to be. For another, the B-theory claims that every moment of time is equally real, and that there is nothing special about the moment we regard as present. These are meant to be significant, indeed revolutionary, claims. So how can it be a *trivial* fact that what we experience we take to be present and to have a special metaphysical status? If the B-theory is correct, these are significant metaphysical errors, so how can the nature of the experience that leads us into these purported errors be any less significant? (Cf. Craig 2001, pp. 135-36.)

What is this experienced present *like*? Benjamin Curtis and Jon Robson identify three aspects of its phenomenology (2016, pp. 160-62). First, the present appears to have some brief *duration* rather than being a durationless instant. Writers like William James (1890) famously refer to this as the "specious present." Second, it is taken to be an experience *of change*. Even when one is not experiencing a change in something *external* to consciousness (a bird taking flight, a dog barking, a sudden chill in the air), there will be a change within consciousness itself as one thought, mental image, or sensation gives way to another. Third, the experienced present seems to have a *stream-like*, *seamless*, or *continuous* quality rather than being a succession of discrete parts. A fourth aspect that should be added to those noted by Curtis and Robson is that the present is experienced as *directed forward* into the future (Dainton 2010, p. 115). Time is experienced as flowing in one direction only. (Whether time *can* in fact be traversed in one direction only is a question to which we will return. Again, the point for the moment is just to describe how time is *experienced*.)

Now, few would deny that our experience of time does *seem* to be the way I have just described it. But one topic of controversy is whether that experience *really is* the way it seems to be. Notice that I am not at the moment talking about whether the physical world external to experience is the way experience makes it seem, as I was when a noted that a star we observe in the night sky may no longer exist. That is a separate question, to which we will return. The question is rather whether *temporal experience itself* is the way it seems to be. As Curtis and Robson rightly note (2016, p. 157), this is bound to strike some readers as an odd or even bogus distinction. How could there be a distinction between the way experience seems and the way experience really is? Is there anything more *to* an experience than the way it seems?

But some philosophers would call attention to the way we are sometimes confused about, or fail to notice, aspects of an experience. Curtis and Robson cite the example of the way that a drop of hot water might for a split second be experienced as cold. (Suppose you were *expecting* it to be cold, and for an instant thought it was until realizing it was hot.) The question on the table, then, is whether the commonsense phenomenological description of temporal experience is confused about, or fails to notice, important aspects of that experience. So, putting aside for the moment the question of what the physical world *beyond* experience is really like, are we wrong about what experience *itself* is really like?

Here there is a dispute between two positions sometimes labeled "extensionalism" and "retentionalism." Extensionalism holds that temporal experiences really do have duration or extension through time, and thus involve change and continuity in a straightforward way. Hence it takes the commonsense phenomenological description of temporal experience to be more or less correct. Retentionalism holds that temporal experiences do *not* really have the duration they seem to have, but occur in durationless instants, so that change and continuity must be experienced in a less straightforward way than it seems. The appearance of duration, on this view, results from a retention or representation of immediate previous experiences (for example, in memory, on some versions of the view). Hence this view takes the commonsense phenomenological description of experience to be wrong in significant respects.

The appeal to memory is the most obvious way retentionalism might be spelled out, but it is also obviously problematic. Suppose you see an apple fall from a tree branch to the ground. This seems like a single experience of brief duration, but according to retentionalism it is really a succession of durationless experiences. You first have a durationless experience of seeing the apple on the branch, then a durationless experience of seeing it in midair, then a durationless experience of seeing it on the ground. The appearance of duration results from your remembering the immediately previous experiences when you have the experience of seeing the apple on the ground. The problem with this analysis is that it cannot account for the difference between actually perceiving *change itself* and merely perceiving *that* a change has occurred (Broad 1923, p. 351; Curtis and Robson 2016, p. 166). To perceive an apple on the ground while remembering that the apple was previously on the tree and in midair would account for your knowing that it *has* changed its position, but actually *perceiving such change occurring* involves more than that.

A more subtle retentionalist analysis is associated with Edmund Husserl (1991). In the case of the example under consideration, Husserl would hold that the apple's being on the tree and its being in midair are retained, not as *memories* of the earlier experiences, but as the experiences *themselves*, *traces* of which still exist even as one has the experience of the apple being at the ground. Why is this perceived as *duration* rather than as a group of durationless experiences stacked on top of one another? The reason, on Husserl's analysis, is that the apple's being at the ground is experienced as *being now*, whereas the apple's being in midair and being on the tree are experienced as *having just been*.

The stream-like quality of temporal experience, on the retentionalist analysis, is to be accounted for in terms of overlaps between successions of instantaneous experiences. Suppose that after falling from the tree and hitting the ground the apple rolls an inch or two. You have an instantaneous experience of the apple being in midair while retaining a trace of the immediately preceding experience of its being on the tree. Then you have an instantaneous experience of the apple's being on the ground while retaining a trace of the immediately preceding experience of its being in midair. Then you have an instantaneous experience of its being an inch from where it landed while retaining a trace of the immediately preceding experience of its having hit the ground. And so on. These overlaps between successive sets of experiences generate the appearance of continuity or seamlessness in temporal experience.

But there are serious problems facing even more subtle retentionalist analyses. For one thing, it is far from clear that the notion of a durationless experience even makes sense. What would it be, for example, to have a strictly durationless experience of a sound (Dainton 2010, p. 106)? For another, as Dainton argues (2010, pp. 111-12), there are at least three difficulties facing attempts to account for the appearance of duration and continuity in terms of successions of durationless experiences. First, if the retentionalist analysis were correct, we would expect consciousness to be "clogged" or "choked" (to borrow some language from Dainton) with fading traces of recent experiences. Yet it is not. Second, the fadedness or decreased intensity of purported traces of immediately previous experiences simply does not suffice to account for why they seem past while a current experience does not. To borrow an example from Dainton, if we consider a visual image some parts of which are faded, the faded parts do not seem any less present or more past than the more vivid parts of the image do. Third, as Dainton puts it, the retentionalist analysis "atomizes" experience and cannot account for the "phenomenal binding" of the succession of instantaneous experiences it posits into an overall experience that appears continuous and durational. The problem is analogous to the one facing attempts to analyze a continuous physical object as a collection of unextended parts.

As Dainton writes, "it is one thing to have a succession of experiences, another to have an experience *of* succession" (2010, p. 107). Now, if retentionalism has difficulty accounting for the latter in terms of the former, so too will extensionalism *if* it is interpreted in a similarly "atomizing" way. In particular, suppose an extensionalist replaces the retentionalist's notion of durationless experiences with the idea of experiences of some minimum non-zero duration or extension through time – "atoms" of time, as it were (or "chronons," as they are sometimes called). Suppose too, though, that the rest of the retentionalist account is more or less preserved, in that the continuity and duration of experiences longer than a chronon are accounted for in terms of memory of, or fading traces of, immediately preceding experiences. Some of the problems facing the retentionalist account will clearly arise in a similar way for such a version of extensionalism.

The lesson, I would suggest, is that it is a mistake to suppose that experience of the specious present can be analyzed in terms of more fundamental experiences either of zero duration or of some minimal non-zero duration shorter than the specious present itself. The shorter durations exist in the specious present only in something like the way shorter spatial distances exist in a continuum, or the way parts exist in a single continuous physical object. Just as the parts of a continuous object exist in it only potentially rather than actually, so too the shorter durations exist in the specious present only potentially rather than actually. What actually is experienced is the specious present itself, not some succession of briefer experiences that somehow add up to the specious present. The correct view of temporal experience, then, is an extensionalist account that takes such experience *really* to be essentially just the way it *seems* to be.

But might not the *directedness* of experience toward the future yet be illusory, even if the continuous or stream-like character of experience is not? Consider a scenario in which we have precognition of future events that is analogous to memory of past events (Newton-Smith 1980, pp. 207-8). Wouldn't time seem to lack direction in such a case? In fact that doesn't follow from the scenario at all. To borrow an analogy from Dainton (2010, p. 412), if you are watching a television show you have seen before and therefore know what you will be seeing over the next few minutes, that doesn't affect the way the experienced present seems to be moving *forward towards* that anticipated outcome. Similarly, having a precognitive knowledge of the future wouldn't by itself suffice to prevent temporal experience from seeming to be directed forward. Nor would it change anything if we imagined that you had no memory of the past but *only* this precognition of the future (Dainton 2010, p. 116). That is to say, even if the status of our *knowledge* of past versus future events were reversed, the *direction of experience toward* the future would not thereby be reversed.

Notice that the situation would still remain unchanged even if time travel were possible (which, as I will argue below, it is not). Suppose you had the experience of stepping into a time machine, followed by an experience of exiting it and encountering Abraham Lincoln. Even if you judged that you had as a matter of objective fact traveled back to 1865, your experiences would the whole time still *seem like* they were directed toward the future. As the machine conveyed you to the nineteenth century, and as you stood there talking to Lincoln, each moment would *appear* to be flowing forward toward something that has not happened yet rather than backward toward something already past. (Again, what this tells us about reality outside experience is not something I am yet addressing. I am for the moment still only talking about the phenomenology of experience itself.)

It is sometimes claimed that evidence from neuroscience and cognitive science casts doubt on the commonsense phenomenological description of experience. For example, James Harrington (2015, pp. 138-44) cites cases of subjects whose judgments about how much time has passed are badly wrong, evidence that the brain integrates into a single experience stimuli that actually occur milliseconds apart, and experiments in which a series of flashing lights is wrongly perceived as the movement of a single light. But the most any of this would show is that experience does not always correspond to objective reality – something we already knew, and which has nothing to do with the perception of *time*, specifically. The evidence Harrington cites in no way shows that the commonsense description of the phenomenology of temporal experience *itself* is mistaken. An experience of a physical object always presents it as something extended and colored. This *phenomenological* point is in no way undermined by the fact that we sometimes hallucinate objects that are not really there, or by the thesis that there is nothing in the objects themselves that corresponds to colors as they appear to us in experience. Similarly, the fact that we sometimes misjudge *how much* time has actually passed in objective reality casts no doubt at all on the phenomenological point that experience itself always seems to involve *some* passage of time. The fact that there can be an objective temporal gap between the stimuli that generate an experience casts no doubt at all on the phenomenological point that an experience always seems to be of something that is present. The fact that what exists in objective reality might be a series of distinct things rather than a single thing changing its position casts no doubt at all on the phenomenological point that an experience always seems to be an experience of change.

Simon Prosser claims to show with his "detector argument" that there could be no such thing as an experience of temporal passage (2016, pp. 33-38). The argument goes as follows. Consider the idea of a physical device we might call a "passage-of-time detector," atop which there is a light that flashes when the device detects the passage of time. Could there actually be such a device? There could not be. For despite the different ways in which they are described, the A-series and the B-series contain the same physical events in the same order. Moreover, despite their disagreement about whether temporal passage is real, the A-theory and the B-theory "share a common fragment, embodied in the mathematical structure of the laws of physics" (p. 34). So, the two theories agree that the same sequence of physical events would occur regardless of the whether the A-theory or the B-theory were correct. But this would include the events involving the detector, which means that the light will either flash or not flash in exactly the same way whether the A-theory or the B-theory is correct. But in that case, the purported detector would not really be capable of detecting temporal passage after all, since it would still flash even if there were no temporal passage (as there is not according to the B-theory). So, in fact there could be no such thing as a "passage-of-time detector." Now, experience, Prosser holds, is no less physical than the purported detector would be. Hence experience could not really detect the passage of time any more than the physical device in question could.

Obviously, this argument presupposes the materialist view of the mind – which, however popular these days, is in my view demonstrably false (Feser 2013a). But even if we put that point aside, the argument still fails for a more fundamental reason. It presupposes that to detect temporal passage would be to detect some *extra* ingredient in nature, *over and above* the series of physical events. But that is precisely what the Aristotelian view denies, since it regards the series of events as a series of changes, and time as the measure of these changes with respect to their succession. Hence to detect change just is to detect temporal passage. Whereas Prosser says that there could not be a "passage-of-time detector," the truth is actually that *everything* that changes is *already* a "passage-of-time detector." You don't need to look for a flash of light in some special machine to know that time is passing. You can simply take note of *any* change that occurs in *anything* in nature. Of course, Prosser would reject the Aristotelian position, but the point is that the detector argument simply begs the question against it. Prosser would beg the question in another way if he were to try to get around this problem by denying that change as the Aristotelian understands it (viz., the actualization of potential) exists in nature, on the grounds that the "mathematical structure of the laws of physics" recognized by A-theory and B-theory alike makes no reference to it. For as I have argued in earlier chapters, the very methods of physics prevent it from capturing everything that there is to physical reality. Hence the fact that some feature does not appear in the *description* of nature provided by physics does not by itself show that it isn't in nature *itself*.

Prosser (2016, pp. 42-51) offers a second argument, labeled the "multi-detector argument," to buttress the first, but it is no better. Once again we are asked to consider the idea of a physical device which detects temporal passage, only in this case the device also has several other lights the flashing of which would detect other features of nature. For example, one light might indicate the presence of a certain color and another the presence of a certain temperature, in addition to the one indicating that time has passed. Now, it is easy to see how such a device might indicate temperature, for example. We can imagine that it contains a tube of mercury which expands when the temperature reaches a certain point, causing a circuit-breaker to trip which in turn results in a certain light flashing. By contrast, Prosser says, there could be no causal mechanism by which temporal passage would cause some *particular* light to flash. For any causal relation holding between temporal passage and one particular light would also hold between temporal passage and every other light. So, there could be no such thing as a "multi-detector" that would detect temporal passage. But experience is relevantly like such a "multi-detector" in that it too detects multiple features of nature. So, experience cannot detect temporal passage either.

The trouble with this argument is not its premise, for as I have just said, *every* event that occurs in nature involves temporal passage, so that Prosser is correct to hold that any causal relation that temporal passage bears to one light is one that it will also bear to every other. The trouble is rather that Prosser's conclusion does not follow. It's not that *no* light would detect temporal passage, but rather that *every* light would. To see the fallacy Prosser is committing, consider the idea of a "multi-detector" that would indicate that a physical object has the attribute of extension in space. Now, any causal relation that extension bears to one particular light is a relation that extension will also bear to every other. For whether an object is red or green, hot or cold, or distinguished by any other feature, it is going to be extended. Hence, the flashing of any light will always indicate the presence of an extended object just as much as it indicates the presence of a red object, a hot object, or what have you. So, there can be no *particular* light that indicates the presence of extension. Should we conclude that extension cannot be detected by any experience?

Of course not. On the contrary, *every* experience of a physical object involves the detection of extension, because (unlike redness and hotness, say) extension is a universal feature of physical objects. Now, by the same token, temporal passage is a universal feature of events, so that it will be detected in *every* case (rather than in no case) in which an event is detected. Prosser would of course disagree with this, but once again the point is that his argument gives no *non-question-begging reason* for disagreeing with it.

Some further arguments from Prosser face similar problems. For example, he claims that the passage of time cannot even be *mentally represented*, much less experienced, so that the A-theory is unintelligible (2016, pp. 51-58). His basis for this claim is that no purported mental representation of temporal passage could be accounted for in terms of any of the currently popular materialist theories of mental representation, such as causal theories and informational theories. For example, any causal relation that temporal passage might bear to one mental representation is a causal relation it would bear to any other, so that there could be no way to account, in causal terms, for why a particular mental representation represents temporal passage, specifically.

Once again, Prosser is making the tendentious assumption that materialism is true, but here too the argument fails even if we put that objection aside. For one thing, if applied consistently, Prosser's reasoning would have implications he would not welcome. Consider, for example, that what Prosser says about temporal passage is also true of other features, such as being located in space. In particular, any causal relation that space bears to any one mental representation is a causal relation it would bear to any other. Hence, by Prosser's reasoning, we would have to conclude that space cannot be mentally represented and is therefore unintelligible. Consider also that theories of mental representation like the ones deployed by Prosser face notorious *indeterminacy* problems (Feser 2011b and 2013a). That is to say, the causal relations, informational roles, etc. to which such theories appeal never suffice to determine that a given mental representation has one specific content to the exclusion of others. To use an example made famous by Quine (1960), they can never determine that the content of a mental representation is the proposition that *a rabbit is present* as opposed to the proposition that *an undetached rabbit part is present* or the proposition that *a temporal stage of a rabbit is present*. Hence, if the failure of causal relations and the like to determine content is a reason for us to deny that we can mentally represent *temporal passage*, it would also be a reason to deny that we can mentally represent *anything*. But since Prosser does not draw that more extreme latter conclusion, he cannot consistently draw the former conclusion.

Then there is the problem that Prosser's argument, like the theories of meaning put forward by Hume and by logical positivists, is simply dogmatic and question-begging. The A-theorist will, of course, insist that we have the concept of temporal passage and that Prosser himself could not so much as disagree with the A-theorist unless he too had it. Hence, if some theory of mental representation cannot account for our having that concept, the problem is with the theory and not with the concept. Prosser would deny this, but once again, the problem is that he has given no non-question-begging reason for denying it.

Since Prosser affirms that "time passes if and only if change is dynamic" (2016, p. 165), he thinks that if his arguments show that temporal passage cannot be experienced and is indeed unintelligible, so too do they show thereby that change in the ordinary sense cannot be experienced and is unintelligible. The banana of our earlier example is green at one time and yellow at a later time, but for Prosser, as for B-theorists in general, all this means is that the banana's being green and its being yellow occupy different positions in the B-series. It "changes" from green to yellow only in the *static* way that a road "changes" from being smooth at one segment to being bumpy at another. Nothing changes in the *dynamic* sense of going from potential to actual, or so Prosser thinks he has shown. But since his arguments against temporal passage fail, so too does his case against dynamic change.

Robin Le Poidevin argues that the claim that we experience such change conflicts with the claim that what we experience we experience as present (2007, p. 87). For suppose there is some change from A to B and I experience B as present. Then A must be past. But if A is past, then I cannot experience the change from A to B, but only the outcome of the change. This would follow, however, only if the experienced present is *durationless*, and as I argued above, it is not (as Le Poidevin himself concedes earlier, at p. 80). So long as the experienced present has at least a very brief duration, both A and B can be included within it.

In any event, I have already argued, in chapter 2, that any attempt to deny that change occurs at least *within* conscious experience is ultimately incoherent. When we add to this the thesis that time is the measure of change with respect to succession, it follows that any attempt to deny that *time* exists at least within conscious experience is also incoherent. Taking these considerations together with those adduced in this section, we have a rebuttal to the charge that temporal experience is other than it appears to be – durational, *of* change, continuous or stream-like, and directed forward. Common sense gets the phenomenology right. What can we conclude from this?

Huw Price (1996, pp. 14-15) seems to think that nothing of metaphysical interest follows from it. The phenomenology tells us only how the world *seems*, not how it really *is*, and the way it seems would be the same whether the A-theory or the B-theory is true. Others compare the features of temporal experience to such secondary qualities as color or sound, which (as the mechanical world picture holds) the mind *projects onto* the external world but which are not really there that world (Bardon 2013, pp. 102-10; Le Poidevin 2007, pp. 93-96).

One problem with this view, as Richard Gale points out (1968, p. 228), is that A-determinations are not relevantly analogous to secondary qualities. With a secondary quality like redness or sweetness, there is a specific quale associated with it, a rough correlation between the having of that quale and certain specific neural processes, a physical feature of external objects that also correlates with the quale, and a causal link between that physical feature and the neural processes correlated with the quale. There are also aberrant cases where the external physical feature fails to trigger the having of the quale in question, or where the quale is triggered by something other than the physical feature. As our discussion of Prosser indicates, no such detailed causal story exists in the case of features like temporal passage. Hence the standard considerations appealed to as reason to treat colors, tastes, sounds, and the like as secondary qualities are not available in the case of temporal features.

Le Poidevin responds to Gale by proposing moral values as a better analogy for temporal passage than color, taste, or other secondary qualities. Unlike colors, etc., moral values are not sensible properties, but we can explain how the mind projects them onto the world in terms of "causal considerations, such as the emotional effect that certain natural properties of events have on us" (2007, p. 96). Temporal passage can be treated as a projection in a similar way. But this analogy too won't work. For one thing, it assumes that goodness and badness are entirely subjective and not features of the objective world, and I would argue that that assumption is false (Feser 2014a). But the analogy would fail even if they were subjective. For here too we can identify specific neural processes correlated with the "emotional effects" Le Poidevin refers to, and specific "natural properties" of external events with which they are causally correlated. Again, there is nothing like this kind of causal story available in the case of temporal passage.

A deeper problem with the claim that the phenomenology has no metaphysical significance, however, is that even if the temporal features did not exist in the world external to experience, their ineliminability from experience itself shows that they exist at least *there*. That suffices to refute the B-theory, since the B-theory denies that there is any dynamic change or temporal passage *anywhere*. Price is simply wrong, then, to claim that the phenomenology would be the same whether the A-theory or the B-theory is true. Since the phenomenology itself manifests dynamic change and temporal passage, it would not exist were the B-theory true. Certainly Price gives no non-question-begging for supposing otherwise.

The B-theorist might respond by alleging that this kind of argument commits a fallacy. To treat features like redness and sweetness as secondary qualities is to say that physical objects are not really red or sweet in the ordinary sense, but are merely *represented as* being red or sweet by our sense-data. But it doesn't follow that sense-data themselves are red or sweet. Similarly, the B-theorist maintains, the external physical world does not really exhibit dynamic change or temporal passage, but is merely *represented as* exhibiting these features by the mind. But it doesn't follow that the mind itself has these features. (Cf. Le Poidevin 2007, p. 83.) But no such fallacy is being committed, because this secondary quality model is, again, not the *A-theorist's* way of understanding the situation in the first place. Nor is he committed to any dubious premise to the effect that representations themselves have the properties that they represent the thing represented as having. Rather, the A-theorist simply notes that we know from introspection that the mind exhibits dynamic change and temporal passage – as we go from one thought or experience to the next – and that this would remain true even if it turned out that dynamic change and temporal passage did not exist in objects and events *outside* the mind.

As Craig argues, there really is no coherent way to deny that dynamic change and temporal passage exist at least within the mind itself (2001, pp. 199-200). You read the beginning of this sentence, and then you read the end of it. Insofar as the first experience gives way to the second, your mind appears to undergo change and time thus seems to pass. The

B-theorist says this is an illusion, but the purported illusion would *itself* be an instance of the very phenomenon claimed to be illusory. For it is only because one and the same thing, namely you, first has the one experience and then the other that you are inclined to judge that change has occurred and time has passed, and can then go on to wonder whether this was illusory. Indeed, the sequence consisting of the thought that change really occurred, then the thought that perhaps this was merely illusory, is itself an instance of change and thus of temporal passage. Were there no *loss* of the one thought or experience and *gaining* of the other, there would be nothing to generate even the illusion of change. Were there no *persistence* of you, the subject of these lost and gained attributes, there would be no one mind which entertains this sequence of thoughts. The very spelling out of a scenario in which change and temporal passage are illusory surreptitiously smuggles them back in again. (Recall that in chapter 2 I argued that four-dimensionalism or temporal parts theory fails to provide a plausible alternative analysis of psychological changes of this sort.)

Nor, once we recognize that dynamic change and temporal passage cannot be eliminated from the mind, is it easy to confine them there. Recall what was said earlier in this chapter about the necessity of thinking in terms of physical objects possessing causal properties and occupying a spatially extended world if we are to make sense of the natural world as mind-independent. Dynamic change and temporal passage, as it turns out, are no less necessary. As Kant famously argued in his *Critique of Pure Reason*, they are prerequisites to our experiences of the external world being intelligible. Hence, suppose I take what I am perceiving to be a dog. Why does it appear to have only two legs? Where is its tail? Why is it not barking? How can it be the same thing as the animal I experienced a moment ago that had four legs and a tail and was barking? Part of the answer is that I take myself to have changed my position in space with respect to the dog. Before I had been looking at it from the side and could see all four legs and the tail, and now I am standing in front of it and cannot see the back legs or tail. Interpreting the sequence of experiences as involving this change in perspective is part of what makes it possible to treat the thing I am perceiving as one and the same object rather than two objects (a four-legged one with a tail one and a two-legged one without a tail). But there is also the fact that I take it that I *had been* perceiving the dog from one point of view and *am now* perceiving it from another, and that the dog *had been* barking and *is now* silent. Furthermore, I take myself to be potentially moving further along still such that I *will be* perceiving the dog from yet another angle if this potential is actualized, and I take the dog to be potentially barking again such that it *will be* doing so in another moment or so if that potential is actualized. Thus am I able both to relate previous experiences to current ones and predict experiences yet to come. Without this conceptual framework which posits persisting physical objects, including myself, causally interacting with one another and also *having been* certain ways and *yet to be* yet other ways, my experiences would lack the coherence they have. The static B-theoretic framework would not suffice, because it cannot account for why certain things and events *are available* to experience whereas others *are not* available and yet others *may yet be* available. Only a framework which distinguishes *what is present* from *what is past* and *what is future* can do that. (Cf. Bardon 2013, pp. 31-38, 104-7.)

Of course, Kant took this conceptual framework to reflect only the way we have to *think about* the world, and not the way the world actually *is* in itself. But there is nothing in the analysis that requires us to adopt this anti-realist interpretation. On the contrary, when we factor in the defense of realism developed in chapter 3, we have a powerful argument for the conclusion that temporal passage is a real feature of the external world no less than of the mind. For the notion of temporal passage to be so extremely useful to organizing experience, and so extremely difficult to eliminate, would be a miracle if it were *not* a real feature.

4.3.3 Aristotle versus Einstein?

4.3.3.1 Making a metaphysics of method

Again, though, it might seem that dynamic change and temporal passage have been banished from nature by Einstein, and in particular by the Minkowski space-time interpretation of the Special Theory of Relativity (STR). Michael Lockwood sums up a common view:

> To take the space-time view seriously is indeed to regard everything that ever exists, or ever happens, at any time or place, as being just as real as the contents of the here and now. And this rules out any conception of free will that pictures human agents, through their choices, as selectively conferring actuality on what are initially only potentialities. Contrary to this common-sense conception, the world according to Minkowski is, at all times and places, actuality through and through: a four-dimensional *block universe*. (Lockwood 2005, pp. 68-69)

Leave aside the question of free will, which is not our concern here. What is relevant is that Lockwood's remarks suggest that there is, just as the Aristotelian holds, an essential connection between time, change, and the actualization of potential, so that to deny one is to deny the others. The idea is that the block universe concept rules out the reality of temporal passage. But temporal passage follows upon change, so that if there is no temporal passage there can be no change either. Change, however (Lockwood's implicit argument continues), is just the actualization of potentiality, so that if there is no change, neither is there any actualization of potentiality. Hence on the Minkowskian interpretation of STR, there is in the natural order no actualization of potentiality; everything in the world, whether "past," "present," or "future," is all "already" actual, as it were. Thus, as Karl Popper (1998) noted, does Einstein recapitulate Parmenides. Physicist Julian Barbour, who also sees in modern physics a return to Parmenides (1999, p. 1), boldly proclaims "the end of time." James Harrington peremptorily declares that "there's simply no way to make any version of A-theory compatible with Einstein's theory of relativity" (2015, p. 93). (Cf. Putnam 1967 and Saunders 2002.)

Yet any argument to the effect that *physics per se* refutes the A-theory would be fallacious. As Craig points out:

> A physical theory is comprised of two components: a mathematical formalism and a physical interpretation of that formalism. Competing theories which differ only in virtue of their divergent physical interpretations can be extremely difficult to assess if they are empirically equivalent in their testable predictions. Considerations which are metaphysical in nature may then become paramount.
>
> The Special Theory of Relativity... provides a case in point... The empirical success of [STR's] testable predictions can... be misleading, dulling us to the truly controversial nature of the correct physical interpretation of the theory's formalism. (2008, p. 11)

Philosopher of physics Lawrence Sklar is perhaps even more dismissive of attempts to draw philosophical conclusions from scientific premises:

> Just as a computer is only as good as its programmer ("Garbage in, garbage out"), one can extract only so much metaphysics from a physical theory as one puts in. While our total world view must, of course, be consistent with our best available scientific theories, it is a great mistake to read off a metaphysics superficially

> from the theory's overt appearance, and an even graver mistake to neglect the fact that metaphysical presuppositions have gone into the formulation of the theory, as it is usually framed, in the first place. (1985b, pp. 291-2. Cf. Markosian 2004)

That you have to read metaphysical assumptions into STR before you can read them out again is only half the problem, though. The other half is that the metaphysical assumptions you have to read into it are bad ones. One indication of this is that to make of STR a premise in an argument for the unreality of dynamic change and temporal passage results in an incoherent position. The reason, as Richard Healey has pointed out (2002, pp. 299-300), is that a physical theory like STR is grounded in empirical evidence in a way that presupposes the reality of dynamic change and temporal passage. The scientist must formulate a prediction, set up an experimental test, and observe the result. He will thereby transition from a mental state of not knowing the outcome of the experiment to a mental state of knowing it. The part of the physical world he is observing and manipulating will also transition from being in one state to being in another. These sequences of thoughts, experiences, actions, and states of the external physical world will all involve change and thus the passage of time. Hence if, in the name of STR, someone denies that any of this is real, then he will implicitly be denying that the empirical evidence for STR itself is real – in which case he will, in the very act of denying the reality of change and temporal passage, be undermining the rational justification for that denial. (Of course, the denier will no doubt claim that he can reformulate our description of the observational and experimental situation so as to make it consistent with this denial – in terms, say, of four-dimensionalism or temporal parts theory – but again, I have already shown in chapter 2 that no such move can succeed.)

So, if STR is true, then any metaphysical premise which, when conjoined to STR, would yield this incoherent result must be false, or at least rationally unjustifiable. As it happens, though, we already have independent grounds to judge the relevant metaphysical premises to be false. The crux of the alleged conflict between STR and the A-theory is the *relativity of simultaneity*. Einstein famously shows that observers in different frames of reference moving relative to one another will disagree about whether two events are simultaneous, and thus about whether they are each occurring *now*. Obviously this has *epistemological* significance. But why suppose that it has *metaphysical* significance? In particular, why conclude that there is *no objective fact of the matter* about whether two events are simultaneous (as opposed to concluding merely that while there *is* an objective fact of the matter, this fact cannot be detected empirically)?

The answer is that the metaphysical conclusion will follow if we make the verificationist assumption that there *are no* objective facts that transcend what can be empirically detected. Now, the influence of verificationism on Einstein's formulation of STR is well known (Brown 1991, chapter 5; Craig 2001b, chapter 7). And though Einstein later abandoned verificationism, it doesn't follow that a metaphysical interpretation of the relativity of simultaneity can survive that abandonment. Sklar writes:

> Certainly the original arguments in favor of the relativistic viewpoint are rife with verificationist presuppositions about meaning, etc. And despite Einstein's later disavowal of the verificationist point of view, no one to my knowledge has provided an adequate account of the foundations of relativity which isn't verificationist in essence. (1985b, p. 303)

The trouble, of course, is that verificationism is false, as we saw in chapter 3. Thus, if STR shows that there is no fact of the matter about simultaneity only given verificationism, then STR does not show that at all. As Yehiel Cohen notes of the problematic relationship between STR and verificationism, "unfortunately, in the voluminous literature on special relativity... this fact has been often neglected' (2016, p. 51). Sklar rightly judges that "acceptance of relativity cannot force one into the acceptance or rejection of any of the traditional metaphysical views about the reality of past and future" (1985b, p. 302).

But there is yet another metaphysical error underlying the grand claims often made about the philosophical significance of relativity. In 1962, philosopher Max Black wrote:

> But this picture of a 'block universe', composed of a timeless web of 'world-lines' in a four-dimensional space, however strongly suggested by the theory of relativity, is *a piece of gratuitous metaphysics.* Since the concept of change, of something happening, is an inseparable component of the common-sense concept of time and a necessary component of the scientist's view of reality, it is quite out of the question that theoretical physics should require us to hold the Eleatic view that nothing happens in 'the objective world'. Here, as so often in the philosophy of science, *a useful limitation in the form of representation* is mistaken for a deficiency of the universe. (1962, pp. 181-2. Emphasis added.)

Black's remark that change is "a necessary component of the scientist's view of reality" is presumably an expression of the point I made above and in chapter 2 that all cognitive and perceptual processes – including those of scientists themselves as they make observations, develop theories, carry out experiments, etc. – presuppose the reality of dynamic change and temporal passage. But it is the other point Black makes that I want to call attention to here. To infer from relativity that the universe is a static four-dimensional block is, he says, fallaciously to make a "gratuitous metaphysics" out of what is really nothing more than "a useful limitation in the [theory's] form of representation." For the absence of dynamic change and temporal passage from relativity's description of the universe is merely an artifact of the representational *methods* of physics, and in no way necessarily reflects the reality studied *by means of* those methods. That is to say, dynamic change and temporal passage could exist in nature even if they don't show up in the physicist's picture of the world, because the methods the physicist uses to paint that picture wouldn't capture them even if they were there.

This is essentially just an application to the specific case of relativity of the point made in chapter 3 about the abstract and idealized nature of physics in general. To return to an analogy from that chapter, despite its usefulness in helping you get around a building, a blueprint is neither an exhaustive nor an entirely accurate representation. It leaves out much that is really there in the building (the furniture, the color of the walls, and so forth) and adds elements that are not really there (blue lines, graphic symbols, simplifications, and so on). Similarly, when an aircraft engineer ignores every characteristic of passengers except their average height and weight, he leaves out much that is really true of them (such as their meal and entertainment preferences) and may introduce features that are not true of them (since there may be no actual individual passenger whose weight or height exactly matches the average). By the same token, despite their predictive success and technological utility, the abstract and idealized mathematical models developed by the physicist are bound to exclude features that exist in nature and to add features that do not exist there. As Jeffrey Koperski writes:

> Continuum mechanics, for example, treats matter as if it were smoothed out and continuous across a region rather than atomic. Aerodynamics treats the airflow over a wing the same way, and these are perfectly good idealizations for the scale at which we normally deal with materials, especially fluids and gases.

> Spacetime theorists make this same move by ignoring [the] mid-scale structure [of the universe]. (2015, p. 136)

As Raymond Tallis points out (2017, p. 30), the method of modern physics when dealing with time has been, first, to treat it as "a quasi-spatial dimension," and then to treat space in turn in terms of "pure quantity." (Recall Strawson's description of physics' conception of space as an "abstract metric" or "abstract dimensionality.") The end result is that time is "shriveled to a number, a quantitative variable signified by a letter – 't' – conceived as a one-dimensional topological space that maps on to a line of real numbers" (Tallis 2017, p. 30). Similarly, John Bigelow has noted the role that mathematics and formal logic have played in modern physics in suggesting the picture of a timeless and changeless world, even apart from relativity (2013, pp. 155-58). The Newtonian analysis of motion represents the speed of a body in terms of the slope of a curve on a graph which measures time on one axis and space on another. Different times and different places are thereby represented in the same way, viz. as points on the graph, all of which exist at once, as it were. Then there is the use of calculus to characterize speed in terms of the limits of infinite sequences. To formulate statements about such sequences in modern predicate logic requires quantifying over past and future bodies and events no less than present ones. Again, this suggests a picture on which all times and places exist at once, as timeless and changeless Platonic mathematical objects. The picture, though, is a byproduct of the use of formal methods in representing nature, just as the color of a blueprint reflects merely the method by which it is drawn up. It no more entails that the world really is timeless and changeless than a blueprint entails that the building it represents is really blue.

Physicist Lee Smolin is another thinker who has emphasized that a timeless picture of nature is a byproduct of the physicist's mathematical manner of representing reality, noting that "the process of recording a *motion*, which takes place in time, results in a *record*, which is frozen in time – a record that can be represented by a curve in a graph, which is also frozen in time" (2013, p. 34). Noting that the concept of four-dimensional spacetime is the result of a thoroughgoing application of this method of representing nature, Smolin warns against too quickly drawing metaphysical conclusions from the successes of the method:

> Some philosophers and physicists see this [method] as a profound insight into the nature of reality. Some argue to the contrary – that mathematics is only a tool, whose usefulness does not require us to see the world as essentially mathematical...
>
> [They] will insist that the mathematical representation of a motion as a curve does not imply that the motion is in any way identical to the representation. The very fact that the motion takes place in time whereas its mathematical representation is timeless means that they aren't the same thing...
>
> By succumbing to the temptation to conflate the representation with the reality and identify the graph of the records of the motion with the motion itself, [some] scientists have taken a big step toward the expulsion of time from our conception of nature.
>
> The confusion worsens when we represent time as an axis on a graph... This can be called spatializing time.
>
> And the mathematical conjunction of the representations of space and time, with each having its own axis, can be called *spacetime*... If we confuse spacetime with reality, we are committing a fallacy, which can be called the fallacy of the spatialization of time. It is a consequence of forgetting the distinction between recording motion in time and time itself. (2013, pp. 34-35)

His label for it notwithstanding, the error Smolin identifies here is not that of reducing time to a kind of spatial dimension – I think that *is* an error, but that is a separate issue to be addressed below – but rather the fallacious inference involved in supposing that the usefulness of the mathematical representation *all by itself justifies* such a metaphysical reduction. That conclusion simply does not follow, any more than the parallel conclusion follows in the case of the blueprint. Or, to take another example, from philosopher Craig Bourne:

> [J]ust because something is represented spatially, we cannot draw the conclusion that it is a spatial dimension or that it is in anyway [sic] analogous to a spatial dimension. For consider... a three-dimensional colour space which illustrates the possible ways in which things can match in colour... [I]t would be misconceived to draw the conclusion that brightness, hue, and saturation were each spatial dimensions, just because they were repre-

> sented spatially. And to go on to conclude that each of these dimensions must be alike just because they comprise the different dimensions of colour space would be equally fallacious, since they're not. We should, then, be equally wary of drawing conclusions from Minkowski space-time diagrams. (2006, p. 158. Bourne attributes the point to Mellor (2005).)

Lowe makes a similar point:

> Physicists often represent spatiotemporal relations graphically by means of (two-dimensional) space-time diagrams, in which one axis represents time, *t*, and the other represents the three dimensions of space, *s*... But it is equally common to use such two-dimensional graphical representations to convey information about relations between, for instance, the pressure and temperature of a gas – and no one imagines that pressure and temperature are literally dimensions of reality in which physical things are extended. So the mathematical representation of time implies nothing, in itself, about the similarity or lack of it between time and the three dimensions of space. (2002, p. 253)

Smolin is not the only prominent physicist to recognize the problem. George Ellis has urged his fellow physicists to "consider... what lie[s] outside the limits of mathematically based efforts to encapsulate aspects of the nature of what exists" and emphasizes that "mathematical equations only represent part of reality, and should not be confused with reality" (Horgan 2014). Nor is the point a new one. It was often made during the first half or so of the twentieth century by thinkers of diverse interests and theoretical commitments. As I noted in chapter 3, process philosopher Alfred North Whitehead characterized the tendency to confuse an abstract mathematical representation of reality with the concrete reality represented as "The Fallacy of Misplaced Concreteness" (1967, pp. 51 and 58), and the error identified by Smolin is a special case of this fallacy. Philosopher of physics Milic Capek criticized what he called "the fallacy of spatialization," and as he notes, scientists Paul Langevin and Emile Meyerson were similarly critical of it (1961, pp. 158-65). Phenomenologist Edmund Husserl (1970) emphasized that the "mathematization" of nature results in an "idealization" which does not capture the whole of reality. Henri Bergson (1998) complained that the mathematician's conception of nature leads us wrongly to think of time as a series of frozen moments, like the still photographs that make up a film strip. (Cf. Canales 2015.)

Historian of science E. A. Burtt lamented the tendency of the modern scientist to "make a metaphysics out of his method" (1952, p. 229). As we have seen, analytic philosopher Max Black made a similar point. Unsurprisingly, Aristotelian-Thomistic philosophers raised such criticisms as well (Bittle 1941, pp. 450-1; Koren 1962, pp. 129-30; Mullahy 1946). Jacques Maritain (1995) argued that mathematics captures only one of three "degrees of abstraction" from concrete physical reality, and other Aristotelian-Thomistic philosophers endorsed Whitehead's "Fallacy of Misplaced Concreteness" objection (Van Melsen 1954, pp. 83-84; Weisheipl 1955, p. 109). But as these various examples illustrate, one hardly need have an Aristotelian ax to grind to suppose that there might be more to physical reality than is captured in the mathematical representation developed by modern physics.

Indeed, the whole point of the epistemic structural realist position developed by thinkers like Russell and defended in chapter 3 is that physics actually tells us relatively little about the nature of physical reality. Hence, the absence from physical theory of tensed notions, or of any other metaphysically significant notions for that matter, *by itself* tells us nothing about whether they have any application to physical reality. What we've seen Strawson argue with respect to space and McGinn argue with respect to motion is true also with respect to time: Physics simply doesn't capture all there is to it. So, the argument from STR against the A-theory is fallacious.

4.3.3.2 Relativity and the A-theory

The skeptical reader will, of course, nevertheless want to know exactly *how* the reality of dynamic change and temporal passage fits into the picture of the world afforded by relativity, given that there at least *seems* to be a conflict. That is a perfectly reasonable question, but it is important to emphasize that answering it is first and foremost a problem *for the physicist*, not for Aristotelians and other defenders of the A-theory. The scientist who pretends that it is the *latter* who have the primary burden here is like the party guest who trashes the house and then demands of the host: "So how do you propose to clean this mess up?" For as we have seen, the physicist himself no less than anyone else ultimately has to affirm the reality of dynamic change and temporal passage. If he puts forward a theory that at least appears to deny them, then, he is the one who has some explaining to do. The predictive and technological successes of relativity are undeniable, but that must not blind us to the fact that *metaphysically* it is something of a mess, and that the physicists are the ones who made the mess.

On the other hand, a metaphysician is better placed to clean up a metaphysical mess, whoever made it. And there is a sizable philosophical literature on the issue of how relativity might be reconciled with the A-theory. Now, presentism is the version of the A-theory which holds that the present alone exists, and thus that past things and events no longer exist and future things and events do not yet exist. This is the version of the A-theory which might seem most obviously at odds with relativity. For if, as STR holds, simultaneity is relative to frames of reference and there is no privileged frame, then whether or not a moment is present would be relative to a frame of reference, and there would be no frame by reference to which we could define an absolute present. Moreover, if the universe is a four-dimensional block, then all things and events (past and future no less than present) would seem to be equally real. So, if presentism can for all that nevertheless be reconciled with relativity, then it seems any A-theory could be. Let's begin with presentism, then.

There are essentially four general approaches to reconciling presentism with relativity. They differ in what they take the physics of relativity to tell us about objective reality. The first and most obvious approach to reconciliation would be to back away from even a structuralist brand of realism and hold that the physics of relativity does not really tell us *anything* about objective physical reality in the first place, but should be given a purely instrumentalist or other anti-realist interpretation. (Cf. Parker 1970, pp. 162-71.) If a model of the universe on which there is no absolute present is merely a useful fiction, then naturally there is no incompatibility with the metaphysical claim that there nevertheless is in reality an absolute present moment. Nor can anti-realism be easily dismissed in this context given relativity's historical and conceptual connections with verificationism, which is itself a kind of anti-realism.

One way to develop an anti-realist approach would be along the lines of Arthur Prior's (1970, 1996) suggestion that relativity has merely *epistemological* significance, and in particular shows only that we cannot *know*, of some events, whether they are absolutely simultaneous with one another, but not that there is no fact of the matter about whether they are. Peter Hoenen (1958) points out that a verificationist inference is obviously invalid when we know independently that the impossibility of verifying a claim about some phenomenon derives not from the nature of the phenomenon itself but rather from the nature of our epistemic access to it. For example, if we cannot observe the far side of some distant planet, we know that this in no way shows that the planet lacks a far side. For the reason we cannot observe it has to do with the fact that light from the far side cannot reach us given our position with respect to it. Similarly, the reason we cannot detect the simultaneity of distant events has to do with the physical laws governing the media by which we know the events (such as light signals) rather than the nature of the events themselves. Hence it would be fallacious to conclude, from this epistemic circumstance, that there is no fact of the matter about whether the events are simultaneous.

Another way to develop an anti-realist approach is proposed by David Woodruff (2011), who suggests that relativity seems inconsistent with presentism only if we think of Minkowski's four-dimensional space-time manifold as a concrete *substance* in its own right. So understood, its future and past components naturally seem no less existent than its present ones. But we can deny the existence of any such substance and hold instead that what exist are only present objects and events of the more ordinary sort. The space-time manifold can then be thought of as a "geometric representation" of "what will happen to accelerated bodies and how it will affect measurements of time and space" as well as "what causal interactions are possible" (2011, p. 118). Since this makes of the manifold a tool for making predictions about the behavior of things, Woodruff allows that his position might be regarded as a kind of instrumentalism, though with an important qualification to be noted presently. (Unlike other presentists, Woodruff does not think it essential to hold on to absolute simultaneity, though that aspect of his position – which seems to me implausible – appears to be detachable from the thesis I'm summarizing here.)

An anti-realist solution has the merit of being simple and straightforward, though of course it inherits the usual difficulties with anti-realism. On the other hand, even if anti-realism is problematic as a completely *general* approach to interpreting physics, it doesn't follow that there aren't *specific* physical theories best given an anti-realist interpretation. And given the metaphysical oddities often claimed to follow from relativity, it is about as good a candidate as any for a physical theory plausibly susceptible of an anti-realist construal. In any event, if an anti-realist approach is rejected, there are still three remaining, more or less realist ways of attempting a reconciliation. This brings us to the second approach to reconciling presentism and relativity, which would be to affirm that the physics of relativity really does capture objective physical reality, but to maintain that Einstein and Minkowski simply got that physics wrong. This is the approach of William Lane Craig (2001b, 2008), who proposes a neo-Lorentzian relativity theory. Lorentz's theory, which is empirically equivalent to Einstein's, affirms absolute simultaneity and thus allows for a privileged present moment. A famous difficulty with it is that it has to posit an empirically undetectable aether by reference to which to define a privileged frame of reference. But there are, Craig argues, various alternatives to the aether as Lorentz understood it (for example, the cosmic microwave background radiation, or quantum non-locality).

A variation on this approach would be to criticize anti-presentist appeals to STR on the grounds that STR is even from an Einsteinian point of view not strictly correct in the first place, but merely an approximation to the General Theory of Relativity (GTR). (Cf. Zimmerman 2011b, though Zimmerman's own position is best understood as a variation on the fourth approach I will describe presently.) Moreover (this line of argument continues) even GTR is not the last word, but is itself merely an approximation to whatever the correct theory of quantum gravity turns out to be (Monton 2006). Unlike Craig, who holds that Einstein got the physics of relativity wrong, this line of argument holds that Einsteinian relativity is not wrong so much as incomplete, that there is a larger framework of still developing physical theory in the context of which relativity must be interpreted, and that this larger framework might turn out to be more favorable to presentism. (Some of the developments potentially useful to presentists are summarized in Koperski 2015, pp. 129-34.)

Of course, such proposals are highly controversial, and a completed physics may turn out *not* to favor absolute simultaneity any more than STR does. But this brings us to the third approach to reconciling presentism and relativity, which suggests that STR as it stands may in fact already be reconcilable with at least a modified form of presentism. Theodore Sider (2001, pp. 48-52) suggests three ways this idea might be developed (though he does not endorse any of them and is not himself a presentist). They all involve the presentist affirming the existence of a part of the Minkowskian manifold. The first affirms the existence of some point in the manifold together with everything in its past light cone; the second affirms the existence of the point and everything in its future light cone; and the third affirms the existence of the point and everything spacelike separated from it. Because each of these options preserves at least part of the manifold it is to that extent realist. Because each denies other parts, it preserves, to that extent, the presentist idea that not all points of time are equally real. Sider quite rightly notes that the first two options nevertheless depart considerably from presentism as usually understood, since the first allows (to use Sider's examples) that dinosaurs are still part of reality, and the second that future Martian outposts are part of reality. The third option, he suggests, most closely preserves the basic thrust of presentism.

Sider not implausibly objects to these sorts of proposals that they privilege a particular point in space-time in a way that has no justification in relativity physics itself. An even more extreme privileging along these lines is represented by a fourth proposal considered by Sider (2001, pp. 45-47), according to which a certain space-time point – by itself, without its future light cone, past light cone, or spacelike separated points – is all that exists. (This idea is also entertained, without being endorsed, in Sklar 1985b and in Hinchliff 2000.) These proposals also privilege a particular point in a way *presentists* typically would not. For presentism, as usually understood, it is a *class* of points (or things and events) that exist in the present. Fitting this idea into relativity would require positing, within the Minkowskian manifold, a privileged hyperplane of simultaneity relative to some frame of reference, and taking that alone to be real. The problem, Sider argues (2001, pp. 47-48), is that there is nothing in the geometry of Minkowskian space-time to justify taking any hyperplane to be privileged in this way. (Another problematic way of trying to marry presentism to STR as it stands would be to allow that what is present is relative to the observer. The idea is discussed, without being endorsed, by Howard Stein (1968). As Koperski points out (2015, p. 119), this would seem to make of presentness a mere subjective or observer-dependent secondary quality, which is hardly what the presentist intends.)

That brings us to the fourth approach to reconciling presentism and relativity. Suppose nothing in the geometry of space-time either as understood within STR, or even within the correct theory of quantum gravity, reveals any point, region, or privileged slicing of the manifold with which the present might be identified. One could still argue – as I have been arguing, and as Dean Zimmerman does in the context of discussing presentism and relativity (2008, 2011a, 2011b) – that physics, while correctly describing objective physical reality as far as it goes, nevertheless does not provide an *exhaustive* description and needs to be supplemented by metaphysics. Philosophical arguments for presentism, in Zimmerman's view, provide us with independent evidence that there must be a privileged slicing of the manifold described by physics, even if physics itself cannot tell us what it is. Zimmerman takes the Minkowskian manifold to represent "the set of locations at which events could happen," with the present amounting to a "wave of becoming" that moves through the manifold, as it were (2011a, p. 140). The presentist can in his view admit the existence of non-present points in the manifold and simply deny that there are any objects and events that occupy them (Zimmerman 2008, p. 219). The presently occupied points constitute a privileged slicing of the manifold, and while Sider rejects such a slicing as too "scientifically revisionary," Zimmerman responds that his position does not *revise* the physics of relativity – he is not claiming that it is wrong as far as it goes – but simply supplements it.

Another variation of the approach of supplementing physics with metaphysics might be to argue that relativity theory simply isn't using the word "time" in the same sense as the presentist metaphysician is. Quentin Smith draws a distinction between *metaphysical time* and *STR time*, which he develops as follows:

> Something "exists in time" in the broadest sense if temporal predicates are required to describe any of the object's states, including such relational states as the exemplifications of relational properties of *being referred to.* Let us call this broadest sense of time *metaphysical time*... Narrower senses of "exists in time" correlate to less complete descriptions of an object, such as a description that mentions only states involving real (rather than Cambridge) changes in an object or only states involving luminal or subluminal physical relations with other objects. One of these narrower senses of "exists in time" is the sense that "time" has in the STR, and time in this sense may be called *STR time*. (1993, p. 230)

One could then argue that since STR and the presentist metaphysician are not really talking about the same thing, they do not contradict one another. Smith proposes that "the STR shows not that [metaphysical] time is relative but merely that certain light-connectibility relations are relative" (p. 231). (Though as Smith's own views have developed, he has come to defend a variation of the neo-Lorentzian view that Einstein got the physics wrong. See Smith 2008.)

Insofar as Zimmerman's position allows for some kind of reality even to non-occupied points of the Minkowskian manifold, it is plausibly realist. But it is worth noting that even Woodruff's view can, as he notes (2011, pp. 120-4), be read as a kind of realism rather than instrumentalism. For one thing, the idea that the Minkowskian four-dimensional space-time manifold exists as a kind of concrete substance is not, Woodruff argues, actually part of the physics of relativity or strictly implied by the physics. Rather, it is a metaphysical interpretation one may (or may not) wish to give the physics. Hence, to deny its existence as a concrete substance is not to take an anti-realist interpretation of the actual *science*. Second, in taking the manifold to represent what will happen to accelerated bodies, how they can interact causally, etc. he regards relativity as telling us something about *the real features of actual things themselves*, not merely about how reality appears to us or how we represent it. As Koperski suggests (2015, p. 129), we can think of Minkowski space-time as a kind of *phase space* representing the possible states of the evolving universe. And to affirm even that much is to go beyond a purely instrumentalist positon, at least as instrumentalism is usually understood. Accordingly, Zimmerman's position too would arguably remain at least minimally realist even if he discarded the notion of the reality of non-occupied points of space-time.

In summary, then, the four general approaches to reconciling presentism and STR are: (1) Interpret STR in an anti-realist way, so that the question of its consistency with presentism is moot; (2) Interpret STR in a realist way, but argue that its consistency with presentism is still moot insofar the physics is either mistaken or incomplete; (3) Interpret STR in a realist way and concede the correctness and completeness of the physics, but argue that a modified form of presentism corresponds to something already there in the physics; and (4) Interpret STR in a realist way, concede the correctness and completeness of the physics, leave presentism unmodified, and argue that while it doesn't correspond to anything already there in the physics, the physics doesn't rule it out either and can therefore be supplemented with a presentist metaphysics.

But an A-theorist could also depart from presentism and mitigate the apparent conflict with relativity by conceding a kind of reality to things and events other than present ones. One way of doing this is by opting for a "growing block" theory on which past things and events are as real as present ones, with the present constituting the growing edge of a four-dimensional universe (Broad 1923; Tooley 1997; Ellis 2014). "Growing block" theorist Michael Tooley proposes modifying STR by replacing the thesis that the one-way speed of light is the same in all inertial frames with the thesis that the average round-trip speed of light is the same in all frames, and adding postulates that would entail that events stand in relations of absolute simultaneity (1997, Chapter 11). The theory that results is, Tooley argues, superior in three respects to standard versions of

STR. First, spatiotemporal regions are *contingent*, and their existence thus stands in need of an explanation. Standard versions of STR offer no explanation, but Tooley's "growing block" version does insofar as it holds that later regions are *caused by* earlier ones. Second, the standard supposition that the one-way speed of light is the same in all inertial frames lacks experimental support and may be untestable in principle. Tooley's position thus avoids what amounts to a "gratuitous metaphysical assumption" (p. 358).

The third advantage concerns quantum mechanics, which, as I have indicated, other A-theorists have also appealed to in defense of absolute simultaneity. As Tooley notes:

> [T]he Einstein-Podolsky-Rosen thought experiment shows... [that] either particles have determinate states prior to measurement, in which case quantum mechanics does not provide a complete description of physical reality, or else correlated particles must acquire corresponding determinate properties simultaneously, in the absolute sense – or at least without there being an intervening temporal gap – in which case the Special Theory of Relativity does not provide a complete description of the spatiotemporal relations between events. (1997, p. 361)

That much reinforces the point that STR *by itself* does not suffice to show that there is no such thing as absolute simultaneity, since though the EPR thought experiment does not force on us the conclusion that STR is incomplete, it allows for it. But we get a stronger conclusion, Tooley argues, when we factor in Bell's theorem:

> John S. Bell... showed that the quantitative predictions generated by quantum mechanics logically *preclude* there being properties that make it the case, for example, that an electron possesses determinate spins along various possible axes before any measurements are made. So the thrust of the Einstein-Podolsky-Rosen thought experiment is no longer merely that either the Special Theory of Relativity, or else quantum mechanics, is incomplete. It is rather that either the Special Theory of Relativity is incomplete, or quantum mechanics is *false*. (1997, p. 361)

Now, the empirical evidence for quantum mechanics is very strong, so that it is not plausible to opt for taking it to be false. In that case, though, we have to conclude that STR is incomplete and that absolute simultaneity is real after all.

Yet another way to reconcile the A-theory with relativity is to go even further in a realist direction and concede the existence of future events no less than past ones, while maintaining that present events are unique in being illuminated by the "moving spotlight" of the *now*. Though not himself a moving spotlight theorist, Bradford Skow (2015, Chapter 9) has defended the possibility of such a reconciliation. In Skow's view, the most plausible way for the moving spotlight theorist to carry it out is to abandon absolute simultaneity and a single "now," and instead posit multiple "spotlights" each of which "shines" on a different part of spacetime. (Though this would seem open to an objection similar to the one Koperski levels against the view entertained by Stein, alluded to above.)

Now, these various approaches to reconciling relativity and the A-theory are not all mutually exclusive (though of course some of them are). For example, recall that structural realism as a general approach in philosophy of science constitutes a middle ground position between other forms of scientific realism on the one hand, and instrumentalism on the other. In the specific case of STR, then, one could adopt a structural realist position that affirms the mathematical formalism without committing to realism about any particular physical interpretation of the formalism. As Craig writes:

> The fact is that the only version of [STR] which is experimentally verifiable, as Geoffrey Builder points out, "is the theory that the spatial and temporal coordinates of events, measured in any one inertial reference system, are related to the spatial and temporal coordinates of the same events, as measured in any other inertial reference system, by the Lorentz transformations"... But this verifiable statement is underdeterminative with regard to the radically different physical interpretations of the Lorentz transformations given, respectively, by Einstein, Minkowski, and Lorentz. (2008, p. 11. The Builder quote is from Builder 1958.)

Similarly, Bourne argues:

> We can, with Einstein, reject the existence of the aether as redundant, and maintain that light does not need a medium in order to be propagated. Nevertheless, we can still hold on to an absolute frame. The Lorentz transformations are, then, to be regarded more as recipes for relating the measurements made by some inertial observer to the measurements made by another inertial

> observer, given a particular well-defined measurement procedure. This is essentially Einstein's way. It follows that the observable content of the theory remains intact and that there is nothing in terms of the physics of the situation which will tell for or against a privileged frame. (2006, p. 178)

(To be sure, Craig and Bourne are not talking about structural realism. The point is just that their observations about the distinction between the Lorentz transformations and alternative physical interpretations of the Lorentz transformations are grist for the structural realist's mill.)

Furthermore, some of the arguments summarized above (such as those that appeal to quantum mechanics in defense of absolute simultaneity) are open to any version of the A-theory. There are various alternative ways, then, that insights from the approaches we've been examining may be combined, and thus even more potential approaches to spelling out an A-theoretic interpretation of relativity than has been indicated already.

Of course, much more would have to be said in order adequately to defend any particular approach to reconciling relativity and the A-theory. Nor, as I have indicated, are all of them equally plausible. The point for the moment, however, is not to endorse any particular approach, but simply to note that a wide variety of approaches *has* indeed been developed in contemporary philosophy – and, moreover, developed *entirely independently of any concern with upholding an Aristotelian philosophy of nature*. They are already there "on the shelf," as it were, awaiting deployment by the Aristotelian. In no way, then, would a reconciliation of Aristotle and Einstein be *ad hoc*.

As I noted above, even a static four-dimensional block conception of the universe wouldn't *entirely* undermine the Aristotelian theory of actuality and potentiality, since the block itself qua contingent would be merely potential and thus require actualization from outside it. (Cf. Feser 2017, p. 50) That at least the core of Aristotelian philosophy of nature would survive even this worst case scenario reinforces the lesson of this section, viz. how little the physics of relativity by itself actually tells us about metaphysics and how wide open is the range of possible interpretations. Given this fact, and given also the incoherence of any attempt entirely to expunge dynamic change and temporal passage from our picture of reality, the question of how to work out the metaphysics of relativity is hardly the *Aristotelian's* problem. It is *everyone's* problem. The idea that Einstein has put paid to Aristotle, like the idea that Newton has done so, simply does not withstand scrutiny.

4.3.4 Against the spatialization of time

As we have seen, both relativity and the B-theory have encouraged a tendency to think of time as something like a spatial dimension. But that is simply not what time is. I have already noted some of the differences, but let us examine the topic in greater depth. (Cf. Huggett 2010, pp. 103-6; Lawrence 1971; Reichenbach 1957, pp. 109-13; Smart 1978; Tallis 2017, pp. 34-59.) We can begin by noting several major differences between our pre-theoretical or commonsense notions of time and space.

First, different regions of space exist all at once, whereas different moments of time exist only successively. Of course, B-theorists would deny that, but again, that is part of the pre-theoretical or commonsense understanding of the difference between time and space.

Second, the spatial dimensions *length*, *width*, and *height* differ profoundly from *time* considered as a dimension. For example, rotating an object ninety degrees will change its width to its height, but no such rotation could turn a spatial dimension into a temporal one. You can use a ruler to measure length, width, or height and a clock to measure time, but you cannot use a ruler to measure time or a clock to measure length, width, or height. There is a unity to the three spatial dimensions that does not exist between any of them and the temporal dimension. For instance, it would be odd to group *length*, *height*, and *time* together while setting *width* off by itself, the way we naturally group *length*, *width*, and *height* together and set *time* off by itself. The spatial dimensions cannot exist in concrete reality apart from one another. A line embodied in ink or in a stretch of thread (as opposed to the abstraction studied in geometry) would have at least some width and height in addition to its length. But they could at least in principle exist apart from time, viz. in a three-dimensional object that happened in no way to be undergoing change.

Third, time appears to have a direction and flow that space lacks. We speak of the passage of time, but not of the passage of width. We have some control over our position in space, and can move in any spatial direction. But we have no control over our position in time. We can only move forward in time, and never back to some earlier point. Moreover, we can't jump forward to just any point in the future we like. We must passively take each moment as it comes to us. In short, we cannot travel through time the way we travel through space. (More on this below.) One thing can be said to come *before* or *after* another both in space and in time, but in the case of space these descriptions are relative whereas in the case of time they are not. For example, whether Boston comes before New York depends on whether you are traveling from the north or the south. But that 1945 came before 1975 does not depend on perspective in this way.

Fourth, things occupy space and time in very different senses. A region of space can be occupied by one thing, then vacated by it and occupied by another thing. Time is not like that. For example, the year 1945 cannot be emptied out so that it might come to be occupied by some different events than the ones that actually occurred that year. Events have a fixed location that things in space do not. You can rearrange the pieces on a chessboard, but you cannot rearrange World War II, the first moon landing, and 9/11 so that they occur in some different order. Events always also stay the same distance apart in time, whereas physical objects can get closer or farther apart in space. An object located at a certain region of space exclusively occupies that region. Two physical objects cannot be in the same place at once. By contrast, an event located at a certain point in time is not the exclusive occupant of that point in time. Many events are occurring at any particular moment.

To be sure, there are also commonsense ways of speaking about time that seem to attribute spatial properties to it. We say things like "My college years are behind me" and "Christmas is just around the corner." But these are obviously intended as mere metaphors. We also sometimes represent the past and the future in ways that might seem to suggest that they are as real as the present, just as other regions of space are as real as the region one currently occupies. For example, a calendar represents future days, weeks, months, and years as if they already exist and are waiting for us to arrive. A timeline in a history book represents past events as if they too still exist and we have simply moved away from them. But it would be fallacious to suppose that any metaphysical conclusion about the reality of past and future events follows from this mode of representation, or that common sense takes such a conclusion to follow from it.

Now, it might be claimed that whether common sense regards time and space as radically different in these ways is moot, on the grounds that the physics of relativity and the B-theory have shown that time is in fact much more like space than common sense supposes. The first thing to say in response to this is that the extent to which even modern physics spatializes time should not be exaggerated. Physics does not in fact treat the three spatial dimensions and the time dimension as *exactly* on a par. If we describe the state of the physical world at some particular point in time in terms of the three spatial dimensions, the laws of physics will tell us the state of the physical world at some other point along the time dimension. But if instead we describe the state of the physical world in terms of two spatial dimensions and the time dimension, the laws will *not* necessarily tell us what is going on along the remaining spatial dimension (Huggett 2010, pp. 104-5). Like common sense, then, physics treats the spatial dimensions as naturally related to one another in a way time is not. (As Tallis puts it (2017, p. 36), "the natural place of time is, at best, a D'Artagnan to the Three Musketeers of space.") Even the Minkowskian model doesn't mathematically represent the time dimension in *exactly* the same way as the spatial dimensions (Reichenbach 1957, p. 112). So, physics cannot be said *entirely* to abandon the commonsense differentiation between time and space.

Still, as noted earlier, the physicist's representation of time as a further axis on a graph alongside the spatial axes does tend to foster the idea that physics has shown time to be far more like space than common sense supposes. But here we need to reiterate, and develop further, the point that it is a fallacy to try to read off the intrinsic nature of physical reality from a mathematical representation of physical reality. In this chapter and the previous one I have already had much to say about the highly abstract character of physics' representation of nature, and thus about how much of nature's concrete features are necessarily left out of it. It is the character of the mode of representation, and not the character of what is represented by means of it, that accounts for the absence of some features. Hence, if physics' description of time leaves out much of what common sense would attribute to it, it simply does not follow that time lacks those commonsense features.

But it is important to keep in mind too that, as we noted earlier in this chapter, the physicist's description of *space* is *also* highly abstract. It is only by marrying a highly desiccated conception of time to this highly desiccated conception of space that physics gives the illusion of having spatialized time. As Tallis has noted (2017, Chapter 3), the latter desiccation proceeds in two stages. In the first, physical space is conflated with geometry. Space is treated as if it were composed of points and physical objects as if they were composed of lines and planes. This puts the cart before the horse, because points, lines, and planes are in fact *abstractions from* concrete physical reality rather than the constituents of concrete physical reality. But that is only the beginning of the confusion of abstractions with reality. For the second stage of desiccation is to conflate geometry, in turn, with a Cartesian coordinate system. Points are defined in terms of numbers, and relations between points in terms of numerical intervals. Length, height, and width are conceived of in terms of the *x*, *y*, and *z* axes, with each axis originating from a point designated 0. Their convergence point 0, 0, 0 is the point by reference to which every other point in space is identified. Motion through space is conceived of as the transition from one point in the coordinate system to another. Geometrical properties and relations in general are thereby reconceived in terms of numbers, and novel manipulations of numbers are interpreted as revelations of unexpected geometrical possibilities, and thus of unexpected spatial possibilities. Hence, when we go beyond operations like squaring and cubing numbers, it seems as if we thereby discover possible higher spatial dimensions. Time, then, is finally introduced as an additional axis *t*, and values along the *t* axis seem to be on all fours with values along the *x*, *y*, and *z* axes. Time is thereby spatialized, but only because time and space together have first been mathematicized.

For purposes of physics this is entirely unobjectionable, and the epistemic structural realist can happily acknowledge that the predictive and technological successes that the mathematicization of time and space has afforded are evidence that it captures something real in nature. However, once we carry out this mathematicization, we are not really talking about space and time *themselves* anymore, but only about certain very abstract *relations* between objects and events located within space and time. It is only the reality of these abstract relations that the success of physics gives us any reason to believe in, and as the epistemic structural realist emphasizes, the equations describing such abstract relations are susceptible of a variety of possible physical interpretations. The success of physics thus gives us no reason to prefer one interpretation over another, and it certainly gives no warrant whatsoever to the conclusion that the nature of time and space is *exhausted* by physics' mathematical description. As Tallis writes:

> To conclude from the fact that, if mathematics greatly extends our gaze into the world it must be constitutive of that world, would be in some respects analogous to citing the success of astronomy as evidence that the stars are made of telescopes. (2017, p. 198)

We have already seen in chapters 2 and 3 why there *must* be more to physical reality than the abstract mathematical description afforded by physics, but Tallis notes some specific conceptual problems facing the thesis that the natures of space and time are exhausted by the abstractions of geometry and its Cartesian mathematicization. First, there is ambiguity in the use that is made of numbers in scientific descriptions. "1," for instance, can correspond to a point, which is unextended, or to a unit of length, which is extended, or to some magnitude that bears no relation to extension at all, such as the intensity of pain (Tallis 2017, pp. 111, 123). It is precisely because of their abstractness or lack of concrete physical content that mathematical descriptions are susceptible of such ambiguity. Now, when we work back from the mathematical descriptions to conclusions about concrete physical reality, there is a danger that the ambiguity of the former will lead to fallacious conclusions about the latter. Obviously, if we were to conclude that there is some relation between the intensity of pain and extension in space, on the grounds that they are susceptible of similar mathematical descriptions, we would be committing such a fallacy. But a similar fallacy would committed if we concluded that time must be space-like, on the basis of the similarity in *their* mathematical descriptions. Now, it is what we know about pain and about extension *independently* of their mathematical representation that tells us that the former inference is fallacious. But by the same token, there is no non-question-begging reason to deny that we can have knowledge about time and space independently of *their* mathematical description that can justify us in rejecting the latter inference. Note that this judgment would still stand even if it turned out that time really was after all a kind of spatial dimension. For the point is that the mathematical description *itself, given its inherent ambiguity*, could not justify this thesis. Those who would spatialize time, no less than those who reject such a spatialization, would have to appeal to something outside the description in order to resolve the ambiguity. We simply cannot read off the nature of space or time from the mathematical description alone.

A second problem is that there are serious questions about how coherent is the description of space and its occupants that results when we conflate physical space with geometry, and geometry in turn with a system of numbers. Neither points (since they lack any extension at all), nor lines (since they lack width and depth), nor planes (since they lack depth), can be said to *occupy* space. In that case, however, it is difficult to see how they can coherently be said to be *located* in space (Tallis 2017, p. 120). (If I told you that there is an apple on the table, but that there is no part of the space of the table or of the space just above it that is actually occupied by the apple, you would have a hard time understanding exactly what I mean.) In that case, though, it is hard to see what sense there is to be made of the notion of there being a distance between two points (since there can hardly be a distance between two things that have no location in space), or of the idea that physical objects can be constructed out of points, lines, and planes (since what is extended cannot be made out of what is unextended, and what has depth or width cannot be made out of what lacks depth or width). As Tallis points out, diagrams and other pictorial representations foster the illusion that we can make sense of all this, because we tend unthinkingly to read the concrete features of these representations into the mathematics that the diagrams and other representations represent (2017, p. 122). For example, we imagine black dots and lines of the kind we would see on the page of a textbook. When we correct for this fallacy and subtract from our conception these features of the representations, there is no concrete reality *left* to the mathematics. All these puzzles disappear when we realize that the mathematics *just is* an abstraction rather than anything concrete. In particular, it is *abstracted from* a concrete physical reality whose nature outruns anything captured by the mathematics, rather than being exhaustively *constitutive of* concrete physical reality.

Similarly, there are difficulties with notions such as that of *curved space*, considered as a concrete physical reality rather than an abstraction (Tallis 2017, pp. 141-4). Ordinarily, we attribute either curvature or straightness to the surface of a physical object, to a boundary, or to the trajectory of an object's motion, all of which *occupy* space. We don't attribute curvature or straightness to space *itself*. So what does it mean, exactly, to say that space is curved? We commonly say that an object or trajectory curves *into* one part of space from another. So what does space *itself* curve into? With respect to *what* is it curved? The answer cannot be that it is curved with respect to its surface, trajectory, or the like, because these are, again, features of the things that occupy space, not of space itself. Nor do the usual analogies help. For example, suppose it is said by way of explanation of non-Euclidean space that such space is like a ball and a triangle in curved space is like a triangle drawn on the surface of the ball. This invites the naïve response that the ball has a surface whereas space does not, that you can tunnel in a straight line through a ball whereas there are not supposed to be straight lines in curved space, that what is drawn on the surface of a ball would not count as a triangle in the ordinary sense in the first place precisely because its sides are curved, and so forth. The retort, of course, would be that this is merely an analogy and that these features of a ball don't carry over to curved space itself. The trouble is that once we delete these features of the analogy, we seem to be back where we started. The analogy only seems to make the notion of curved space intelligible insofar as it *includes* features such as the surface of the ball. When we subtract them, what we are left with is just the idea of curved space *itself* rather than any illuminating analogy *for* curved space, which is supposed to be what we were being offered.

Now, if the notion of curved space is mysterious, the notion of a collapse of the very distinction between space and the things that occupy it, and the notion of curvature in *space-time* – two further aspects of GTR – are hardly less mysterious. As Tallis emphasizes, the point is *not* to suggest that there is something wrong with GTR itself. On the contrary, no one who has studied GTR can fail to be struck by its brilliance and elegance, and its empirical successes are, needless to say, striking and undeniable. The point is rather that the notions in question are clearly intelligible only at the level of mathematics, and the mathematics simply doesn't force on us any particular physical interpretation. In particular, interpreting it in either an instrumentalist or epistemic structural realist manner is perfectly consistent with the empirical evidence, and hard to avoid once the philosophical considerations are factored in. Nor is it being claimed that space is really, after all, Euclidean. As Tallis argues, the point is rather that to think of space itself as either non-Euclidean *or* Euclidean "is just one example of the wider tendency to confuse the way we represent what is there with the intrinsic properties of what is there" (2017, p. 147).

A third conceptual problem Tallis identifies in the attempt to mathematicize space and time is that the coordinate system in terms of which space-time is characterized is supposed to describe reality as it exists objectively and apart from any observer, yet this coordinate system *itself* cannot be made sense of without reference to an observer:

> $x = 2, y = 3$, and $z = 4$ does not generate a point in real space unless I have located the axes in a place defined non-mathematically, with myself or someplace I have chosen as the point of origin, at the heart of "egocentric" space. Without an audit trail leading back to egocentric space, mathematical space is not really space at all...

> The mathematical account of what is there seems to be stand-alone and self-sufficient only because the individuals or the communities that generate and apply the axes – enabling them to get a grip on the real world and to have a concrete reference to real places – are off-stage. There is in reality no frame of reference without reference outside the activity of framing; something (more precisely some*one*) has to plant 0, 0, 0 in space otherwise empty of coordinates. (Tallis 2017, p. 124)

The idea that reference to such human observers can be eliminated and replaced by reference to measuring or recording instruments is, as Tallis notes, completely fallacious, because what such instruments do counts as "measuring," "recording," or the like, *only relative to human observers* who *interpret* their deliverances as measurements and recordings, and who *designed* them for the purposes of measuring and recording (2017, pp. 152-7).

As Tallis notes, somewhat acidly but entirely correctly, "it is because all of this is so obvious that it is overlooked" (2017, p. 157). And it reinforces the point that the mathematical representation of space and time afforded us by modern physics is something that *the intellect abstracts from* concrete physical reality, rather than an exhaustive description of concrete physical reality. This brings us to one further problem with the attempt to spatialize time. As with other abstractions – universals, propositions, possible worlds, etc. – the mathematical representation of time and space has a timeless or eternal quality. But that is precisely because it is an *abstraction*, and not because of anything to do with time and space themselves. It has no tendency to show that time and space don't really have the qualities that common sense attributes to them, any more than the timeless or eternal character of universals like *dogginess* shows that individual dogs are timeless or eternal.

We saw earlier that a similar point has been made by Smolin, but there is a further implication that has not yet been considered. When we identify space and time with the static four-dimensional space-time of mathematical physics, we are essentially *collapsing the distinction between time and eternity*. Or rather, we are changing the subject, and talking about some eternal Platonic object rather than talking about space and time. As E. J. Lowe has argued (1999, pp. 95-98), the problem afflicts the B-theory or tenseless approach to time even apart from the mathematical models of physics. The relations between points in time, when described in exclusively tenseless or B-series terms, seem to be as timeless as mathematical relations are. (Cf. Craig 2001, p. 193; McTaggart 1908, pp. 461-62.) Nor, Lowe argues, can the B-theorist plausibly argue that, unlike numbers, the elements related in the B-series are *events*, and bear *causal* relations to one another. For once we drain tense out of our description, we are no longer plausibly talking about either events or causation in the ordinary sense, but rather about some timeless ersatz. The B-theory thus seems "open to the charge that it offers not merely a *tenseless*, but a *timeless*, view of time, and thereby eliminates the very phenomenon which it is supposed to explain" (Lowe 1999, p. 98). Nor will it avail the B-theorist to consider biting this particular bullet and acquiescing to the elimination of time, since, as I have argued, even if time could be banished from the external world, it *cannot* coherently be banished from the conscious subject himself.

4.3.5 The metaphysical impossibility of time travel

The notion of *time travel* poses a problem for the spatialization of time and for the B-theory that merits special attention. (Cf. Koons and Pickavance 2015, pp. 196-7.) If time is like space, or if past and future events are as real as present ones, then it should be possible at least in principle to travel to other times just as we can travel to other places. But (contrary to what some philosophers and physicists claim these days) time travel is *not* possible, not even in principle. Therefore, time is not like space, and past and future times do not exist.

What would time travel be if it *were* possible? The standard answer appeals to a distinction made by David Lewis (1976) between *personal time* and *external time*. Personal time is time as measured by the purported time traveler. It reflects changes taking place within him and his immediate environment (e.g. the time machine by means of which he travels), such as the movement of the hands on a wristwatch he is wearing. External time reflects changes taking place in, and time as measured within, the world beyond the time traveler and his immediate environment. It would be reflected, for example, in the movement of the hands of a watch worn by an observer who watches the time traveler depart in his time machine.

A straightforward but inadequate way of defining time travel would be to characterize it as a discrepancy between personal time and external time. Take Doc Brown's journey from 1985 to 2015 at the end of the movie *Back to the Future*. Suppose that between the moment he left 1985 and the moment he arrived in 2015, the second hand on his wristwatch moved forward by one marker. In other words, though thirty years had passed in external time, only one second had passed in Doc Brown's personal time. Though this might seem to suffice for time travel, on closer inspection we can see that it does not. For suppose that Doc Brown had instead been cryogenically frozen in 1985 and awoke in 2015. Suppose his bodily processes were suspended so that he aged almost not at all, and suppose the wristwatch he was wearing when frozen ticked forward only one second during his thirty years in suspended animation. We could say that there was a discrepancy between his personal time and external time, but we couldn't plausibly say that he had traveled through time. (Cf. Le Poidevin 2003, pp. 175-6; Wasserman 2018, pp. 10-11.)

So, a discrepancy between personal time and external time would not suffice for time travel. There would have to be an additional element. A standard proposal is that the needed additional element is *discontinuity* in external time. Consider the *Back to the Future* scenario once again. From the point of view of his personal time, Doc Brown and his time-traveling DeLorean would continue to exist from the moment he leaves 1985 to the moment he appears in 2015. But from the point of view of external time, Doc Brown and the DeLorean would go out of existence in 1985 and reappear in 2015. Time travel, on the view under consideration, would be a discrepancy between personal time and external time that is associated with such a discontinuity in external time (Le Poidevin 2003, p. 176).

However, even this is not enough, for given this discontinuity in external time, we ought to ask what makes it the case that the Doc Brown and the DeLorean that appear in 2015 are *identical to* the Doc Brown and the DeLorean that went out of existence in 1985, rather than mere *duplicates* of the latter. For if they are mere duplicates, then Doc Brown and his time machine will not really have traveled into the future at all, but rather been annihilated and replaced by these duplicates. Consider also Marty McFly's journey, in the same movie, from 1985 back to 1955. In this case too we have a discrepancy between personal time and external time, and we have discontinuity between the Marty that exists in 1955 and the Marty that exists in 1985. But this isn't because Marty traveled from 1955 to 1985. Rather, according to the movie, he traveled in the other direction. Merely noting a discrepancy between personal time and external time and a discontinuity in external time would not suffice to explain what would make it true that Doc Brown traveled *forward* in time whereas Marty traveled *backward*.

The notion of *causation* might seem to provide a solution to both of these problems. Doc Brown's departure in 1985 *causes* his arrival in 2015, and this causal relation, it might be proposed, suffices to guarantee that the man who arrives is identical to the man who left. Furthermore, it might be suggested that the fact that it is Doc Brown's actions in 1985 that cause his presence in 2015 (rather than the other way around) entails that he has traveled forward in time, whereas the fact that Marty's actions in 1985 cause his presence in 1955 (rather than the other way around) entails that he has traveled backward in time. So, this might seem to give us, at last, an adequate characterization of time travel. Time travel, on this revised analysis, involves a discrepancy between external time and the personal time of the time traveler, where the stages of the traveler are discontinuous in external time but causally related. (Cf. Wasserman 2018, p. 8.)

But even given these qualifications, the whole idea is incoherent, and for reasons that have to do precisely with causation. Consider first the notion of a discontinuous discrepancy between personal time and external time. In the case of Doc Brown's journey from 1985 to 2015, we are supposed to believe that thirty years of external time pass whereas only one second passes in the personal time of Doc Brown, or more precisely, in the personal time of everything within his time machine. But how exactly do we draw the line between the part of the universe described in terms of external time and the part described in terms of personal time? Presumably the idea is that everything from the surface of the DeLorean inward is in personal time and travels to 2015, and everything beyond that is in external time and does not travel. But why does that surface mark the boundary? Why doesn't the boundary extend to, say, everything within six feet of the DeLorean, so that if you happen to be standing right next to it when it travels to 2015, you'll go with it? Or why does it extend even as far as the surface of the DeLorean? Why don't some of the outer parts of the car – the stainless steel outer skin, the rearview mirrors, and the tires say – get left behind in 1985 while the rest travels to 2015? Is it because they are *fastened* to it? Why does that matter? Suppose Doc Brown does not buckle his seat belt. *He* won't be fastened to the machine either, but merely touching it. So why doesn't he get left behind? Is mere *contact* with the machine sufficient to travel back with it, then? In that case, all the molecules of air making contact with the outer surface of the car will travel to 2015 too. But then it seems that the molecules in contact with *those* molecules will also travel, as will the molecules making contact with those *further* molecules, as will the physical objects making contact with those molecules, and so on until the *entire universe* travels forward to 2015! (Note that it is no good merely to *stipulate* that the time machine just works in such a way that only the car and its occupants travel, because what we need to know is precisely how something *could* work that way. We need to know how the idea of a time machine could in principle be *more* than just some sheer stipulation for the purposes of fiction.)

The problem is that, as Tallis points out (2017, pp. 84-85), a time machine would on its departure have to *sever* all causal connections to the rest of the universe. For example, at the moment of its departure, Doc Brown's DeLorean would have to cease being affected by the friction of the road under it as that road existed in 1985, by the gravitational pull exerted by the Earth in 1985, by the temperature of the air that surrounded the car in 1985, and so forth. Yet as the questions I've just raised indicate, every part of the purported time machine would be so connected to every other part and to the wider universe around it that any line we might try to draw between some region of the world that breaks all causal connections with the wider universe and the regions that do not seems arbitrary. But that is only half the problem, for as Tallis also notes (2017, p. 87), when a time machine arrives at its destination, it would have to *reestablish* causal connections to the rest of the universe. For example, the DeLorean would have to start being affected by the gravitational pull exerted by the Earth in 2015, by the friction of the road as it exists in 2015, by the temperature of the surrounding air in 2015, and so on. If it didn't reestablish causal connections with the universe as described by external time, it couldn't *stop* traveling. But how *could* something which has completely severed all causal connections to the world reestablish them? Even a ship dropping its anchor is not beforehand *entirely* disconnected from the sea bottom below, since the ocean waters, through which the anchor falls, connects them. But a time machine would lack *all* causal connection to the world to which it would try once again to anchor itself.

Moreover, even *before* the DeLorean tries to stop its journey so as to arrive at 2015, it seems that it would already have to have reestablished some causal connections to the world of external time. For it has to set course, as it were, for *2015, specifically*, as its destination (as opposed to 2014, 2016, or some other date). The driver of a car can set course for a specific destination only because he is causally connected to it. He can see where he is going, or can rely on GPS guidance, or at least knows that his destination lies at the end of a road or series of roads that link his current location to it. But nothing like this exists for the time machine that has severed all causal connections to external time on its departure. So how can it so much as *aim* at a certain destination, much less arrive at it? (Cf. Tallis 2017, p. 87.)

So, it is hard to see how to make sense of the notion of a discontinuous discrepancy between personal time and external time. Then there is the problem of making sense of the idea that Doc Brown's getting into his time machine in 1985 is the *cause* of his arrival in 2015, or that Marty's getting into the time machine in 1985 is the cause of his arrival in 1955. Of course, there is no difficulty in understanding how something happening in 1985 could be *indirectly* causally linked, by way of a series of intermediate causes, to something happening in 2015. That happens all the time. For example, if a father begets a son in 1985 and that son goes on to beget a son of his own in 2015, we can say that the father's act of begetting in 1985 was among the causal factors responsible for his grandson's being begotten in 2015. What is mysterious is how something happening in 1985 could cause something to happen in 2015 *directly, without* any intermediate causal links. When time travel is not in view, everyone would agree that such direct causal influence between temporally separated events is not possible. Now, the defender of the possibility of time travel might respond: "Yes, direct causal influence between temporally separated events is not possible – *apart from time travel*, that is. But once we factor time travel in, such direct causal influence between temporally separated events *is* possible." The trouble with such a response, though, is this. The idea of direct causal influence between temporally separated events was introduced as a way to make sense of the possibility of time travel. But now we are appealing to the possibility of time travel as a way of making sense of the possibility of direct causal influence between temporally separated events! So we have gone around in a circle, in which case the attempt to elucidate the notion of time travel in causal terms fails.

The problem only gets worse when we consider the *backward* causation entailed by travel to an earlier moment of time, such as Marty's journey from 1985 to 1955. A well-known objection to the very possibility of backward causation is known as the "bilking argument" (Horwich 1987, pp. 91-105; Dainton 2010, pp. 131-33). Suppose it is claimed that some earlier event *E* is caused by a later event *L*. Then, the argument suggests, we can refute this claim in the following way. When *E* occurs, we can try to prevent *L* from occurring, and when *E* does not occur, we can try to ensure that *L* does occur (a procedure labeled "bilking"). Suppose we succeed. Then we will have falsified the claim that *L* causes *E*, because we will have shown that *E* can still occur even when *L* does not, and that *E* can fail to occur even when *L* does occur. But suppose we fail. In that case too we will have falsified the claim that *L* is causing *E*, because if we are unable to prevent *L* once *E* has occurred or to bring about *L* when *E* does not occur, what that really shows is that it is *E* that is causing *L* rather than the other way around.

Now, some philosophers (such as Dainton and Paul Horwich) argue that the bilking argument does not really refute the claim that backward causation can occur, because the defender of backward causation can interpret the bilking scenario in a way consistent with the reality of such causation. For example, he can suggest that the specific *way* we bring about *L* after *E* fails to occur itself ends up disrupting the causal process by which *L* normally brings about *E*, and that had we not so acted then *L would* have brought *E* about. But even if we were to concede that the bilking argument does not prove the *impossibility* of backward causation, it does seem to undermine any attempt to prove the *possibility* of backward causation. For it seems that any scenario the defender of backward causation might appeal to in order to demonstrate the possibility is a scenario that could instead be given a "bilking" interpretation. Merely describing a scenario in which *L* is said to cause *E* will not suffice, because such a scenario will also be susceptible of an interpretation in which it is really *E* that causes *L*. Some additional element must be added to the scenario in order to rule out the bilking interpretation.

Now, it seems that the only plausible candidate for such an additional element will be an appeal to time travel. That is to say, in order to rule out the bilking interpretation, the defender of backward causation will have to hold that in some way or other, causal efficacy extends backward in time from *L* to *E*, but not forward in time from *E* to *L*. After all, as Horwich points out (1987, p. 96), if the causal factors antecedent in time to *E* are sufficient to produce *E*, then *L* would not be strictly necessary for the occurrence of *E* and we would have a case of causal overdetermination. So for *L* to be a necessary condition for *E* it seems that some causal factor would have to travel backward in time from *L* to *E*. But the problem with this should be obvious. The notion of backward causation only entered our discussion in the first place because it was supposed to provide a way of making time travel intelligible. Yet now we are appealing to the notion of time travel in order to make backward causation intelligible!

Once again we have gone in a circle, and once again we have thereby failed to elucidate the notion of time travel in causal terms.

It might seem that some of the problems we've been considering derive from the fact that we are conceiving of time travel as a kind of *jump* from one point in time to another. But we might instead think of time travel as a kind of *time-slide* rather than a time-jump (Dainton 2010, p. 122). For example, instead of thinking of Doc Brown's journey from 1985 to 2015 as a matter of *bypassing* all the intervening years and arriving directly in 2015, we can think of it as a matter of him journeying through all the intervening years, but at a faster rate than usual. And if we think of it this way, then it might seem that we can avoid problems such as how to explain what makes the Doc Brown who arrives in 2015 identical the Doc Brown who left 1985, and how to explain the possibility of severing all causal connections with external time and then reestablishing them.

But on closer inspection, this proposal will not work, because the notion of a time-slide is ambiguous. What exactly does traveling through all the years between 1985 and 2015 at a faster rate amount to? One possibility is that it involves a *series* of time-jumps rather than a single time-jump. For instance, it might involve Doc Brown jumping from 1985 to 1986, then from 1986 to 1987, and so on through all the other intervening years until he arrives at 2015. But in this case all the problems we've been discussing reappear for each of these shorter time-jumps and the time-slide scenario is not really an *alternative* view at all. Should we say instead that a time-slide is something *gradual* rather than "jumpy"? In that case, Doc Brown's "sliding" from 1985 to 2015 at a faster rate seems to amount to nothing more than his *aging more slowly* than the rest of the world, and as we saw when considering the suspended animation example, that does not suffice for time travel.

The most famous challenge to the metaphysical possibility of time travel that is grounded in considerations about causation is no doubt the *grandfather paradox*. The idea here is that if time travel were possible, then I could in theory travel back in time and murder my grandfather before he has a chance to beget my father. If I were to do so, however, then I would never be born, in which case I would never be able to travel back in time and commit the murder. Hence the notion of time travel seems to entail a contradiction insofar as this scenario implies both that I am born and that I am never born. A different sort of time travel paradox related to causation is the *bootstrapping paradox*, in which a later event causes an earlier event which in turn causes the later event. A famous example is found in Robert Heinlein's short story "'—All you Zombies – '" (which was made into the movie *Predestination*), in which an intersex person turn out, as a result of a seduction, a pregnancy, a sex change operation, and a series of jumps through time, to be his own father and mother. Whereas objections based on the grandfather paradox claim to show that the notion of time travel leads to self-contradiction, objections based on the bootstrapping paradox claim to show that time travel entails vicious causal circularity.

Now, in the case of the grandfather paradox, defenders of the possibility of time travel respond that all the thought experiment really shows is that since I *have* in fact been born, then even if I travel back in time I *will not* in fact murder my grandfather. (Cf. Lewis 1976.) For one reason or another, I will fail to do so even if I try. (Think of time travel movies like *Timecrimes* or *12 Monkeys*, in which attempts to alter the past inadvertently end up helping to bring about the very events the protagonist was seeking to prevent.) This might seem to entail inexplicable coincidences. Why couldn't I find out, through historical research, exactly what circumstances blocked me from carrying out the murder of my grandfather (such as a defective weapon or whatever), and take steps to ensure that once I arrive in the past I avoid those circumstances? Why would there always have to be some *further* unforeseen circumstance that somehow prevents me from carrying out the murder?

Here the defender of time travel might appeal to the idea that time is really just a further quasi-spatial dimension in a four-dimensional block universe. Think of a block of marble which has been turned into a statue, such as Michelangelo's *David*. Is it a coincidence that the legs of the statue happen to be positioned in such a way that they hold up the torso, arms, and head? No, because they are all parts of a single coherent whole designed by Michelangelo to fit together in precisely this way. Similarly, the set of events that includes my grandfather's begetting of my father, my own birth, my stepping into a time machine with the intention of killing my grandfather, my being prevented from ever actually doing so, etc. are all parts of a single coherent whole, the four-dimensional spacetime block, which has been determined by the laws of physics together with the requirements of logical consistency to fit together in precisely this way. As with the statue's legs, the consistency of the events in question can seem coincidental only if we ignore the larger context. (Cf. Dainton 2010, pp. 128-30.) The defender of time travel might deal with bootstrapping paradoxes in a similar way. When considered in isolation, a scenario in which a later event *L* causes an earlier event *E* which in turn causes the later event *L* seems to entail a vicious explanatory circle. But we can break out of the circle by situating *E* and *L* in the larger context of a four-dimensional block universe, and arguing that the unusual way that *E* and *L* fit together in this universe can be explained by reference to the laws that govern the block.

But a serious problem faces any attempt to defend time travel by characterizing time in spatial terms. Suppose a two-dimensional creature were to defend the possibility of time travel by characterizing time as a third spatial dimension and then describing the way that objects can move through three-dimensional space. The problem with this defense, of course, is that what this two-dimensional creature would be describing is not *time* travel at all, but just travel through *space* as we three-dimensional creatures understand it. Similarly, for us three-dimensional creatures to defend the possibility of time travel by characterizing time as a fourth, quasi-spatial dimension in a four-dimensional block is essentially to commit the same fallacy. What we are describing is not time travel at all, but just an exotic form of travel through space. Moreover, just as (the Aristotelian holds) time as we normally understand it is the measure of change within three-dimensional space, so too, if what we thought was time is really just a fourth spatial dimension, then what that would entail (the Aristotelian would argue) is that time is really the measure of change in this *four*-dimensional space – and that for all the defender of time travel has shown, we do *not* travel through time in *that* sense. "Proving" that time travel is possible by redefining time as a spatial dimension is like "proving" that you are a millionaire by redefining a millionaire as someone who has at least one dollar in his bank account.

(This suggests a further argument against any attempt to spatialize time, which is that it can never be completely carried through. Again, time is the measure of change within space. If we think of space as three-dimensional, then time is the measure of change within three-dimensional space, but if instead we say that what common sense *conceives of* as time is "really" just a fourth spatial dimension, then what that implies – again, for all the defender of the spatialization of time has shown – is that time ought really to be thought of as the measure of change within *four*-dimensional space. If the defender of the spatialization of time now claims that time so understood is really just a *fifth* spatial dimension, then the response will be that in that case time turns out to be the measure of change in *five*-dimensional space. And so on *ad infinitum*. So the reduction of time (as the Aristotelian and as common sense understand time) to space cannot be carried out. But if we have to stop *somewhere* in the regress just described and admit that at that level there is such a thing as time conceived of as non-spatial, then we might as well just admit this from the get-go, and abandon the idea that time as common sense understands it is a kind of fourth spatial dimension.)

It is sometimes thought that paradoxes like the grandfather paradox can be avoided if we think of travel to the past as travel to a parallel universe. (Cf. Deutsch and Lockwood 1994; Dainton 2010, pp. 126-7.) Suppose I am in universe A and I travel back in time and succeed in killing my grandfather. On this proposal, I do not thereby prevent myself from being born in universe A, but do prevent myself being born in a parallel universe *B*, which is otherwise like A but now lacks my grandfather and therefore my own father and my birth. The trouble, though, is that travel between parallel universes also seems really to be an eccentric kind of *space* travel and not time travel at all. As with the four-dimensionalist model, the parallel universe model simply changes the subject.

That time travel presupposes the spatialization of time is commonly acknowledged. (See e.g. Bigelow 2001 for a defense of this view.) But there are some philosophers who claim that the possibility of time travel does not in fact require thinking of time in spatial terms or in B-theoretic terms. For example, Ryan Wasserman (2018, pp. 38-49) argues that the possibility of time travel could be defended even given the presentist view that past and future events do not exist. Wasserman proposes that a presentist might conceive of time travel, not as a matter of traveling to some past or future time, but rather as a matter of doing something now that makes some past-tensed or future-tensed statement true. For instance, he could say that when Marty activates his time machine in 1985, this makes it true to say that Marty arrived in 1955. The trouble with this proposal, however, is that it too seems to "prove" the possibility of time travel merely by redefining it. In the context of discussions of time travel, a statement like "Marty arrived in 1955" is typically understood to entail that 1955 exists no less than the present time does, and that Marty traveled to that time. But that is not what it entails on Wasserman's proposal. Indeed, Wasserman's proposal seems not only to allow for, but even to require, that once Marty activates the time machine, he no longer exists! For, having left the present, he no longer exists in the present. And on the presentist view, there are no past or future times for him to have gone to. So there just is nowhere for him to exist anymore. In that case, it is not clear what the statement "Marty arrived in 1955" *could* mean on Wasserman's "presentist time travel" proposal. In any case, whatever exotic meaning we are to attach to it, it has nothing to do with time travel as usually understood. (Cf. Dainton 2010, p. 139.)

It might be claimed that, whatever philosophical puzzles might face the defender of the possibility of time travel, modern physics has established that it is possible, so that there must be some way to resolve the puzzles. But modern physics has established no such thing. There are three sorts of considerations from physics often said to favor time travel. The first concerns various phenomena related to quantum mechanics which, some propose, can be interpreted as instances of backward causation (Dainton 2010, pp.135-6 and 142). For example, some have suggested that the measurement of a system can determine what state the system was in prior to the measurement; and some have proposed backward causation as a way of avoiding the instantaneous action-at-a-distance that the EPR correlations seem otherwise to entail. The problem here is that backward causation interpretations are *interpretations*, and alternative interpretations are available. Moreover, it is only if we *already* know independently that backward causation is possible that we can reasonably deploy the notion in these interpretations. In that case, though, such interpretations *presuppose* the possibility of backward causation and therefore cannot provide a non-question-begging *argument for* the possibility of backward causation. Here as elsewhere, you cannot read metaphysical conclusions *out of* physics without first reading them *into* the physics.

A second consideration concerns the *time dilation* entailed by STR, illustrated by famous examples like that of the astronaut who travels for ten years at close to the speed of light and then returns home to find that centuries had passed on Earth. It is sometimes claimed that such an astronaut will thereby have traveled to the future. But that is not what is happening, any more than in the suspended animation example. As James Gleick has noted, time dilation is "hardly time travel... It's an antiaging device" (2016, p. 58).

The third relevant set of considerations from physics concern GTR. Kurt Gödel (1970) discovered solutions for the field equations of GTR that allow for the possibility of closed causal chains in a rotating universe, where the "backward" part of such a chain can be interpreted as an object's revisiting its earlier self. Physicist Kip Thorne (1994) argues that GTR allows for a scenario in which Minkowskian spacetime curves back on itself and the two halves of the curve are then joined together by a wormhole. An object which traveled around the curve and then entered the wormhole would pass by its earlier self when it emerged from the other end of the wormhole – thereby, it is proposed, traveling back in time.

Now, as Wasserman points out (2018, pp. 67-68), there is a problem with such proposals (and indeed, with any other defense of time travel that appeals to relativity theory) that derives from the denial of absolute simultaneity. The problem is that without the notion of absolute simultaneity, we cannot make sense of the notion of *external time*. All time becomes *personal time*. But time travel, on the standard analysis, is defined in part in terms of a discrepancy between personal time and external time. Hence, if there is no such distinction, then there can be no discrepancy, and thus no time travel. But Wasserman notes that GTR arguably provides a way to amend the standard analysis. If some event *E* is in the past light cone of an event *L*, then we can say that *E* is objectively earlier than *L* and define this objective ordering as *light cone time*. Time travel could then be defined in terms of a discrepancy between personal time and light cone time.

Whatever one thinks of such a revision, there is a deeper problem with such GTR-based defenses of the possibility of time travel – a problem Gödel and Einstein saw, though many who appeal to their work seem not to. Gödel presented his results, not as an argument for the possibility of time travel, but as an argument for the *unreality of time*. As Einstein wrote in response to Gödel:

> If, therefore, *B* and *A* are two, sufficiently neighboring, world-points, which can be connected by a time-like line, then the assertion: "*B* is before *A*," makes physical sense. But does this assertion still make sense, if the points, which are connectable by the time-like line, are arbitrarily far separated from each other? Certainly not, if there exist point-series connectable by time-like lines in such a way that each point precedes temporally the preceding one, *and if the series is closed in itself*. In that case the distinction "earlier – later" is abandoned for world-points which lie far apart in a cosmological sense, and those paradoxes, regarding the *direction* of the causal connection, arise, of which Mr. Gödel has spoken. (1970, p. 688).

The problem, as Einstein indicates, is that in a causal chain that loops around back on itself, there is no reason to regard any member of the chain as *objectively* "earlier" or "later." If *E* is part of a chain that leads to *L* and *L* in turn leads back to *E*, then while you could regard *E* as the earlier event and *L* the later one, you could just as well regard *L* as the earlier event and *E* the later one. There is nothing in the scenario itself that forces either interpretation. But the notions of "earlier" and "later" are essential to time, as even the B-theorist (who preserves *that* much of our commonsense notion of time) acknowledges. Hence, in a scenario in which there are no objective "earlier than" and "later than" relations, there is no *time* either. But in that case, neither can there be any *travel* through time.

Once again, then, we have a purported defense of time travel that really just changes the subject. The GTR-based scenarios are not about *time* travel, strictly speaking, at all, but rather about something else – such as an eccentric kind of space travel, or even (given that the operative notion of space is as desiccated as the operative notion of time) about a static mathematical abstraction that doesn't count as travel or change of any kind. But mightn't the defender of time travel still argue that physics has shown this abstraction correctly to describe reality, whether or not we want to label it "time travel"? No. For one thing, the fact that physicists can come up with such abstractions by itself shows far less than some breathless science popularizers (and indeed, some breathless scientists and philosophers) suppose. As Horwich, who is much more sympathetic to the notion of time travel than I am, acknowledges:

> [We] might rule out Gödel's solutions in the way that we often reject unacceptable mathematical solutions to physical problems. (For example, using the equation, "distance (in feet) = 16 x time (in seconds) squared," to find out how long a stone would take to fall, say, 64 ft, we obtain $t^2 = 4$, and one of the solutions, minus 2 secs, is dismissed out of hand.) (1987, p. 111)

As in other contexts, the fact that certain mathematical solutions are possible does not *by itself* tell us anything about concrete physical reality. But nor do the solutions in question tell us anything about concrete physical reality even when supplemented by empirical considerations. Neither Gödel nor those who appeal to his scenario think the empirical evidence is actually consistent with the rotating universe model the scenario presupposes. Nor does the empirical evidence confirm Thorne's scenario. These are, for all anyone has shown, at *best* mere abstract possibilities. It would be fallacious, then, to appeal to models like Gödel's or Thorne's as evidence that physics has established the possibility of time travel or anything analogous to it. But even that is too generous. For more fundamental than the empirical considerations are the *metaphysical* considerations, and as I have argued, the thesis that time is entirely illusory simply cannot coherently be maintained. *We know* that time is real, and thus we know that any mathematical model a physicist concocts that leaves time out is either mistaken or incomplete.

Palle Yourgrau (1991, 1999, 2005) has long rightly complained of the failure of commentators on Gödel's argument to perceive its true, radical implications. This failure consists, in part, in their not seeing that what Gödel's model implies is not time travel, strictly speaking, but rather the unreality of time. But it also involves their failure to realize that Gödel's claim is not merely that his solutions to the equations of GTR imply the unreality of time, but rather that *GTR itself* implies the unreality of time. As Yourgrau puts it, Gödel's model is essentially a "limit case" for GTR's spatialization of time, which reveals what follows from pushing that spatialization through consistently. It is true that whether a universe is of the rotating kind operative in Gödel's model depends on the distribution of matter within it, which is a contingent matter that GTR itself says nothing about. But in Gödel's view, a fundamental metaphysical question such as whether time is real cannot hinge on such contingent matters of fact. For Gödel, if GTR implies the unreality of time under even one interpretation, it implies it full stop. As Yourgrau writes:

> This is something that even the "friends of Gödel," who in recent years have stepped forward to defend his account of time travel as logically and physically coherent, have failed to note. For Gödel, if there is time travel, there isn't time. The goal of the great logician was not to make room in physics for one's favorite episode of *Star Trek*, but rather to demonstrate that if one follows the logic of relativity further even than its father was willing to venture, the results will not just illuminate but eliminate the reality of time. (2005, p. 134)

Or at least the results will imply this *given* the assumption that GTR is a complete description of physical reality, an assumption Gödel was willing to make. It is not an assumption *we* should make, however. Since (as I have been arguing) the reality of time cannot coherently be denied, what we should conclude from Gödel's argument (if it is otherwise sound) is precisely that GTR does *not* give us a complete description of physical reality.

In any event, the upshot of our discussion in this section is that there is no coherent sense to be made of the notion of time travel. But there *would* be if time were as the B-theorist or the defender of the spatialization of time say it is. Therefore, time is *not* as they say it is.

4.3.6 In defense of presentism

Though most of the argumentation developed so far has been generically A-theoretic, since I have now suggested that past and future times do not exist, I have also implicitly defended presentism, specifically, at the expense of other versions of the A-theory. It is now time to make that defense explicit. The basic idea is this. As I have argued, the objections to the A-theory (such as those that appeal to relativity theory or to other attempts to spatialize time) all fail. Meanwhile, the ineliminability of tense, the incoherence of denying that dynamic change and temporal passage exist at least within conscious experience, and the impossibility in principle of time travel, give positive reason to conclude that the A-theory is true. So, some version of the A-theory is true. But, I want now to argue, the only version of the A-theory that does not face insuperable objections is presentism. So, presentism is true.

Let's consider, then, the deficiencies of the other A-theories of time, starting with the "moving spotlight" theory. Like the B-theory, the moving spotlight view holds that past and future events exist no less than present events do. What makes it an A-theory rather than a B-theory is that it also holds that only present events fall under the "spotlight" of the *now*, which "moves" along, and successively illuminates, the B-series of events. Now, one problem with this view is that it inherits some of the problems facing the B-theory because of *its* affirmation of the reality of past and future events. For example, like the B-theory, the moving spotlight view would seem to imply that time travel is possible at least in principle. (Sara Bernstein (2017) proposes that this would involve the time traveler's somehow being able to move the "spotlight" to his desired location in the series of events.) Since, as I have argued, time travel is *not* possible even in principle, we have reason to reject the moving spotlight theory. In affirming the reality of past and future events, the moving spotlight theory also at least partially spatializes time, which is problematic in the ways described above.

But the theory also faces problems of its own. McTaggart (1908, 1927) famously argued against the reality of time with reasoning that, as is now widely acknowledged (Bardon 2013, p. 85; Craig 2001, p. 148; Dainton 2010, p. 18), has no force against presentism, but does have force against a moving spotlight conception of time. Suppose, as the moving spotlight theory does, that past, present, and future times are all equally real, and consider some event *E* which occurs at time t_2. Now, nothing can be past, present, and future all at once. However, argues McTaggart, *E is* past, present, and future all at once on the supposition in question. For at t_2, *E* is present. But at the earlier time t_1, *E* is future, and at the later time t_3, *E* is past. Because these past and future times are no less real than the present, *E* is future and past no less than it is present. So, we have a contradiction. Now, suppose McTaggart's critic responds that there is no contradiction insofar as no event is past, present, and future *at the same time*. For example, *E* is *not* past or future *at* t_2, but *only* present at that time. The problem with this response, McTaggart argues, is that *E* will count as absolutely present (and thus in no sense past or future) at t_2 only relative to some *higher-order* time (or *hyper*-time). But then a similar contradiction will arise at the level of this higher-order time, insofar as *E* will be past and future no less than it is present in this higher-order time. Resolving this contradiction will require appeal to a yet *higher*-order time, leading to a vicious regress. Whatever one thinks of this argument (and it has generated an enormous literature), it has force only given the thesis that past, present, and future times are all real, which the moving spotlight theory accepts but presentism does not.

Now, McTaggart's argument emphasizes the idea that to affirm the equal reality of past, present, and future times entails a *contradiction*, but in my view that is not the most interesting or important part of the argument. For suppose the moving spotlight theorist can find a way to avoid a contradiction without resorting to higher-order times. He will still be led to posit such times, and thus be led into a vicious regress, for another reason. The series along which the spotlight of the "now" is said to move is supposed to contain all events. But the arrival of the spotlight on this event, then the next, then the next one after that, and so on, is *itself* a series of events. Hence we have a higher-order series of events, and given the assumptions of the moving spotlight theory, this would seem to entail a higher-order "spotlight" which moves along this higher-order series. But that will entail a yet *higher*-order series of events, and thus a third "spotlight," and so on *ad infinitum*. (Cf. Broad 1923, p. 60; Prosser 2016, pp. 4-5.)

Responding to McTaggart, Ross Cameron (2015, pp. 60-63) suggests that while the moving spotlight theory entails a regress, it is a *benign* one rather than a vicious one. A benign explanatory regress, Cameron says, is one in which a problem at level n is solved by reference to an explanation that raises a similar problem at level n + 1, but where the success of the explanation at level n does not depend on the success of a parallel explanation at level n + 1. A vicious explanatory regress, by contrast, is one in which a problem at level n is solved by reference to an explanation that raises a similar problem at level n + 1, and where the success of the explanation at level n *does* depend on the success of the parallel explanation at level n + 1. Consider the following examples (mine, not Cameron's). Suppose we explain Smith's death by reference to the hypothesis that he was murdered by Jones, and suppose that Jones too has died in the time since Smith's death. Even if we cannot plausibly explain Jones's death as a further murder – and indeed, even if we don't have any explanation at all of how he died – that will not undermine our explanation of Smith's death. For Jones could have murdered Smith whether or not he was himself murdered. This would be a benign explanatory regress. But suppose we explain how a certain chandelier is being held aloft by reference to the chain from which it hangs, but have no explanation for how the chain itself is being held aloft. For example, suppose there is no hook or ceiling or anything else above the chain, yet somehow the chain and chandelier are suspended in midair. Then our explanation of how the chandelier is being held up would be undermined. For a chain cannot hold up a chandelier or anything else unless it is itself being held up. This would be a vicious explanatory regress.

Again, Cameron claims that the regress that the moving spotlight theory entails is benign. Perhaps this is true of the specific regress *McTaggart* was concerned with, where a contradiction is resolved at one level in a way that appears to generate a similar contradiction at a higher level. Or perhaps not. Either way, I submit that what I have characterized as the more interesting and important regress entailed by the moving spotlight theory is clearly *vicious* rather than benign. The moving spotlight theory accepts the claim of the B-theory that past, present, and future events are all equally real. This seems to imply a static universe, viz. one from which change and temporal passage are absent. The moving spotlight theory purports to explain how change and temporal passage can exist in such a universe by positing the moving spotlight of the "now." But it turns out that this entails a higher-order series of events – the successive "illuminations" of one time after another by the spotlight – which, for all the moving spotlight theory shows, are *also* all equally real, whether past, present, or future. In this case, though, it seems that we are once again faced with the implication that change and temporal passage are unreal. And the attempt to avoid this result by appealing to a yet higher-level "spotlight" will just generate the same problem again. Now, this is clearly analogous to the chandelier and chain example. Since a chain cannot hold anything aloft unless something holds *it* aloft, we haven't really explained how the chandelier is held aloft if we don't know how the chain is. Similarly, since, on analysis, the series of illuminations by the spotlight is as static as the series we started out with, the former cannot explain how real change and temporal passage can exist in the latter.

Consider now the "growing block" theory, which allows that *future* events do not exist, but holds that *past* events exist no less than present ones do. Like the moving spotlight theory, this version of the A-theory also inherits some of the problems of the B-theory. In particular, since it treats at least past events as real, it too at least partially spatializes time, and it too implies that at least some kinds of time travel are possible. The growing block theory also faces a problem analogous to the regress problem facing the moving spotlight theory. The theory holds that past events and present events are all the events that exist, until the coming about of new event. But it seems that the addition of this new event to the series would *itself* be an event, over and above the new event itself together with all the past ones (Prosser 2016, p. 5). In that case, then as with the moving spotlight theory, it seems we have a higher-order series of events (in this case, the series of additions to members of the first-order series) leading to an infinite regress.

In addition, both the moving spotlight theory and the growing block theory face an epistemological problem (Curtis and Robson 2016, p. 74). On both theories, past events are as real as present ones, and therefore *people* who exist in the past are just as real as present ones. Now, past people, no less than present ones, *believe* themselves to be present, though of course their belief is false. But how do we know that *we* are correct in believing that *we* exist in the present? For it would *seem* to us as if we were even if we are not. Indeed, since there is only one present moment but a vast number of past moments, it is far more probable that we are actually in the past than that we should just happen to be in the one *seemingly* present moment that really *is* present! In defense of the growing block theory against this objection, Peter Forrest (2004) proposes that consciousness exists only at the growing edge of the block and not at earlier moments. But as Chris Heathwood (2005) has noted in response, this is essentially to fall back to a *presentist* position vis-à-vis consciousness while maintaining a growing block position with respect to everything else, which undermines the point of the growing block theory.

Now, there are further objections to the moving spotlight and growing block theories, though of course there are also attempts to reply to the various objections I've been surveying, and there are yet other variations on the A-theory as well. I will not attempt to deal with all of the enormous body of literature devoted to these topics. Suffice it for present purposes to note that the main problems facing non-presentist versions of the A-theory seem to arise precisely because of the concessions they make to the B-theory. That is to say, they run into trouble because they are less pure versions of the A-theory than presentism is. As the purest version of the A-theory, presentism must surely be at least the *default* position for anyone convinced that some form of A-theory is true. And the problems that arise when concessions to the B-theory are made tell against moving away from that default position. Hence it is no surprise that presentism appears to be the most widely accepted version of the A-theory.

Of course, presentism too faces various objections. I have already responded to objections grounded in relativity theory, but let us now turn to the others. It is often claimed that presentism faces a "truthmaker" problem. If some statement is true, says the objection, then there must be something in the world that *makes* it true. For example, the statement that the *cat is on the mat* is true because there actually exists a cat who is on a mat. Now, among the true statements there are are statements about past events. For example, it is true that *Julius Caesar was assassinated on the Ides of March*. So, there must be some truthmaker for such statements. But the presentist claims that there are no past events, only present ones. So, it seems that if presentism were true, then there would be nothing to serve as the truthmakers for such statements.

Now, there has been a lot of ink spilled about this objection, but it seems to me unimpressive. In response to it, some presentists hold that the world presently has the property of *having contained Julius Caesar's assassination on the Ides of March*, and that its having this property is the truthmaker for the statement that *Julius Caesar was assassinated on the Ideas of March* (Bigelow 1996; Zimmerman 2008, pp. 217-18). This seems to me correct as far as it goes, though whether this is the best way to put things depends on how much metaphysical baggage one reads into talk of such "properties." (Some will object that considerations of parsimony tell against the existence of such properties, but no one who claims that past and future events are as real as present ones really has any business appealing to parsimony!) A better way to respond, in my view, is to say that the truthmaker for the statement that *Julius Caesar was assassinated on the Ides of March* is simply the fact that Julius Caesar actually was assassinated on the Ides of March, and nothing more need be said. In particular, we needn't cash out claims about what *was* the case in terms of some claim about what presently *is* the case, or in terms of something (a "property" or whatever) that exists *now*. Facts about what *was* are as primitive as facts about what *is*, and irreducible to the latter. (Cf. Baia 2012; Sanson and Caplan 2010; Tallant 2009.) The whole point of presentism, after all, is that the past and future don't have the kind of reality that the present does. Hence it shouldn't be surprising if the truthmakers for statements about the past and future are unlike the truthmakers for statements about the present. The non-presentist is bound to disagree with this, but to insist that all truthmakers have to be of the same kind simply begs the question.

A similar response should be given to another objection to presentism, to the effect that it cannot account for the *relations* that present things and events bear to past and future ones. Consider, for example, the fact that Caesar's assassination millennia ago is among the *causes* of my writing about it now. For a causal relation, or any other relation, to hold between two things, it seems that both relata have to exist. You can't bear any causal relations to unicorns, for example, because there are no unicorns. But the presentist holds that past things and events do not exist. Hence (so the objection claims) presentism cannot account for the causal relation between Caesar's assassination and my writing about it.

Now, the trouble with this objection is that it assumes that for a relation to hold between two things, they both have to exist *now*. But that is just what the presentist denies, or should deny. The presentist should say that it is sufficient that the relata *did* exist or *will* exist. I cannot be causally related to unicorns, because they not only do not exist, but never did. But I can be related to Caesar's assassination, because even though that event does not exist *now*, it *did* exist in the past. Again, facts about what *was* or *will be* the case are simply irreducibly different from facts about what now *is* the case, and it begs the question against presentism to insist that all of these facts must really be of the same kind.

Finally, there is the objection that the present moment lasts for merely an *instant*, and that this is too brief for anything to exist. Hence, the objection claims, presentism implicitly denies that *anything* exists, and not merely that the past and the future exist. To see what is wrong with this objection, it is important to recall the Aristotelian definition of time as *the measure of change with respect to succession.* As we have seen, this is a middle-ground position between the absolutist view that time exists entirely independently of changing things and of minds, and the idealist view that time is entirely mind-dependent. We also noted that this view is analogous to the Aristotelian realist position vis-à-vis universals. An abstraction like *triangularity* exists only insofar as there is an intellect that abstracts it, but it is abstracted *from* mind-independent concrete particulars rather than being a free creation of the intellect. Similarly, time as the Newtonian absolutist conceives of it exists only insofar as there is an intellect that abstracts it, but it is abstracted *from* the world of mind-independent changing things rather than being the intellect's free creation.

Hence, what exists in mind-independent reality is, say, a green banana. The banana has the potential to be yellow and the potential to be brown, but neither potential happens to be actualized. Rather, what it actually is, is green. If this were the end of the story, there would be no change and thus no time. With the actualization of the banana's potential yellowness (as it ripens), we have change, and thus time; and with the subsequent actualization of its brownness (as it rots) we have further change and thus further moments of time. When an intellect measures these changes by judging that they occurred over the course of several days, we have units of time, and these units can be divided into hours, minutes, seconds, and so forth. Once we have this system of units of measurement we can go on to form concepts like the concept of an instant. But it would be a fallacy to consider the banana at the stage it is actually at, and look for a second, or half-second, or instant as the duration of this present stage. That is somewhat like looking in the world of concrete material objects for a geometrical point, or a perfect line, or a perfect triangle. It is to confuse an abstraction *from* concrete reality with concrete reality itself. The assumption that the presentist must be committed to holding that the present lasts for an instant, or some specific brief duration, is rooted in this confusion.

Similar confusion underlies the objection sometimes raised against the A-theory in general to the effect that it is committed to there being a *rate* at which time passes and that this entails a vicious regress (since that rate would presuppose some higher-order time by reference to which the rate is measured). This is like saying that if you measure other things by reference to the standard meter, then you are committed to measuring the standard meter by reference to some yet further standard. In both cases, we have a category mistake. The standard meter is not some particular length alongside the others, standing in need of measurement the way they do. Rather, it just *is* the standard by which lengths are measured. Similarly, time is not some further kind of change alongside other kinds, standing in need of measurement the way they do. Rather, it just *is* the measure of change. Once again, the problem arises from a confusion of an abstraction with the concrete reality from which it is abstracted. It is when we treat time as some entity in its own right, existing over and above both changing things and the minds that measure change, that it comes to seem like something which itself undergoes change, and therefore must do so at a certain rate.

4.3.7 Physics and the funhouse mirror of nature

It is time to bring this long chapter to an end and to draw some general conclusions. I argued in chapter 2 that science cannot in principle eliminate from our picture of reality the point of view of the conscious and rational subject, and that we cannot make sense of this subject without attributing to it real change and embodiment. I also argued that this in turn entails that the key Aristotelian notions of actuality and potentiality, substantial form and prime matter, and efficient and final causality, retain their indispensability *at least* within the realm of this conscious and embodied rational subject, whatever we say about their applicability to the world external to the subject. In this current chapter I have argued that a further lesson that follows from all this is that *temporal passage* and the *now* cannot coherently be eliminated from our conception of the conscious and rational embodied subject. In short, *we know that the A-theory of time* must *be true of at least one portion of the natural world, namely us*.

I also argued in chapter 3 for an epistemic structural realist interpretation of physics, according to which the mathematical depiction of nature physics affords us gives us at most a highly abstract *blueprint* of nature which, though capturing real features of the world, is nevertheless far from an exhaustive description, and also adds elements that merely reflect the mathematical mode of representation rather than anything really there in nature itself. In this current chapter I have argued that this epistemic structural realist perspective should inform our understanding of what modern physics tells us about space, time, and motion. Insofar as physics represents the world as a static four-dimensional block, we should regard this as an artifact of the mathematical methods of physics, whose predictive successes no more show that the world is devoid of real change or temporal passage than the utility of a blueprint shows that a building is really a two-dimensional blue and white expanse. In short, *properly understood, physics gives us no reason to doubt that the A-theory of time also applies to the natural world* beyond *the conscious and rational embodied subject.*

To change metaphors, and to modify an arresting phrase from Richard Rorty (1979), what modern physics' picture of time, space, and motion gives us is something like a *funhouse mirror of nature.* The image in a funhouse mirror is by no means entirely inaccurate, and it can even allow you to see things you wouldn't otherwise see. You will see your face, arms, legs, etc. in the mirror, and these correspond to features you really have. If you see someone in the mirror approaching from behind, you can correctly predict that he will indeed be there if you turn around. If the funhouse mirror is of the magnifying kind, it will also allow you to see details (pores, blemishes, and the like) you wouldn't otherwise notice, and this may have utility outside the context of the mirror. All the same, the mirror does give a greatly distorted image, even a grotesque one. The portrayal of nature as a four-dimensional block universe is like that. Much of what is in the picture corresponds to what is really there in nature, and the picture also allows us to see features of nature we wouldn't otherwise see. All the same, the picture distorts nature, and some of what is in its portrayal really just reflects the mode of portrayal rather than the thing portrayed, just as some aspects of your image in the funhouse mirror are there because of the nature of the *mirror*, and not because of anything to do with you.

How distorted is the image physics gives us? That depends on which A-theory of time one endorses. If, as I have argued, presentism is true, then the image is highly distorted indeed. The character of the distortion, in any event, is that it drains the world of potentiality and represents it exclusively in terms of actuality. It is a kind of "Parmenideanization" of nature. As I have already noted, this has been a general trend in modern physics, from Newton's treatment of inertial motion as a kind of stasis to Einstein's spatialization of time. (As we will see in the next chapter, the "many worlds" interpretation of quantum mechanics reflects the same tendency, though in other ways quantum theory reflects an opposite "Heracliteanizing" tendency, to drain the world of actuality and represent it as pure potentiality.) To the extent that this is all intended merely as *methodologically* useful, there is no problem. Potential problems arise only when we start drawing *metaphysical* conclusions from it.

Readers who have not been paying careful attention are bound at this point to accuse me of letting philosophical considerations trump the findings of empirical science. But that is not at all what I am doing. To be sure, I would not for a moment concede that philosophy must always yield to what empirical science purports to show. As I have been arguing, philosophy is no less rational and objective than empirical science, and some of what it has to tell us is more fundamental than anything science can tell us, precisely because any possible empirical science would have to presuppose it. Hence it is certainly possible for philosophy to yield results to which empirical science has to conform itself.

But again, that is not what is going on in the present case. For there is nothing in the first place in the actual results of physics *per se* that tells us that dynamic change and temporal passage are illusory, that past and future events are as real as present ones, or the like. Physics could tell us these things only if *combined with* certain background philosophical assumptions about the nature of time, space, and motion and about how to interpret the results of science. And it is these latter assumptions that I am challenging. I am not pitting philosophy of nature against physics. I am pitting one philosophy of nature against another philosophy of nature.

Nor need one be an Aristotelian to see this. That claims made in the name of relativity theory often reflect philosophical assumptions as well as empirical results is widely noted. As we saw earlier, the role of verificationism in Einstein's formulation of STR is often remarked upon. It is also a commonplace in philosophy of physics that there are conventional elements in both STR and GTR. In particular, that spacetime appears curved could be interpreted as evidence that it really is curved, but it could also be interpreted instead as evidence that some force is affecting our measuring devices (Kosso 1998, pp. 102-3; Rickles 2016, pp. 83-90; Sklar 1992, pp. 53-69). And while the round-trip speed of light is empirically measurable, the one-way speed of light is not (Kosso 1998, pp. 105-6; Lowe 2002, pp. 268-69; Tooley 1997, pp. 357-58). The curvature of space and the constancy of the one-way speed of light are matters of convention, and while it does not follow that these conventions are wrong, their existence underlines the point that there are *already* philosophical considerations at work in the standard interpretation of relativity, long before the Aristotelian shows up. Then there is the fact that with relativity no less than any other physical theory, the question of whether an instrumentalist or epistemic structural realist interpretation is preferable to a standard scientific realist interpretation would arise for the usual reasons, whatever one thinks of Aristotelianism.

Once again, the confident claim that modern science has empirically refuted Aristotelian philosophy of nature disintegrates on careful inspection.

5. The philosophy of matter

5.1 Does physics capture all there is to matter?

As we saw in the previous chapter, the epistemic structural realist interpretation of physical science that I have been defending entails that there is more to space, time, and motion than meets the eye of the modern physicist. There is more to matter too. Indeed, philosophers sensitive to the highly abstract character of physics' description of the world often put special emphasis on how little physics tells us about the nature of matter. In light of modern physics, writes Bertrand Russell:

> It has begun to seem that matter, like the Cheshire Cat, is becoming gradually diaphanous until nothing of it is left but the grin, caused, presumably, by amusement at those who still think it is there...
>
> All that I know about matter is what I can infer by means of certain abstract postulates about the purely logical attributes of its space-time distribution. *Prima facie*, these tell me nothing whatever about its other characteristics. (1995, pp. 135 and 144)

Similarly, Michael Lockwood argues that physics' characterizations of material phenomena are "topic-neutral" in the sense that they "represent physical attributes only in the mode of... whatever it is that occupies the relevant positions within a certain causal structure" (1989, p. 160). (Cf. Chalmers 1996, p. 153; Strawson 2008.) If I tell you the rules of checkers, you will be able to figure out that a checkers game board will have squares of different colors on it, that the game pieces will be movable from one square to another, and so forth. But the rules will not tell you whether the board will be made out of cardboard and the pieces made of plastic, what color the board and pieces will be, etc. Similarly, physics tells us about different types of particles, their spatiotemporal relations, their causal roles, and the like. But it does not tell us the intrinsic nature of the entities that bear these relations to one another and play these roles.

The abstract mathematical character of physics is part of the reason for this. Simon Blackburn elaborates on the point by noting that when formulating conservation laws, physics must make use of predicates that apply across different possible realizations of the feature being predicated of a system (1990, pp. 63-64; 1991, pp. 205-6). For example, the physicist must characterize temperature in a way that applies across changes of state from solid to liquid to gas. This applicability across different possible realizations requires that a predicate be formulated in terms of the *functional role* of the feature being predicated, rather than its intrinsic nature, the thing that *plays* the role. Hence, in seeking unifying features or patterns that persist across the evolution of systems, "physics deals only in roles; it is role seeking 'all the way down'" (1991, p. 206).

As John Foster argues (1982, pp. 64-67), physics' inability to reveal matter's intrinsic nature also reflects the nature of the empirical testing of physical theories. Spatiotemporal relations, causal roles, and the like are all that *can* be tested by way of making empirical predictions. Hence, suppose we are trying to decide whether to affirm the existence of a particle of type *A* versus a particle of type *B*, where *A* and *B* have exactly the same causal roles, spatiotemporal relations, and other "topic-neutral" features and differ only in their intrinsic nature. Then there will be no empirical test to determine, for any particle we discover that exhibits those features, whether it is of type *A* or type *B*. Thus, concludes Foster, beyond its topic-neutral features, "matter is empirically inscrutable" (p. 66).

Unsurprisingly, then, the concept of matter as such does not actually play much of a role in modern physics. As Ernan McMullin has written:

> [T]he concept of matter faded out of science, leaving only its grin, so to speak, behind... [It] plays no *direct* part in the doing of science today. It still plays a part, albeit a tenuous and difficult-to-define one, in *talking about* science and its implications. (1963, p. 2)
>
> In the seventeenth century, the term 'matter' lost much of its significance for the working scientist because (1) from Descartes' time onwards, it was used more and more as an empty but convenient general term to denote the objects of physical science, without any sort of commitment as to the nature of these objects; (2) Newtonian mechanics substituted mass for matter in its analysis of the causes of motion. The term 'matter' does not occur, consequently, in any physical theory today; one finds no symbol

> corresponding to it in the equations of twentieth-century physics...
>
> [T]he result [is] that in most modern languages 'matter' is a vacuous term denoting simply that which is accessible to empirical investigation. To say that something is "matter" or "material" in this broad usage, therefore, is to say no more than that it is part of the physical world; nothing whatever is conveyed about the *properties* of the entity so labeled. (1978, pp. 295-96)

Noam Chomsky has also often noted how ill-defined "physical" and related terms are in contemporary academic discourse, conveying little more than the idea of whatever it is the physical sciences happen to investigate:

> There is no longer any definite conception of body. Rather, the material world is whatever we discover it to be, with whatever properties it must be assumed to have for the purposes of explanatory theory. Any intelligible theory that offers genuine explanations and that can be assimilated to the core notions of physics becomes part of the theory of the material world, part of our account of body. (1988, p. 144)

But might this not suggest that matter should simply be *identified with* whatever it is that physics describes or is capable of describing? Why suppose there is anything more *to* matter than that? One problem with such a suggestion is that, since physics gives us only abstract structure, the claim that there is nothing more to matter than what physics describes amounts to the claim that there is nothing more to the material world than abstract structure. And we have already seen, in chapter 3, that that thesis is incoherent. It makes no sense to suppose that there is nothing more to matter than what can be stated in topic-neutral or functional role terms, any more than it makes sense to suppose that there is nothing more to any game of checkers than its rules. (Cf. Robinson 1982, pp. 113-21; Blackburn 1990, p. 64.)

Another problem is that the attempt to identify matter with mass, or energy, or some other ersatz from physical theory simply doesn't work. Take the thesis that matter is identical with mass. As Colin McGinn points out (2011, pp. 37-38 and 63), there are several problems with this proposal. First, the idea of mass is the idea of a *quantity* of matter, and it would be circular to define matter as a quantity of matter. Second, neutrinos, though material, were once thought to have no mass. Though this thesis has now been abandoned, the fact that it was once held suffices to show that even modern physics supposes that there is more to matter than mass. Third, physics characterizes mass in terms of its *functional role* insofar as mass is understood in inertial terms, viz. in terms of the amount of force it would take to change an object's state of motion. But to explain what matter is would require determining what *plays* the role in question, rather than describing the role itself. As McGinn points out, this last problem also faces any attempt to identify matter with energy, since energy too is defined in physics in functional role terms (2011, pp. 63-64).

The upshot of these considerations is that the absence of some feature from physics' description of matter simply does not entail that the feature is absent from matter itself, since the methods of physics preclude it from providing an exhaustive description of matter in the first place. This is a point often emphasized by contemporary philosophers influenced by Russell (such as Michael Lockwood, David Chalmers, and Galen Strawson), who deploy the point in defense of the proposal that *mental* properties give us a model for what the intrinsic nature of matter might be like. But the Aristotelian may also deploy the same point to his own, very different, ends. In particular, the absence from physics of notions like substantial form, prime matter, and other concepts in Aristotelian philosophy of nature does not entail that these notions do not correspond to anything in physical reality, any more than the absence of certain features from a blueprint entails their absence from the building represented by the blueprint. Physics and philosophy of nature are simply concerned with different aspects of matter, just as a blueprint and a photograph of a building capture different aspects of the building.

5.2 Aristotle and quantum mechanics

That suggests that modern physics' fundamental theory of matter, quantum mechanics, is as neutral between Aristotelian and non-Aristotelian interpretations as (I argued in the previous chapter) relativity theory is when rightly understood. But an even stronger claim can be made, because there is a sense in which quantum mechanics actually points *toward* Aristotelianism, at least obliquely, rather than pointing away from it or even merely being neutral.

Needless to say, caution is called for in making such a claim, so let me immediately qualify it. As Peter Lewis emphasizes (2016, Chapter

1), when discussing quantum mechanics, we need to be careful to distinguish between (1) quantum phenomena, (2) quantum theory, and (3) alternative possible interpretations of quantum theory. As Lewis notes, the two main quantum *phenomena* are *interference* phenomena of the kind famously illustrated by the two-slit experiment, and *entanglement* phenomena of the kind famously illustrated by correlations holding between particles that are inexplicable in terms of properties of the individual particles themselves. Quantum *theory* comprises the mathematical representation of physical systems like the ones central to quantum phenomena, a law describing how such systems change over time, and a measurement postulate by which states of such systems are related to the results of measurements. The matrix mechanics of Werner Heisenberg, Max Born, and Pascual Jordan, and the wave mechanics of Erwin Schrödinger, are alterative but mathematically equivalent formulations of quantum theory. *Interpretations* of quantum mechanics are accounts of how the mathematical representation of quantum theory relates to objective physical reality – for example, the "Copenhagen interpretation" of Niels Bohr, the "pilot wave" interpretation of Louis de Broglie and David Bohm, and the "many worlds" interpretation of Hugh Everett.

Now, I am certainly not claiming that *all* of this points in an Aristotelian direction. Indeed, some of it is decidedly un-Aristotelian (such as the "many worlds" interpretation, on which I will comment presently). The idea is rather this. As we have seen, on the most *natural* interpretation of relativity (even if, as I have stressed, this is not the *only* possible interpretation or the *correct* interpretation), the theory describes a world that is entirely actual and devoid of potentiality. By contrast, on the most *natural* interpretation of quantum mechanics (even if, here too, this is not the only possible interpretation), the theory describes a world that is merely potential until actualized with the collapse of the wave function, and where parts exist virtually rather than actually in the wholes of which they are parts. In other words, it can be read as recapitulating Aristotelian hylemorphism and the theory of actuality and potentiality that is at the core of hylemorphism. Quantum mechanics arguably does for Aristotle what relativity theory is sometimes said to do for Parmenides.

Or it does so, at any rate, if we are talking about the nature of matter. But there are other aspects of physical reality to which quantum mechanics is relevant; and one of them is causality, which quantum mechanics is sometimes alleged to undermine. In *that* respect the theory might seem at odds with Aristotelianism. But here too, properly understood, quantum mechanics if anything supports a distinctively Aristotelian position. Let's consider these issues in turn.

5.2.1 Quantum hylemorphism

Recall that hylemorphism holds that a physical substance is a composite of prime matter and substantial form. Prime matter on its own is wholly indeterminate. By itself it is not an actual particular physical thing of any kind, but rather the pure potentiality to be a particular physical thing of some kind. If we think of matter on the analogy of the position of a needle on a dial and the values on the dial as representing the various specific kinds of material thing that might exist, prime matter is like a needle that is flitting wildly all across the face of the dial. It has no intrinsic tendency to stop at any particular value, though potentially it could be made to stop at any of them.

Substantial form is what actualizes that potential and makes of otherwise indeterminate prime matter a substance of some determinate kind – a molecule, a rock, a tree, a dog, or what have you. It also grounds a substance's properties and causal powers (for instance, a dog's four-leggedness and capacity to reproduce itself). Like prime matter, though, the substantial form of a purely corporeal substance does not subsist on its own. What subsists as a concrete particular thing in nature is a *dog*, not the substantial form of a dog or prime matter of a dog. Just as prime matter exists only as informed by a substantial form, so too a substantial form exists only as inhering in prime matter.

The fundamental existents in nature are therefore physical substances conceived of as composites of substantial form and prime matter. The mark of a true substance, and thus of a substantial form, is the presence of irreducible properties and causal powers. A true substance is, accordingly, to be distinguished from an *aggregate*, the properties and causal powers of which are reducible to the sum of the properties and causal powers of its parts; and from an *artifact*, the properties and causal powers of which are reducible to the sum of the properties and causal powers of its parts together with the intentions of its designers and users. Water is plausibly a true substance insofar as its distinctive properties and causal powers cannot be reduced to the sum of the properties and causal powers of hydrogen and oxygen. A random pile of stones is not a true substance, because its properties and causal powers are reducible to the sum of the properties and causal powers of the individual stones that make it up. A stone wall is not a true substance, because its properties and causal powers are reducible to the properties and causal powers of the individual stones that make it up, together with the intention of its designers and users that it function as a barrier.

The form of being a pile of stones and the form of being a wall are, accordingly, *accidental forms* rather than substantial forms. Accidental forms inform matter that is already informed by a substantial form, so that aggregates and artifacts are metaphysically less fundamental than true substances. The matter they inform, since it is already informed by a substantial form, is *secondary matter* rather than prime matter. Secondary matter, matter already informed by substantial form and bearing various accidental forms in addition to that, is the kind of matter we actually encounter in nature. Substantial change – the transition from one substance to another, as when water gives way to hydrogen and oxygen in hydrolysis – involves prime matter losing one substantial form and gaining another. Other kinds of change (of quantity, quality, or location) involve a substance gaining or losing an accidental form.

Whereas the parts of an aggregate or artifact exist within it actually, the parts of a true substance exist within it only virtually or potentially rather than actually. For example, in a pile of stones or a wall made of stones, the stones are all actual. By contrast, the hydrogen and oxygen in water are virtual or potential rather than actual. This does not mean these parts are *unreal*, for one of the central claims of Aristotelian metaphysics is that reality is not exhausted by what is actual. What is potential or virtual is not actual, but it is not nothing either. It is real, but a middle ground kind of reality, between actuality or full reality on the one hand and nothingness or unreality on the other. That hydrogen *really is* in water at least virtually or potentially is the reason you can draw hydrogen out of the water by hydrolysis, which you could not do with other substances. That it is in the water *only* virtually or potentially rather than actually is the reason you cannot burn the hydrogen in water, which you could do with actual hydrogen.

What is true of hydrogen and oxygen relative to the water of which they are constituents is true also of the more fundamental particles of which they are composed. In general, the particles of which any true physical substance is composed exist within it virtually or potentially rather than actually. For example, if a stone is a true substance, then while the innumerable atoms that make it up are real, they exist within it virtually or potentially rather than actually. What *actually* exists is just the one thing, the stone itself.

Hylemorphism is therefore anti-reductionist. Fundamental particles are not more real than the substances composed of them. Indeed, in a sense it is the substances (water, stone, etc.) that are more real, since they exist in an actual way and the particles exist in only a virtual or potential way. Hence it cannot be said that the substance is reducible to its constituent particles, both because it has properties and causal powers that are not reducible to those of the particles, and because this would make the particles metaphysically prior, whereas in fact the particles do not exist apart from the substance. But of course, there is also a sense in which substances depend for their existence on their constituent particles. Hence, hylemorphism is a kind of holism. Prime matter depends for its existence on substantial form and substantial form in turn depends for its existence on prime matter; a whole substance depends for its existence on its parts but the parts in turn depend for their existence on the whole. The dependence of matter on form and of parts on the whole is what Aristotelians call *formal causation.* The dependence of form on matter and of whole on parts is what they call *material causation.*

Now, if prime matter is like the needle flitting wildly across a dial's face, a fundamental particle, considered apart from any substance it might partially constitute, is like a needle which has narrowed its flitting somewhat to a certain range of possible values. Fermions do not have the indeterminacy of prime matter, for they are matter of a certain *kind*, with properties and causal powers distinctive of that kind. However, they do maintain a very high degree of indeterminacy insofar as there is an extremely wide variety of more complex kinds of matter that they might constitute. They do not flit back and forth past *every* possible value on the dial, but they do still flit past most of them. A fermion qua fermion can be a constituent of water, a stone, a dog, or what have you. Water and stone, by contrast, are like a needle that has settled down to flitting only across a very narrow range of possible values. Water may take a liquid, solid, or gaseous state; stone may be arranged in a pile or used to construct a wall. Compared to a fundamental particle, though, there is relatively little transformation they can undergo consistent with remaining what they are (viz. water or stone). Whereas prime matter is the pure potentiality to be any material thing, fermions have a somewhat narrower range of potentiality, and water a much narrower range.

It cannot be emphasized too strongly that the basic claims of hylemorphism need to be carefully distinguished from the specific examples used to illustrate it. Hence, suppose it turned out that water was more plausibly thought of as an aggregate than a substance in the Aristotelian sense, with properties and causal powers that were, after all, reducible to the sum of the properties and causal powers of its microphysical parts. That would not suffice to show that hylemorphism is mistaken. The most it would show is that the traditional hylemorphic analysis of the nature of *water*, specifically, is mistaken. Aristotelian philosophy of nature is, in any event, not primarily concerned with the natures of water, stone, fermions, dogs, or any other particular kind of material thing, but rather in the most general metaphysical features of physical reality. The Aristotelian would be the first to insist that philosophical claims about particular kinds of physical substance have to be informed by, and evaluated in light of, empirical considerations drawn from physics, chemistry, and biology.

So far this just recapitulates some of the points made in chapter 1, which I have elsewhere developed and defended at length on philosophical grounds (Feser 2014b). When we factor in what quantum physics tells us about matter, the broad outlines of this Aristotelian picture of nature are given further support. Indeed, the neo-Aristotelian character of quantum mechanics was recognized by Heisenberg. Echoing the hylemorphic account of the relationship between prime matter and fundamental particles, Heisenberg writes:

> All the elementary particles are made of the same substance, which we may call energy or universal matter; they are just different forms in which matter can appear.
>
> If we compare this situation with the Aristotelian concepts of matter and form, we can say that the matter of Aristotle, which is mere "potentia," should be compared to our concept of energy, which gets into "actuality" by means of the form, when the elementary particle is created. (2007, p. 134)

Regarding the "statistical expectations" quantum theory associates with the behavior of an atom, Heisenberg says:

> One might perhaps call it an objective tendency or possibility, a "potentia" in the sense of Aristotelian philosophy. In fact, I believe that the language actually used by physicists when they

> speak about atomic events produces in their minds similar notions as the concept "potentia." So the physicists have gradually become accustomed to considering the electronic orbits, etc., not as reality but rather as a kind of "potentia." (pp. 154-5)

And again:

> The probability wave of Bohr, Kramers, Slater... was a quantitative version of the old concept of "potentia" in Aristotelian philosophy. It introduced something standing in the middle between the idea of an event and the actual event, a strange kind of physical reality just in the middle between possibility and reality. (p. 15)

To be sure, Heisenberg's way of expressing the point needs tidying up. For one thing, he appears in these passages to *contrast* "potentia" with "reality." But as I have said, for the Aristotelian, potentiality is itself a kind of reality, albeit distinct from the reality that is actuality. Indeed, Heisenberg himself clearly agrees that "potentia" have a kind of reality, for elsewhere he notes that they "are completely objective, [and] do not depend on any observer" (2007, p. 27). Heisenberg's identification of "mere 'potentia'" or pure potentiality with energy also needs qualification. On the one hand, as David Oderberg has suggested (2007, p. 76), *if* we think of energy as having "no determinate form in itself," then one *might* be able to argue that it is identical to prime matter (though Oderberg refrains from endorsing such a proposal). On the other hand, as Stanley Grove has pointed out (2008, pp. 282-83), energy is *quantifiable*, whereas prime matter, being wholly indeterminate, is not – in which case, energy would be a kind of secondary matter that is "at least one structural level above that of prime matter."

Robert Koons (2018a, 2018b, and unpublished) identifies several further respects in which quantum mechanics might be said to support an Aristotelian conception of matter (yielding a synthesis Koons labels "quantum hylomorphism"). First of all, the Copenhagen interpretation treats microphysical phenomena, and in particular the position and momentum of a particle, as merely potential apart from interaction with macro-level systems (such as observers who measure the micro-level phenomena). Hence, like hylemorphism, the Copenhagen interpretation implies that the microphysical level is not metaphysically more fundamental than macro-level objects, nor sufficient by itself to ground all facts about the macro-level. Rather, the micro- and macro-levels are mutually interdependent, just as the Aristotelian claims.

Second, there is the holism implied by quantum entanglement phenomena. The properties of a system of entangled particles are irreducible to the properties of the particles considered individually or their spatial relations and relative velocity. The whole is more than the sum of its parts, as it is on the hylemorphic account of physical substances. Third, quantum statistics treats elementary particles of the same kind as indiscernible and essentially *fused* within a larger system, thereby losing their individuality and "merging into a kind of quantum goo or gunk" (Koons 2018a, p. 163). As Dean Rickles puts it, in quantum mechanics, such particles "are really excitations of one and the same basic underlying field" and best thought of "as 'dollars in a checking account' rather than 'coins in a piggy bank': they can be *aggregated* but not counted and distinguished" (2016, pp. 161-3). This echoes the Aristotelian position that parts exist in a substance virtually or potentially rather than actually.

Grove (2008, pp. 252-59) proposes that the wave-particle duality famous from quantum interference phenomena reflects precisely the greater indeterminacy that matter exhibits the closer it is to the level of prime matter. A photon can readily flit back and forth between wave-like and particle-like manifestations in a way that a cow cannot readily flit back and forth between cow-like and hamburger-like manifestations, because particles, being closer to prime matter, are (to use my analogy from earlier) like the needle on a dial flitting across a wide range of possible values. By contrast, a cow is far from the level of prime matter insofar as there are several intermediate levels of kinds of physical substance between it and prime matter (e.g. purely vegetative substances, middle-sized inorganic substances, particles of greater or lesser complexity). A cow is like a needle that flits only across a very narrow range of possible values on the dial. As Grove notes (pp. 263-66), the probabilistic nature of quantum events also reflects proximity to the level of prime matter. Again, the closer a substance is to the level of prime matter, the greater its indeterminacy (its tendency to flit across the dial, as it were), and thus the greater its unpredictability.

Grove also points out (pp. 259-63) that the abrupt or discontinuous changes described by quantum physics (such as an atom gaining an electron, or an electron being boosted to a higher energy level within an atom) are reminiscent of hylemorphic substantial change, which is also abrupt or discontinuous. The gain or loss of a substantial form is all or nothing, unlike the gain or loss of an accidental form, which can be continuous. Now, Democritean atomism, as we have noted in earlier chapters, essentially reduces changes that the Aristotelian regards as substantial to accidental changes, and thus interprets what the Aristotelian would take to be discontinuous transitions as in fact continuous ones. Insofar as quantum physics affirms abrupt or discontinuous transitions, it confirms an Aristotelian rather than Democritean understanding of matter, at least in a very general way.

Heisenberg's famous uncertainty principle is also, in Grove's view (pp. 266-72), something that should not be surprising on a hylemoprhic conception of matter. A particle's momentum implies a potentiality toward a range of possible positions, and its position implies a potentiality toward a range of possible momenta. But since momentum and position yield only potentialities rather than actualities, we should expect that knowledge of the one will yield less than certainty vis-à-vis the other.

I noted in chapter 1 that among the potentialities Aristotelian philosophy of nature attributes to natural substances are causal powers or dispositions. Physicist Ian Thompson argues that whereas modern physics had hoped that there might be only "a minimum number of these peculiar dispositions or potentialities, which seem like 'occult powers,'" in fact:

> Quantum physics shows, however, that this hope is not satisfied. In the quantum world there are in fact *more* kinds of dispositions than in Newtonian physics. For the properties of position and velocity, previously thought quite definite, now may or may not have definite values. Position and velocity seem to behave more like dispositional properties... In the quantum world, it turns out, there are very few *non*-dispositional properties. (2010, p. 37)

Now, this is not to endorse every claim made by any of the familiar alternative interpretations of quantum mechanics. Bohr's version of the Copenhagen interpretation is anti-realist with respect to the quantum level, and smacks of verificationism (even if Bohr regarded his position as motivated by scientific rather than philosophical considerations). In chapter 3, I argued against both anti-realism and verificationism.

The notion of observation or measurement in the Copenhagen interpretation is also notoriously problematic. On the one hand, we have the *wave function* of a quantum system or object, such as a particle – a mathematical description of the different possible states it might be in, together with the probabilities of those states. Apart from observation, the system or object is said to be in a *superposition* of these alternative possible states. On the other hand, we have an act of observation or measurement of the quantum system or object. This act is said to *collapse* the wave function, i.e. to determine which of the alternative possible states the system or object is actually in. For the Copenhagen interpretation, there is a sharp boundary between the observer or measurement apparatus and the system being observed, and the former is governed by classical or non-quantum principles. But how exactly is the boundary to be drawn? How exactly does the act of measurement collapse the wave function? Why do measurements bring about a collapse whereas other events do not? What counts as an observer? Are human beings alone able to perform measurements? Or can non-human animals do so? Which ones? Can an inanimate object count as performing a measurement? Weren't there wave function collapses before observers existed?

In the famous *Schrödinger's cat* thought experiment, an unobserved radioactive particle is in a superposition of both decaying and not decaying, and an unobserved cat whose being poisoned depends on whether or not the decay occurs is, accordingly, in a superposition of both being dead and being alive. Given its claim that a superposition collapses only when an observation is made, the Copenhagen interpretation seems to have the bizarre implication that there is no fact of the matter about whether the cat is dead or alive until it is observed. Indeed, the implications are even stranger, as Eugene Wigner showed with what has come to be called the *Wigner's friend* paradox. Suppose not only that Schrödinger's cat is in a box waiting to be observed, but also that the box is in a larger box together with a scientist who waits to open the first box and observe the cat. Then, just as the particle is in a superposition of being decayed and not being decayed and the cat is in a superposition of being dead and being alive, so too the scientist is in a superposition of both observing the cat and not observing it. And of course, we could also imagine a scenario in which whoever observes the scientist is himself in some yet larger box, waiting to be observed by someone who is in a larger box still, and so on *ad infinitum*. (Cf. Gribbin 1984, pp. 205-8; Rickles 2016, pp. 155-56.)

There are several problems illustrated by examples like these. Again, the Copenhagen interpretation claims that there is a sharp divide between quantum systems on the one hand, and classical systems on the other, and it puts the observer or measurement apparatus on the classical side of the divide. But as the Wigner's friend example shows, it is difficult to find a non-arbitrary way to draw such a line. A further problem is that it is not plausible to suppose that there is no fact of the matter about whether macro-level objects are in one state rather than another. Even if one were to bite the bullet and draw such an odd conclusion in the case of the cat, it is harder to draw it in the case of the scientist who observes the cat. Suppose, after opening the box and observing what the scientist is doing, we ask him whether he was observing the cat or not before we opened the box. He will tell us either that he was or that he wasn't. Whatever answer he gives, it follows that there *was* a fact of the matter about what he was doing in the box even before we observed him. (Cf. Barrett 1999, pp. 51-55; D'Espagnat 2006, pp. 228-36.) Wigner himself drew an essentially Cartesian conclusion to the effect that the observer's mind must be outside the physical world altogether. This would certainly give us the sharp divide between a quantum system and the observer of the system required by the Copenhagen interpretation, but as I argued in chapter 2, the Cartesian conception of the observer is false.

On the "many worlds" interpretation of quantum mechanics, observation of the cat famously leads to a splitting of the world into parallel and equally real branches, in one of which the particle has decayed and the cat is observed to be dead and in the other of which the particle has not decayed and the cat is observed to be alive. And the same is true for every other superposition, on this interpretation. Each possible outcome turns out to be equally real, so that there are parallel worlds in which the Allies lost World War II, parallel worlds in which Einstein was a bus driver rather than a physicist, and so on. Or rather, there is just one reality, a multiverse governed by a universal wave function, where the myriad parallel worlds reflect the various possible outcomes represented by the wave function.

Naturally, a common objection to this view is that it is difficult to see how it could be empirically tested, since our branch of the purported multiverse is the only one to which we have access. Another objection is that no one would take such a bizarre view seriously except on the supposition that the nature of physical reality can be read off from the mathematics of the universal wave function. But as I have argued, it is simply a basic metaphysical error to identify the physical world with what a mathematical representation of it can capture. It is also still not clear even on the "many worlds" interpretation whether measurements alone lead to a branching of the universe, or if any interaction does so – and if so, why (Kosso 1998, p. 169). A fourth problem is that if all possible outcomes are equally real, then there is no sense to be made of the probabilities that quantum mechanics assigns to the outcomes – *every* outcome has the same probability, namely 1 (Koons 2018c; Putnam 2012a). This reflects what the Aristotelian must regard as a deeper problem with the "many worlds" interpretation, which is that it is essentially a variation on Parmenidean monism, in which the world is regarding as a single substance and there are no unactualized potentialities.

There are variations on the many worlds interpretation that are even more bizarre. The "many minds" interpretation attempts to deal with the problem that probabilities raise for the many worlds interpretation by positing an infinite number of minds associated with each observer, where each mind has a certain probability of observing one or the other outcome described in a superposition (Albert and Loewer 1988). The "traveling minds" interpretation posits only a single mind associated with each observer, and holds that when branching occurs, that mind travels down one of those branches and not the other (Barrett 1999). These already strange proposals entail yet stranger further problems (Pruss 2018, pp. 108-11). For example, an apparent implication of the traveling minds view is that the mind now associated with your brain is likely to be surrounded by other people whose bodies do not really have minds, but are among the mindless bodies that resulted from earlier branching events. You inhabit a world of "mindless hulks" or zombies (Albert 1992, p. 130). Like Wigner's proposal, the "many minds" and "traveling minds" proposals also entail a Cartesian form of dualism that I have already argued against (Lewis 2016, pp. 131-32).

Then there is Bohm's "pilot wave" interpretation of quantum mechanics, which deals with wave-particle duality by positing *two* things, particles *and* waves (rather than one thing that exhibits properties of both), with waves guiding or "piloting" particles. As Koons points out, Bohm's position, like the many worlds interpretation, amounts to a kind of monism insofar as it treats the universe as a single fundamental substance, of which particles are passive components rather than having active causal powers of their own. (Koons unpublished; cf. Lewis 2016, pp. 169-70 on the holistic character of Bohmian mechanics.) Interpreted in a deterministic way, Bohm's position also has difficulty accounting for the probabilities described by quantum mechanics (Koons 2018c, p. 93).

From an Aristotelian point of view, then, the traditional alternatives among possible interpretations of quantum mechanics are all seriously problematic in one respect or another. This is no embarrassment for the Aristotelian, because quantum mechanics is, notoriously, a mess from *anyone's* point of view. A famous remark of Feynman's has become a cliché, but it is true: "I think I can safely say that nobody understands quantum mechanics" (1994, p. 123). What physics has given us is a mathematical description and technical jargon the application of which has yielded unprecedented predictive and technological success. And yet it leaves the question of how to interpret the mathematics and the jargon fairly wide open. As Hilary Putnam writes, "mathematically presented quantum-mechanical theories do not wear their ontologies on their sleeve... the mathematics does not transparently tell us what the theory is *about*. Not always, anyhow" (2012b, p. 161). A term like "superposition," as David Albert notes, is "just a name for something we don't understand" (1992, p. 11). What the physics tells us about a quantum object in a superposition of states *A* and *B* is that it is not in *A* and it is not in *B* and it is not in both and it is not in neither, but rather in a superposition of *A* and *B*. "And what that means (other than "none of the above") we don't know" (Albert 1992, p. 11).

What the Aristotelian proposes is that while in principle there are a number of philosophical glosses one could put on this, the most *natural* way to interpret a notion like that of a superposition is as a rediscovery of the Aristotelian concept of potentiality, and the most *natural* way to account for the mysteriousness of quantum phenomena is in terms of their proximity to the indeterminacy of prime matter.

As Koons notes, there are at least two directions in which the quantum hylemorphist could go from here. The first would be to allow that there is a single wave function that describes the whole of quantum reality, even though there is more to physical reality than the quantum realm. This is the approach of what Alexander Pruss (2018) calls the "traveling forms" interpretation of quantum mechanics. This is similar to the traveling minds interpretation, except that it is Aristotelian substantial forms, rather than minds, that "travel" down the branches. It is a form together with the branch it travels down that constitutes a macro-level substance.

But more in the spirit of traditional Aristotelian philosophy of nature (and of the general philosophy of science defended in earlier chapters of this book) is what Koons calls *pluralistic* quantum hylemorphism, which is informed by Nancy Cartwright's (1999, Chapter 9) pluralistic interpretation of quantum mechanics. (Cf. also Grove 2008; Smith 1999 and 2005; and Wallace 1997.) As we saw in chapter 3, for Cartwright, laws of nature are not universal regularities, but rather describe how natural phenomena behave only under certain conditions. She also takes laws to form a patchwork rather than a pyramid. There are the laws governing the behavior of phenomena in *this* domain, and the laws governing the behavior of phenomena in *that* one, but there is no single most basic set of laws to which all the others are reducible. Hence, while in Cartwright's view we "should take the quantum state seriously as a genuine feature of reality and not take it is as an instrumentalist would," at the same time:

> Nor should [we] insist that other descriptions cannot be assigned besides quantum descriptions. For that is to suppose not only that the theory is true but that it provides a complete description of everything of interest in reality. And that is not realism; it is imperialism.
>
> But is there no problem in assigning two different kinds of descriptions to the same system and counting both true?... Problems are not there just because we assign more than one distinct property to the same system. If problems arise, they are generated by the assumptions we make about the relations among those properties: do these relations dictate behaviours that are somehow contradictory? The easiest way to ensure that no contradictions arise is to become a quantum imperialist and assume there are no properties of interest besides those studied by quantum mechanics. In that case classical descriptions, if they are to be true at all, must be reducible to (or supervene on) those of quantum mechanics. But this kind of wholesale imperialism and reductionism is far beyond anything the evidence warrants. (1999, pp. 232-33)

As Koons characterizes pluralistic quantum hylemorphism, it combines Cartwright's patchwork conception of nature with certain elements of both the Copenhagen interpretation and what are called "objective collapse" interpretations of quantum mechanics, while rejecting other elements of these latter interpretations. Like objective collapse theories and unlike Bohr's version of the Copenhagen interpretation, it affirms realism about the quantum realm. Like objective collapse theories and unlike the Copenhagen interpretation, it does not make wave function collapse depend entirely on observers. Rather, it takes wave function collapse to result from the interaction of the irreducible macro-level properties of natural objects and systems with their micro-level quantum properties. Like the Copenhagen interpretation and unlike objective collapse theories, pluralistic quantum hylemorphism is ontologically pluralistic insofar as it affirms that the classical domain is real and irreducible.

Indeed, it is more pluralistic than the Copenhagen interpretation, since, following the traditional Aristotelian view that a variety of irreducibly different kinds of substance exist in nature, it allows for further pluralism *within* the classical domain. As Koons summarizes the view:

> [T]he world consists of a variety of domains, each at a different level of scale. Most of these domains are fully classical, consisting of entities with mutually compatible or commutative properties. At most one domain is accurately described by quantum mechanics... Interaction between quantum properties and classical properties, including those of experimenters and their instruments, precipitates an objective collapse of the quantum object's wavefunction because of the joint exercise of the relevant causal powers of the object and the instruments and not because of the involvement of human consciousness and choice. (2018b)

As with the treatment of relativity in chapter 4, I do not claim to have done more here than scratch the surface, but what has been said suffices for present purposes, viz. to give the lie to any suggestion that modern physics is incompatible with a broadly Aristotelian philosophy of nature.

5.2.2 Quantum mechanics and causality

This is true even where quantum mechanics touches on matters of causality. Indeed, Adam Schulman (1989) has argued that quantum mechanics has recapitulated the Aristotelian conception of change as the actualization of potential, which is central to the Aristotelian understanding of causality within nature. An object's movement from *A* to *B* is, for Aristotle, the actualization of its potential to be at *B*. Now, as Schulman reads Aristotle, it is only when the object arrives at *B*, and the motion is completed, that the object is *actually* anywhere other than *A*. In between, it is no longer actually at *A*, but not yet actually at *B* either – it is still only potentially at *B* or anywhere else. Moreover, there are various alternative trajectories the object could take on its journey from *A* to *B*. The object's position and trajectory are *indeterminate* insofar as it is still in motion, and it is only after the motion has stopped that it can be said that the object was determinately on such-and-such a path and at such-and-such particular points between *A* and *B*. This is, in Schulman's view, why Aristotle famously defines motion in Book III of the *Physics* as "the actuality of the potential, as potential." Something in motion is *actualizing* a potential, but until the motion is complete, the actualization is not complete, and so there is still potentiality. Hence motion entails the actualization of potential *as still potential*, rather than as actualized.

In Schulman's view, the indeterminacy Aristotle attributes to an object still in motion is echoed in Heisenberg's uncertainty principle, and especially in Feynman's "sum over histories" approach to quantum mechanics. In Feynman's famous model, there are an infinite number of possible paths a particle can take when traveling from *A* to *B*, each path associated with a certain probability. If we think of the particle as taking every possible path, the paths will interfere with each other, some reinforcing others and some canceling others out, until a single path emerges at the end as the one the particle actually took. But each possible path makes a contribution to the outcome. Schulman proposes that the indeterminacy in the particle's path from *A* to *B* on Feynman's model is a recapitulation of Aristotle's notion that an object's positon and trajectory are still potential and indeterminate so long as the motion is incomplete.

Be that as it may, in other respects quantum mechanics might appear to conflict with the Aristotelian understanding of causality – and in particular, with the *principle of causality*, according to which any potential that is actualized is actualized by something already actual. There are three main arguments to this effect. The first is that the non-deterministic character of quantum systems is incompatible with the principle of causality. The second appeals to the Bell inequalities (named for physicist John S. Bell), which have to do with measurements made at distant locations between which there are correlations that appear not to have a common cause. The claim is that these correlations lack a causal explanation, and thus conflict with the principle of causality. The third argument is that quantum field theories show that particles can come into existence and go out of existence at random.

As to the objection from indeterminism, it is sometimes pointed out in response that the de Broglie-Bohm hidden variable interpretation of quantum mechanics provides a way of seeing quantum systems as deterministic. But from an Aristotelian point of view it is a mistake to suppose in the first place that causality entails determinism. For a cause to be sufficient to explain its effect it is not necessary that it cause it in a deterministic way. It need only make the effect intelligible. And that condition is satisfied on a non-deterministic interpretation of quantum mechanics. As Koons writes:

> According to the Copenhagen version of quantum mechanics, every transition of a system has causal antecedents: the preceding quantum wave state, in the case of Schrödinger evolution, or the preceding quantum wave state plus the observation, in the case of wave packet collapse. (2000, p. 114)

As to the objection from the Bell inequalities, it is sometimes suggested that one could respond to it by denying that causal influences never travel faster than light (Koons 2000, p. 114), or by allowing for either backward causation, or an absolute reference frame, or positing a law to the effect that the correlations in question take place (Pruss 2006, pp. 166 and 169). As to the objection that particles can come into or go out of existence at random in a quantum vacuum, Alexander Pruss (2006, pp. 169-70) suggests that here too one might propose a hidden variable theory, or, alternatively, propose that the system described by the laws of quantum field theory is what causes the events in question, albeit indeterministically. Of course, all such proposals raise questions. My aim, however, is not to defend any of them, but rather to note that their very existence illustrates the point that quantum mechanics *by itself* does not say anything one way or the other about causality. What lessons we draw depend on what further assumptions we bring to bear.

This brings us to the deeper point to be made in response to objections to the principle of causality that appeal to quantum mechanics, which is the epistemic structural realist point that quantum mechanics, like mathematical physics more generally, doesn't give us anything close to an exhaustive description of nature in the first place. Hence the absence of causality from a quantum mechanical description of a system does not by itself entail that causality is absent from the system itself. Recall, from chapter 4, James Weisheipl's remarks about Newton's principle of inertia:

> [T]he nature of mathematical abstraction... must leave out of consideration the qualitative *and causal* content of nature... [S]ince mathematical physics abstracts from all these factors, it can say nothing about them; it can neither affirm nor deny their reality. (1985, p. 48, emphasis added)

This is as true of quantum mechanics as it is of Newtonian mechanics and relativity. Recall also, from chapter 2, Russell's argument to the effect that causation must be illusory because it does not appear in the mathematical description of nature afforded by physics. Though, as I there argued, the conclusion does not follow, the premise is true enough. As Grove writes, "efficient causality is at best implied, *never described*, by the quantitative spatiotemporal relations studied in physics" (2008, p. 293).

Now, the *ways* these theories leave out causation are different. As I have said, quantum mechanics' description of matter at least approximates the pure potentiality of Aristotelian prime matter; and the measurement problem illustrates the difficulty quantum theory has had in making it clear exactly how that potentiality gets actualized. By contrast, and as I argued in chapter 4, Newtonian mechanics, and to a much greater extent the four-dimensional block universe interpretation of relativity, at least approximate a Parmenidean description of nature as purely actual. Now, causation, the Aristotelian argues, involves the *actualizing* of a *potential*. Hence to leave out either one of these two elements is to leave out causation. That is exactly what relativity does to the extent that it suggests a picture of the world as entirely actualized and devoid of potentiality, and it is exactly what quantum mechanics does to the extent that it suggests a picture of matter as purely potential and leaves it murky exactly how the wave function is collapsed and thus exactly what actualizes the potential. In both cases what is missing is missing, not because it is absent from reality, but because it is bound to be absent from a consistently mathematicized description of reality.

It must also be emphasized that objections to the principle of causality which appeal to quantum mechanics are, ultimately, appeals to laws of physics. But as we saw in earlier chapters, for the Aristotelian, a law of physics is essentially a shorthand description of the way a thing will behave given its nature or substantial form. Thus, to explain something in terms of the laws of physics is hardly an *alternative* to explaining it in terms of the actualization of a potential. For the substantial form of a physical substance relates to the substance's prime matter as actuality to potentiality. In particular, the substantial form is what actualizes the potentiality of prime matter to be a substance of such-and-such a kind. In that way, the operation of any laws of nature, including the laws of quantum mechanics, *presupposes* the actualization of potential.

Hence, consider radioactive decay, which is usually regarded as indeterministic, and thus often claimed to pose a challenge to the principle of causality. Specifically, consider an example given by philosopher of science Phil Dowe:

> Suppose that we have an unstable lead atom, say Pb^{210}. Such an atom may decay, without outside interference, by α-decay into the mercury atom Hg^{206}. Suppose the probability that the atom will decay in the next minute is *x*. Then
>
> $$P(E|C) = x$$
>
> where C is the existence of the lead atom at a certain time t_1, and E is the production of the mercury atom within the minute immediately following t_1. (2000, pp. 22-23)

Now, from an Aristotelian point of view, what is going on here is that Pb^{210} simply behaves, like all other physical substances do, according to its nature or substantial form. Copper, given its nature or substantial form, will conduct electricity; a tree, given its nature or substantial form, will sink roots into the ground; a dog, given its nature or substantial form, will tend to chase cats and squirrels. And Pb^{210} is the sort of thing which, given its nature or substantial form, is such that there is a probability of *x* that it will decay in the next minute. The decay is not deterministic, but that does not entail that it is unintelligible. It is grounded in *what it is to be* Pb^{210} as opposed to being some other kind of thing – that is to say, it is grounded, again, in the nature or substantial form of Pb^{210}. This is what in the Aristotelian calls the *formal cause* of a thing. There is also a generating or *efficient* cause, namely whatever it was that originally generated the Pb^{210} atom at some point in the past (whenever that was).

There is a parallel here with Aquinas's views about local motion or change with respect to location or place, which we discussed in chapter 4. Aquinas took the view that a substance can manifest certain dispositions in a "spontaneous" way in the sense that these manifestations simply follow from its nature or substantial form, and that a thing's natural tendencies vis-à-vis local motion would be an example. Because such motions simply follow from the thing's form, they do not require a continuously conjoined external mover. Now, that is not, in Aquinas's view, to say that the motion in question does not have an efficient cause. But the efficient cause would just be whatever generated the substance, and thus gave it its substantial form which (qua formal cause) accounts for its natural local motion. (It is commonly but erroneously thought that medieval Aristotelians in general thought that all local motion *as such* required a continuously conjoined cause. In fact that was true only of *some* of these thinkers, not all of them. For detailed discussion of this issue, see Weisheipl 1985, from which I borrow the language of "spontaneity.")

Now, Aquinas elaborated on this idea in conjunction with the thesis that the "natural place" toward which heavy objects are inclined to move is the center of the earth, and he supposed also that *projectile* motions *did* require a conjoined mover insofar as he regarded them as "violent" motions rather than natural ones. Both of these suppositions are scientifically outmoded, but the more general thesis summarized in the preceding paragraph is logically independent of them and can easily be disentangled from them. One can consistently affirm (a) that a substance will tend toward a certain kind of local motion simply because of its substantial form, while rejecting the claim that (b) this local motion involves movement toward a certain specific place, such as the center of the earth. Indeed, as we saw in chapter 4, some Aristotelians have proposed that affirming (a) while rejecting (b) is the right way to think about *inertial* motion: Newton's principle of inertia, on this view, is a description of the way a physical object will tend to behave vis-à-vis local motion given its nature or substantial form.

The point for present purposes, though, is that the idea just described also provides a model – not necessarily the only model, but *a* model – for understanding what is going on metaphysically with phenomena like radioactive decay. We can say that the decay described in Dowe's example is "spontaneous" in something like the way Aquinas thought the natural local motion of a physical substance is "spontaneous." In particular, given the nature or substantial form of Pb^{210}, there is a probability of *x* that it will decay in the next minute. The probability is not unintelligible, but grounded in *what it is to be* Pb^{210}. The decay thus has a cause in the sense that (i) it has a *formal* cause in the nature or substantial form of the particular Pb^{210} atom, and (ii) it has an *efficient* cause in whatever it was that originally generated that Pb^{210} atom.

Dowe makes a further point which reinforces the conclusion that examples like that of Pb^{210} show merely that not all causality is deterministic, but not that there is no causality at all in radioactive decay:

> If I bring a bucket of Pb^{210} into the room, and you get radiation sickness, then doubtless I am responsible for your ailment. But in this type of case, I cannot be morally responsible for an action for which I am not causally responsible. Now the causal chain linking my action and your sickness involves a connection constituted by numerous connections like the one just described [in the passage quoted above]. Thus the insistence that C does not cause E on the grounds that there's no deterministic link entails

> that I am not morally responsible for your sickness. Which is sick. (2000, p. 23)

Dowe also points out that "scientists describe such cases of decay as instances of *production* of Hg^{206}... [and] 'production' is a near-synonym for 'causation'" (p. 23). This sounds paradoxical only if we fallaciously conflate deterministic causality and causality as such.

5.3 Chemistry and reductionism

Again, for the Aristotelian, the mark of a true physical substance is the presence of irreducible properties and causal powers. A genuine substance thus contrasts with an aggregate, whose properties and powers are reducible to the sum of those of its parts; and with an artifact, whose properties and powers are reducible to the sum of those of its parts, together with the intentions of the artifact's designers and users. The parts of a true substance, unlike the parts of an aggregate or artifact, exist within it virtually rather than actually. Traditionally, the Aristotelian takes nature to be filled with genuine substances in this sense – water, stone, gold, trees, fish, birds, dogs, cats, and so on.

Traditionally, this view contrasts with ancient atomism, according to which these things are all reducible to collections of particles moving through the void. Water, stone, gold, trees, fish, birds, dogs, and cats are, on the atomist analysis, all essentially aggregates (or divine artifacts, if theism is added to the story). Their parts, the atoms, do exist in them actually and not just virtually. The atoms thus turn out to be the only true physical substances. If what is ontologically fundamental is what is irreducible to anything else, then for the atomist, only the atoms are ontologically fundamental. For the Aristotelian, by contrast, water, stone, gold, trees, fish, birds, dogs, cats, and the like are no less ontologically fundamental than the particles spoken of by physics. Indeed, there is a sense in which these ordinary objects are *more* fundamental than the particles that make them up, insofar as the particles exist in them only virtually, only relative to the wholes of which they are parts.

This is what makes the Aristotelian picture of nature what Cartwright calls a "dappled" or "patchwork" conception. There are the nature and causal powers of water, and the laws governing water are just a description of how it will behave given that nature and those powers. There are the nature and causal powers of dogs, and the laws governing dogs are a description of how they will behave given that nature and those powers. And so on. When, mentally, we abstract away what is distinctive about water, what is distinctive about dogs, and what is distinctive about all other particular kinds of physical thing, we can formulate laws that describe how they will behave qua bodies in motion, qua quantum systems, or whatever. But what actually exist concretely in the world outside the mind are *water*, and *dogs*, and all the rest – not "bodies in motion," "quantum systems," or the like. The latter are *abstractions* rather than concrete realities, and the laws describing them have the generality they do precisely because the entities are abstractions. To be sure, the laws really do describe how water, dogs, etc. behave, but because water, dogs, etc. are not *merely* bodies in motion or *merely* quantum systems, their actual behavior will be much more complex than that of the abstractions described by the laws. Hence the laws will only ever approximate the actual behavior of the concrete objects. That the observed behavior of things in the world only ever approximates what the laws predict has *metaphysical,* and not mere epistemological, significance. It reflects, not a mere gap in our knowledge, but the fact that actual, concrete, mind-independent reality is simply not identical to the abstractions of physical theory.

The atomist picture, by contrast, denies that nature is dappled or a patchwork. Because everything is really just the same kind of thing (atoms in motion) there is one set of laws which governs everything (the laws governing the atoms), and since the atomist description of a thing captures the whole truth about it rather than just being an abstraction, the laws are not mere approximations. The atomist view is thus radically reductionist. If we had complete knowledge of every atom and the laws governing the atoms, we would have complete knowledge of everything that is true of water, stone, gold, trees, fish, birds, dogs, cats, etc., because *there is simply nothing more to these things* than what can be stated in terms of atoms and the laws governing them. That what we observe only ever approximates what the laws predict does have mere epistemological significance, on this view, reflecting mere gaps in our knowledge either of the relevant empirical circumstances or of how correctly to formulate the laws.

Now, no modern philosopher or scientist would accept every aspect of the ancient atomist picture, as is evident even just from the fact that neither "atoms" in the modern sense nor any other particles are taken to have exactly the nature the atomists ascribed to the atoms. (Cf. Hoenen 1960 and Van Melsen 1960 for treatments of the differences between Democritean atomism and the modern scientific understanding of particles.) But the basic reductionist thrust of atomism certainly survives in much modern thinking about these matters – for example, in that of philosophers who maintain that there are no stones but only "particles arranged stone-wise" (Van Inwagen 1990; Merricks 2001).

As I have argued elsewhere (Feser 2014b, pp. 177-84), it can be shown on general metaphysical grounds that some form of hylemorphism is true and atomism and its modern descendants are false. Even if a reduction of all ordinary objects to atoms (or some modern surrogate) could be carried out, the atoms *themselves* (or their modern surrogates) would, on analysis, still have to be conceived of as composites of substantial form and prime matter. The dispute would, on analysis, turn out to be over *which things* are hylemorphic compounds, not over *whether there are* hylemorphic compounds. Moreover, there are general metaphysical problems facing attempts to reduce ordinary objects to atoms (or their modern surrogates).

I won't repeat all those arguments here, but an objection grounded in modern physical science needs to be addressed. For it might appear that science has vindicated the reductionist ambitions of atomism at least *to a large extent* even if not completely. For example, it might be claimed that modern chemistry has shown that water is really nothing but hydrogen and oxygen, that modern physics has shown that the facts about elements like hydrogen and oxygen are in turn reducible to quantum mechanics, and that a similar story can be told about every other object or stuff familiar from ordinary experience. Hence, even if hylemorphism gets the big picture right, it might appear that it is still largely wrong about the everyday world. It wins the war, but only after losing most of the battles. (Something in the ballpark of this view was taken by early twentieth-century Aristotelian-Thomistic philosophers who proposed replacing hylemorphism with "hylosystemism," at least as an analysis of inorganic bodies. See Bittle 1941, Chapter XIV for a sympathetic exposition and Greene 1952 for criticism.)

Yet you don't need to be an Aristotelian to reject this characterization of the situation. For work in contemporary philosophy of chemistry casts doubt both on the claim that chemistry affords us reductionist accounts of ordinary substances, and on the claim that physics affords us a reductionist account of chemistry. Indeed, in their *Stanford Encyclopedia of Philosophy* article surveying the field (2011), Michael Weisberg, Paul Needham, and Robin Hendry speak of an "anti-reductionist consensus in the philosophy of chemistry literature."

One of the problems is usefully approached by way of John Locke's account, in Book III of the *Essay Concerning Human Understanding*, of the relationship between *nominal essence* and *real essence*. The nominal essence of a substance like water is essentially the collection of attributes common sense groups together under the *term* "water," such as being liquid at room temperature, being clear, boiling when heated, turning to ice when frozen, and so forth. The real essence is the corpuscular structure of water as revealed by physical science, on which the attributes associated with the nominal essence depend. The nominal essence might be said to correspond to the "manifest image" of water, and the real essence to the "scientific image" of water (to borrow Sellars' language).

Now, I certainly wouldn't endorse everything Locke has to say about essence. (For one thing, Locke's overall position, as I have explained elsewhere, tends toward nominalism. Cf. Feser 2007, pp. 56-66.) But among the useful points he makes is that the relationship between nominal and real essences is *reciprocal*. On the one hand, deep and truly scientific knowledge of a thing cannot rest at the level of the nominal essence, but requires penetrating to the underlying real essence. That much might sound like grist for the reductionist's mill. But at the same time, Locke tells us that a real essence is always the real essence of a *species* or *class* of things; and it is the nominal essence that determines what species or class a thing belongs to (*Essay*, Book III, Chapter VI, paragraphs 6-7). Hence, while only a grasp of the real essence of water gives us a deep and scientific knowledge of its nature, we couldn't know in the first place that what we thereby grasp the nature of is *water*, specifically, unless we knew the nominal essence. Real essence, you might say, "piggybacks" on nominal essence, and the scientific image on the manifest image. Our grasp of the former presupposes a grasp of the latter.

J. van Brakel (2000, pp. 73-82) deploys this point in criticism of facile reductionist claims about water and H_2O. To forestall misunderstanding, note that the question is *not* whether modern chemistry is correct to identify hydrogen and oxygen as the constituents of water. Of course it is correct; no one denies that. What is at issue is the idea that water is *nothing but* hydrogen and oxygen in a certain arrangement – that the scientific image of water exhausts its nature and the manifest image is otiose.

To be sure, the chemical facts are actually a bit more complicated than the routine reference to water as H_2O would imply (Chang 2012, pp. xvi and 210; Hoffmann 1995, pp. 32-34; Van Brakel 2000, pp. 80-81), but that is not the main point. As Van Brakel points out, when dealing with water we judge that we are dealing with a substance, and with the same substance over time, from its macroscopic properties and apart from our knowledge of chemistry. For example, we don't need to know anything about the microstructure to judge that the water in a certain glass that is first liquid, then freezes, then melts and is liquid again, is the same stuff throughout. We don't need to know the microstructure to judge that when we stir sand into a glass of water and then filter it out again, we haven't altered the nature of the stuff in the glass but just temporarily made it dirty. And so on. We have a rough and ready "manifest image" conception of what water is, and this guides us when we investigate the microstructure. It is *that stuff*, specifically, the stuff that counts as water in the *manifest* image, that the scientific image is describing when it makes reference to H_2O. It is only because a certain part of nature has first been carved off in the manifest image and labeled "water" that chemistry can go on to investigate it. "[S]uch scientific explanations refer to *water*, where the latter term refers to manifest water (the same water Aristotle speculated about)... [T]he manifest image determines which micro-essences are to be selected" (Van Brakel 2000, pp. 78 and 81).

The reductionist might respond that it is chemistry that *explains* the manifest image attributes, in terms of a microstructure of molecules and their interrelationships. Once we have this microstructural description, he might claim, we can account for the nature and identity of water without making reference to the manifest image description that initially guided our investigation. But as Van Brakel argues, this will not work (2000, p. 79). Do the identity criteria for the water in the glass in our example include the velocity and relative positions of the molecules? If so, then the reductionist cannot account for the fact that the same water persists over time, since the velocities and relative positions are constantly changing. By contrast, the non-reductionist has no problem, because the macroscopic features of the glass of water persist despite these changes at the micro-level. Suppose, then, that the reductionist says that velocity and relative position are not part of the identity criteria. Then he will be unable to appeal to a feature such as temperature as a mark of identity or difference, since the molecular motion that is the micro-level correlate of temperature is constantly changing. Again, the non-reductionist has no problem here, because these variations average out at the macro-level.

Thus, the macro-level manifest image description of water remains both indispensable and irreducible. This is in no way to deny that the manifest image's conception of water can be tightened up or even corrected here and there by the scientific image. As Hendry (2010b) argues, there are cases where we cannot plausibly answer a question about whether water is actually or only virtually present in something (to use the Aristotelian's language, not Hendry's) unless we determine whether there are H_2O molecules in it. But it simply does not follow that the manifest image can be *wholly overthrown* by what chemistry tells us about water. Hendry himself writes:

> [S]cientists are people, and their technical discourse must have its origin in the colloquial. Though scientific concepts are refined and often highly abstract, they are honed for the description of the very same world in which everyday life takes place.

The manifest image is like a ladder on which the scientist stands. He can, even from his position atop it, modify the ladder and move it around a bit, if he does so carefully. But he cannot kick it away without taking himself down with it.

A further problem for the reductionist is that water has properties and causal powers that hydrogen, oxygen, and a mere juxtaposition of hydrogen and oxygen, all lack. This is precisely what we should expect if water is irreducible to the sum of its parts. Yet another problem is that water lacks properties and causal powers that hydrogen and oxygen have. This is precisely what we should expect if hydrogen and oxygen are in water in the sense a hylemorphic analysis would imply, viz. virtually rather than actually. As David Oderberg writes:

> [I]f the water contained actual hydrogen, we should be able to burn it – but in fact the opposite is the case. If the water contained actual oxygen, it should boil at -180°C – but in fact it boils at +100°C (at ground level).
>
> Of course the response is that the oxygen and hydrogen are bonded in water and so cannot do what they do in the absence of such a bond. But that is precisely the point. The combustibility of hydrogen and the specific boiling point of oxygen are *properties* of those elements in the technical [Aristotelian] essentialist sense – they are accidents that necessarily flow from their very essence. Since the properties are absent in water, we can infer back to the *absence* of the essences from which they necessarily flow. Therefore neither hydrogen nor oxygen is actually present in water. Rather, they are *virtually* present in the water in the

> sense that some (but not all) of the powers of hydrogen and oxygen are present in the water (though all properties requiring the elements to be actually present will be gone), and these elements can be *recovered* from the water by electrolysis – not in the way that biscuits are recovered from a jar, but in the way that the ingredients of a mixture can (sometimes) be reconstituted. (2007, p. 75)

It is worth pausing over the notion of something's being in a substance "virtually." As I have indicated, to be in a substance virtually is to be in it *potentially*; here we have another application of the theory of actuality and potentiality. But to be in a substance virtually is to be in it in a stronger sense than being in it in a *purely* potential way. After all, hydrogen and oxygen are in prime matter potentially, but only in the sense that *every* physical substance is in prime matter potentially, since prime matter just is the pure potentiality to be any physical substance. So, to say that hydrogen and oxygen are in water virtually is to say that they are in it in a stronger sense than the sense in which they are in prime matter. It is to say that they are in water in a sense that they are *not* in stone, for example.

As Peter Hoenen (1955, pp. 39-50) explains, to be in a substance virtually is therefore to be in it in a way that is a kind of middle ground between pure potentiality on the one hand, and actuality on the other. Hoenen identifies two criteria for attributing to something this virtual or middle ground kind of reality. The first he calls the "proof... taken from the compounding and resolution of compounds" (p. 42). For example, when we break water down by electrolysis, we reliably get hydrogen and oxygen; and when we combine hydrogen and oxygen in the right way, we reliably get water. The second criterion of something's being in a substance virtually is that the substance still exhibits at least some of the virtual thing's properties. For example, as Oderberg notes in the passage quoted above, "some (but not all) of the powers of hydrogen and oxygen are present in the water." These two criteria indicate that hydrogen and oxygen, though not in water *actually*, are nevertheless in water in a special way that they are not in other substances, and thus in a stronger way than the *purely* potential way in which any physical substance exists in prime matter.

What the Aristotelian has in mind is similar to the "fusion emergentism" of Paul Humphreys (2008). On Humphreys' account, emergent properties exist at a higher level within a system when lower-level properties are "fused" in such a way that they cease to exist as separate entities and lose some of their causal powers, and where this fusion yields novel causal powers at the higher level. This is, I have been claiming, exactly what happens with hydrogen and oxygen when they combine to form water. As Humphreys emphasizes, a failure to perceive that the emergence of higher-level properties involves such fusion bedevils much traditional debate over reductionism. Discussion about whether the higher-level properties supervene upon the lower-level ones, and whether, if so, the higher-level properties are causally superfluous, presupposes that the lower-level properties exist in the emergent system in just the same manner in which they exist apart from it, and that is exactly what the fusion emergentist (and the Aristotelian) deny. The higher-level properties neither supervene upon the lower-level ones nor causally compete with them, because the lower-level ones no longer exist *in the way* that the higher-level ones do.

Where Aristotelians would differ from Humphreys is, first, in urging the metaphysics of actuality and potentiality, and the notion of virtual presence as an application of the idea of potentiality, as a way of elucidating the ontological status of fused properties. For flatly to assert (as Humphreys does) that lower-level properties "no longer exist" after fusion (2008, p. 120) needlessly makes the resulting position sound mysterious. (How can higher-level properties "emerge" from something that does not exist?) Rather, the right way to put the point (so the Aristotelian argues) is to say that the lower-level properties are real, but exist as potentialities rather than actualities. That there is a middle-ground kind of reality between actuality on the one hand and sheer nothingness on the other is, of course, the whole point of the theory of actuality and potentiality; and that theory has independent motivation in considerations about how to avoid the Eleatic and Heraclitean extremes regarding change and multiplicity. It is available, then, as a way to answer a potential objection to Humphreys' analysis.

Second, the Aristotelian resists the language of "emergence" because, despite its anti-reductionism, it gives the impression of conceding to the reductionist the thesis that the micro-level is ontologically fundamental or privileged. It is as if the emergentist allows that the macro-level is problematic in a way that the micro-level is not, so that we should concede the reality of macro-level phenomena only to the extent that we can make sense of them somehow "emerging" from the micro-level. As I have said, the Aristotelian rejects any such privileging of the micro-level.

From an Aristotelian point of view, modern emergentist arguments, though salutary, are at best only partial rediscoveries of the correct, hylemorphist account of nature.

A further well-known example of the failure of reductionist claims about modern chemistry concerns thermodynamics (Sklar 1993; Van Brakel 2000, pp. 123-28; Koons 2018a). The thesis that temperature is molecular motion, like the claim that water is H_2O, is routinely cited in philosophical literature as an example of a successful chemical reduction. But as with water, the reality is more complicated. For one thing, molecular motion is not by itself sufficient for a system's having temperature; the system must also be in thermodynamic equilibrium. But equilibrium is a macro-level concept, so that for the reductionist to appeal to molecular motion together with equilibrium would not be to *reduce* the macro-level to the micro-level at all. Nor will it work to appeal to statistical mechanics in order to find some micro-level property to put in place of equilibrium, because the relevant processes from statistical mechanics cannot be characterized without reference to the very macro-level concept of temperature that the reductionist is trying to reduce. As Lawrence Sklar writes:

> It is important to note... the degree to which macroscopic knowledge of the nature of a system, framed in the conceptual languages of thermodynamics, is utilized in choosing the basic probabilistic posits necessary for the underlying statistical mechanics... [I]t is essential to have some thermodynamic appreciation of the system in order to even begin to look for the correct initial probability distribution that must be posited in order to get the ensemble dynamics of statistical mechanics underway...
>
> [T]he very structure of the solutions found is guided by our antecedent knowledge at the empirical macroscopic level of the kind of solution we must look for. (1993, pp. 372-73)

There is, of course, nothing necessarily special about water or temperature. These examples are intended merely to illustrate the general point that for chemistry to identify the microstructure of some manifest image phenomenon and the laws that govern it is not, by itself, sufficient to establish reductionism or refute hylemorphism. What the hylemorphist should say about other examples depends on the specific empirical facts in each case. The Aristotelian does not reject *a priori* the possibility of a reductionist account of some macro-level phenomena. For example, there could be cases where what at first appears to be a genuine substance turns out to be an aggregate. But as Weisberg, Needham, and Hendry note:

> While there is no in-principle argument that reductions will always be impossible, essential reference is made back to some macroscopically observable chemical property in every formal attempt of reduction that we are aware of. In the absence of definite arguments to the contrary, it seems reasonable to suppose that chemistry employs both macroscopic and microscopic concepts in detailed theories which it strives to integrate into a unified view. Although plenty of chemistry is conducted at the microscopic level alone, macroscopic chemical properties continue to play important experimental and theoretical roles throughout chemistry. (2011)

These examples have to do with the question of reductionism *within* chemistry, where what is at issue is the relationship of macro-level chemical phenomena to micro-level chemical phenomena. But again, there is also the question of the reduction of chemistry itself to physics, and here too many philosophers of chemistry have expressed skepticism (Scerri 1994; van Brakel 2000, pp. 128-50; Hendry 2006). Eric Scerri (2007a) notes that the question whether chemistry has been reduced to physics is ambiguous. On the one hand, quantum mechanics has provided chemistry with computational methods of undeniable utility. In that sense a reductionist methodology has proved fruitful. On the other hand, "one cannot begin with quantum mechanics alone and predict the configuration of a particular atom," nor deduce the periodic table (pp. 74-75). (Cf. Scerri 2000 and 2007b, pp. 242-48.) In particular, reductionist accounts face difficulty when time is factored in. For example, once we know the structure of a boron atom, it might appear obvious how its properties derive from those of its microstructure. But if we only knew the particles of which boron is composed, we could not have predicted the properties that would arise once they combined to form a boron atom (2007a, pp. 77-78). Hendry (2006, 2010a, 2017) also argues that there are cases of molecules which differ chemically (such as ethanol and dimethyl ether) even though their description at the level of quantum mechanics is the same.

To be sure, Scerri (2016) has in recent years judged that the empirical considerations that have been raised against the reduction of chemistry to quantum mechanics are weaker than he had at first supposed, and he is now more optimistic about reductionism. However, even so, he appears to concede that a reduction has not yet been entirely achieved. More importantly, his current misgivings leave untouched the deeper conceptual problems for reductionism. In particular, with attempts to reduce chemistry to quantum mechanics as with purported reductions within chemistry, identification of the relevant lower-level features presupposes an independent prior grasp of the higher-level features that are supposedly being reduced. As van Brakel observes:

> [A]s computers get bigger, more sophisticated approximation methods become feasible. But the general structure of how practice works remains the same. There are some classically observed data (which have been around for a long time) and with great effort it is possible to have the quantum mechanical formalism approximate these classically observed data better and better until it fits. Using the methods (models, approximations) that have worked, these methods are then extrapolated to similar molecules – similar according to chemical expertise. This may give the impression of ab initio calculations, but at every stage the experimental data steer the development of the models. (2000, p. 139)

5.4 Primary and secondary qualities

To defend an Aristotelian conception of matter naturally invites the question whether the early modern distinction between primary and secondary qualities ought to be reconsidered. Must the neo-Aristotelian hold that color and other so-called secondary qualities really do, after all, exist in the physical world in just the way common sense supposes? Were the early modern critics of hylemorphism wrong about this too? Early twentieth-century Aristotelians were divided on the issue, with some defending the commonsense understanding of the secondary qualities and others essentially conceding the modern Lockean position. (Cf. Bittle 1936, chapter XII for an overview of the debate.) Of course, one could defend a commonsense position with respect to some secondary qualities but not others.

A primary quality, as Locke conceives of it, is a quality of a physical object which produces in us sensations that "resemble" the quality itself; whereas a secondary quality produces in us sensations that do not resemble anything in the object. For example, in Locke's view, the way a shape like *roundness* looks to us in perceptual experience resembles roundness as it exists in a round object, so that roundness and other shapes count as primary qualities. But there is, in his view, nothing in physical objects that really resembles the redness we see when we look at objects we call "red." So redness, and other colors, are secondary qualities. On Locke's account, colors as they exist in physical objects are really nothing but *powers* that objects have to produce in us sensations that do not resemble anything that is really objectively there in the objects. *If* by "redness" we mean a power of this sort, then we can say that redness is a quality that exists in the objects. But if by "redness" we mean the *look* that we associate with redness, then there is nothing in the objects that corresponds to that. It is entirely mind-dependent. Not every early modern philosopher or later thinker who endorses the distinction between primary and secondary qualities puts it exactly Locke's way, but they essentially agree with this main idea. As is commonly done, I will, for ease of exposition, ignore the distinction between the power to produce a sensation in us, and the sensation itself, and just speak of primary qualities as mind-independent and secondary qualities as mind-dependent. Nothing of substance will ride on this usage.

Among the key issues here is whether a sharp distinction between primary and secondary qualities can actually be maintained. Berkeley, of course, argued that it cannot be; and many contemporary philosophers, such as Michael Lockwood (1989, pp. 155-56), agree with him. (Cf. Lowe 1995, pp. 53-59.) Now, if one denies the distinction, there are several alternative directions one could go in next. Berkeley, needless to say, went in the direction of idealism. Since he agreed that secondary qualities are mind-dependent, he reasoned from the further premise that so-called primary qualities collapse into secondary ones to the conclusion that all the qualities of physical objects are mind-dependent, and from there to the further conclusion that physical objects themselves are mind-dependent. Lockwood takes a different position. Like Locke, he does not regard physical objects as mind-dependent. But since, like Berkeley, he takes primary and secondary qualities alike to be mind-dependent, he concludes that sensation does not tell us anything about what physical objects are like in themselves, not even in the case of so-called primary qualities. (J. L. Mackie (1976, chapter 1), who purports to be defending Locke's distinction, nevertheless gives up the idea that primary qualities always "resemble" something in sensation, with the consequence that his positon seems more like Lockwood's than like Locke's.)

However, a third option would be to reason as follows. Suppose that perceptual experience of primary qualities does reveal something about what physical objects are really like in themselves, as Locke and other defenders of the distinction between primary and secondary qualities traditionally do. If we add to this thesis the further premise that the distinction between primary and secondary qualities cannot be maintained, then we can infer that perceptual experience of secondary qualities too tells us something about what physical objects are really like in themselves. Common sense is correct to take so-called secondary qualities to exist in physical objects in essentially the same way that primary qualities do.

That this is a more plausible path to take than Berkeley's and Lockwood's alternatives can be seen when we consider that the main traditional argument for the mind-dependence of so-called secondary qualities – the argument from perceptual relativity – is no good. The starting point of the argument is the observation that the appearance of qualities like color varies from observer to observer. The same object will look bright red or dull red depending on the lighting; a color blind person might not be able to tell it apart from a green object; another person's color experiences could in theory be inverted relative to my own; and so on. The best explanation of these facts, the argument concludes, is that color is not really there in the objects themselves but only in the mind of the observer.

But there are several problems with this argument, which Putnam (1999, pp. 38-41) has usefully summarized (where Putnam is reiterating points that go back to writers like J. L. Austin (1962) and P. F. Strawson (1979)). First, the argument rests on a simplistic characterization of the commonsense understanding of color. Common sense allows that the same color can look different under different circumstances, just as it allows that a round object can appear oval under certain circumstances. Hence the commonsense thesis that color is mind-independent is not undermined by the fact that an object will look bright red in some contexts and dull red in others. Furthermore, color blindness no more casts down on the supposition that color is mind-independent than hallucination casts doubt on the reality of physical objects. In both cases, the defender of common sense can note that a perceiver's faculties are simply *malfunctioning*, and thus not presenting objective reality as it really is. Meanwhile, the inverted spectrum scenario presupposes that the physical facts about both external objects and the brain could be exactly as they are while the way colors look is different. It presupposes, in other words, that color can float entirely free of the way things really are in the material world. But that is exactly what the commonsense view denies, so that to appeal in this context to the alleged possibility of color inversion is to beg the question.

Then there is the objection that treating color as a secondary quality "solves" one problem (the problem of explaining perceptual relativity) at the expense of creating far more serious problems. For one thing, again, the distinction between primary and secondary qualities is difficult to maintain. Hence if we treat secondary qualities as mind-dependent, we are bound to treat primary qualities as mind-dependent too. But once we do that, the reliability of perception is cast into doubt. This in turn undermines the evidential foundation of natural science, in the name of which the distinction between primary and secondary qualities was being defended. For another thing, removing colors from the objective world requires treating them instead as the subjective qualities or qualia of conscious experience, which gives rise to the notoriously intractable qualia problem in the philosophy of mind.

Of course, another argument for the mind-dependence of qualities like color (and the one modern Locke sympathizers like Mackie are more likely to recommend) is the appeal to science. Physical science, the argument says, makes no reference to features like color as common sense understands it, but only to surrogates like the surface reflectance properties of objects. Therefore, the argument concludes, we have no good reason to believe that there is anything in objective physical reality that corresponds to color as we experience it. But the problem with this argument, as Barry Stroud (2000, pp. 61-68) points out, is that the absence of color from physical science's description of the world does not entail that it is absent from the world itself, any more than the absence of height and weight from the economist's description of economic agents entails that human beings lack height and weight. Here Stroud is, of course, making the point I have made several times in this book in other contexts. The presence or absence of some feature from science's representation of physical reality can reflect merely the mode of representation rather than physical reality itself, and thus cannot by itself license any metaphysical conclusions about the latter. In the present case, for all the Lockean has shown, the absence of color merely reflects modern physics' methodological preference for a quantitative rather than qualitative representation of nature.

Accordingly, Keith Allen (2016, Chapter 1) notes that the commonsense or "naïve realist" view of color has two essential components. First, it takes color to exist in nature in a mind-independent way. That it takes color to exist *at all* distinguishes naïve realism from the eliminativist view that color is unreal. That it takes color to exist in a mind-independent way distinguishes naïve realism from the view that color exists only as a disposition of things to produce certain sensations in us. Color, the commonsense or naïve realist view says, would exist in nature whether or not there were minds around to perceive it. But the second essential component of naïve realism, as Allen says, is that it understands color in a *non-reductionist* way. That is to say, it takes color to be something *distinct from* or over and above the sorts of properties to which physical science confines itself, such as surface reflectance properties. Color as *common sense* understands it, and not merely as *physics* understands it, would exist even if there were no minds. Note that the claim is *not* that color is not a physical property, but rather that physical science does not capture all the physical properties that there are.

In defense of the mind-independence component of naïve realism, Allen argues that it best accounts for the *constancy* of our experience of color. (Cf. Decaen 2001, p. 213.) In general, a physical object will appear to have the same color over time under different kinds of natural and artificial illumination and against a variety of different backgrounds. For example, a certain apple will consistently look red as you carry it from brightly lit room to a shadowy one and then outside into the bright sun and whether you see it against the background of your hand, or a nearby wall, or the blue sky. The apple's appearance will change in some respects insofar as it will look darker in one context and shinier in another, and certain features will be easier to see in brighter light than in dimmer light. But it will not seem to be anything but red in color despite all these changes in appearance. Such stability of color allows us to distinguish, identify, and re-identify objects over time under a wide variety of circumstances. It is true that there some cases where color might falsely seem to change, as when an object is seen in colored light. But this is the exception rather than the rule.

Color is in all of these respects comparable to shape and size, which are traditionally regarded as mind-independent primary qualities. Shape and size too are constant under a wide variety of circumstances and allow us to differentiate and identify objects, even though here too there are occasional exceptions. Now, with shape, size, and other properties we take to be mind-independent, there is a distinction between appearance and reality and criteria that enable us to tell them apart. For example, we know that though a stick immersed in water will look bent, this is merely an illusion, and by removing the stick, observing how it looks in other circumstances, etc., we can determine that in reality it is straight. We know that a large object might look small from a distance, but by getting closer we can see that this is an illusion and determine its true height. But there are also criteria for distinguishing appearance and reality in the case of color. What we judge to be the true color of an object seems to be "in" the surface of an object, whereas what we judge to be merely illusory appearances of color (such as a shadow flitting across the surface of an object, or the redness a white object might seem to have in colored light) seem merely to "overlay" the surface. (Cf. Decaen 2001, p. 209.) In the latter case we are inclined to judge that it is merely our *experience* of a color that is changing, just as we judge that it is merely our *experience* of an object's shape that changes as we look at it from a different angle. The phenomenology of such cases is very different from that of cases where we take the object's color *itself* to change, as when we see a wall being painted – just as experiences of an object's shape *itself* changing (as when we see ice cream melt) are phenomenologically very different from the experience of looking at an object's shape from different angles.

The best explanation of this constancy, Allen argues, is that color is no less objective and mind-independent a feature of physical objects than shape and size are. The other essential component of naïve realism about color, the non-reductionist thesis that color is *distinct* from any property described by physical science, is defended by Allen in part by way of what he calls a *modal* argument analogous to Saul Kripke's (1980) modal argument against materialism. The basic idea is that the connection between color on the one hand, and a property like a surface reflectance profile on the other, appears to be *contingent*. In particular, Allen argues that in principle, color could exist in the absence of any particular surface reflectance profile (just as, according to Kripke's argument, pain could in principle exist apart from the firing of c-fibers). The argument is not *exactly* parallel with Kripke's, though, because Allen is willing to allow that the presence of the surface reflectance profile might be *sufficient* for color even though it is not necessary (whereas for Kripke, the firing of c-fibers is neither necessary nor sufficient for pain).

It seems to me, however, that Allen could make the case for distinctness in a way that both puts the Kripkean metaphysical baggage to one side and results in a stronger conclusion. In particular, he could point out that *the reductionist himself* insists that color, understood in the irreducibly *qualitative* way that commonsense conceives of it, is not in apples, oranges, or other physical objects, but only in the consciousness of the observer. This is a consequence of the reductionist's purely *quantitative* conception of the physical world. Hence the reductionist would say that in a world without conscious observers, apples, oranges, and the like would not have color understood in the irreducibly qualitative commonsense way. This is so even though these objects would in that case still have the same surface reflectance properties they have in the actual world. But then, *the reductionist himself* is committed to the thesis that color, in the commonsense qualitative sense, is distinct from any physical property. Moreover, he is committed to the thesis that physical properties are neither necessary *nor sufficient* for color in that irreducibly qualitative sense – since, again, color in that sense would not exist in the absence of observers, even though the relevant physical properties would. So, Allen doesn't *need* some special argument, Kripkean or otherwise, to establish distinctness. He merely needs to call attention to what the reductionist himself already says.

What the reductionist really objects to in naïve realism, then, is not what Allen calls the *distinctness* thesis, but rather the *mind-independence* thesis. To establish mind-independence is, accordingly, sufficient to refute reductionism. For it shows that color properties of the kind the reductionist acknowledges are distinct from any that he is willing to count as "physical" really do after all exist in the physical world, so that there is more to the physical world than the purely quantitative properties the reductionist is willing to recognize.

Stroud (2000, chapter 7) proposes a line of argument that is in one respect even more ambitious. He characterizes the Lockean position as an "unmasking" exercise, seeking as it does to expose the commonsense understanding of color as an illusion. But in Stroud's view, no such unmasking exercise can coherently be carried out. The unmasking project takes physical reality to be utterly devoid of color in the ordinary sense, but it acknowledges that we have a vast number of interlocking perceptual judgments and beliefs to the effect that physical objects *are* colored in the ordinary sense. Now, that means that the unmasker has to attribute to us a vast number of perceptual judgments and beliefs that get objective reality *systematically wrong*.

This, Stroud argues, is incoherent for reasons related to considerations I briefly raised in chapter 2 when discussing Donald Davidson's (1986, 2001) influential account of how the interpretation of linguistic utterances presupposes a shared environment. The basic idea is that in order to interpret your utterances as *language*, I have to regard them as the expression of beliefs and other propositional attitudes, and in order to identify which beliefs those are, I have to be able to relate them to things in our common environment. To use Quine's (1960) famous example, in order for a field linguist to interpret a native speaker's utterance of "Gavagai" as meaning "Lo, a rabbit," he has to be able to attribute to the speaker the belief that a rabbit is present, and relate that belief to the rabbit that actually is present to the both of them as they try to communicate.

But that entails, as Davidson famously argues, that we cannot so much as communicate with others unless we regard most of what they believe as *true*. If I were to regard your beliefs as *entirely* out of sync with the objective world we both share – if I thought that what was going on in your thoughts had *no* connection at all to the world of tables, chairs, rocks, trees, and other people that we both occupy – then I could have nothing to ground any attribution to you of any specific beliefs or other propositional attitudes, and thus no basis for attributing any particular content to your utterances. I would have to regard them as gibberish. Of course, we can and do regard *some* of what other people say and think as mistaken, especially where the utterances and beliefs concern matters remote from directly observable reality. But this disagreement makes sense only against a background of agreement on more basic matters. I can judge that you are *mistaken* when saying "That guy across the street is John" only because I also judge that you *correctly* perceive that there is a street in front of us and a man standing across it, and have simply misperceived exactly who that man is. If I did not think that you were getting things right at least about the street and the presence of a person across it, I could not attribute to you the belief that John is the person across the street and then go on to judge your belief to be false.

Now, Stroud argues that attributions of beliefs about colors are like this. Naturally, we sometimes judge that we or others are mistaken about various *particular* claims we make about the color of an object. You say that a certain object is red and I correct you by pointing out that it is merely being illuminated in red light and that when we move it out of that light we can see that it is actually yellow. But this correction is possible only because we can indeed go on correctly to judge that the object is yellow. And that judgement is precisely a judgement about something that forms part of our common public environment. The "unmasker" claims that colors are properties of sensations, which are private rather than public, but that is not actually how we understand or use color terms. When I say that the pain in my back is dull rather than sharp, I am attributing something to a sensation. But when I say that the lemon on the table is yellow, I am attributing something to the *lemon*, not to some perceptual experience I am having. Even when we make color-related claims about our perceptual experiences, these are parasitic on the claims we make about physical objects. If I say that I have yellowish qualia, what I mean is that there is something about my qualia that is like what we ordinarily attribute to lemons when we say things like "That lemon is yellow."

So, I have to have some objective, mind-independent reference point in order to attribute to you beliefs about the color of objects, and then go on to judge those beliefs to be either true or false. But as with your beliefs about other things, I have to judge most of your beliefs about the colors of objects to be *true*. Otherwise I could not get as far as attributing any beliefs about color to you *at all*, any more than I could attribute to you a false belief about the person across the street without judging you to be correct at least about the fact that a street and a person exist, or any more than the field linguist could attribute to the native speaker a belief about a rabbit without judging the speaker to be able accurately to perceive rabbits. If I suppose that *none* of your beliefs about color correspond to objective reality, then I would have no way of knowing what beliefs about color to attribute to you in the first place. But if I must regard most of your beliefs about color to be true, then that entails that I must take color really to exist in the physical objects of our common environment.

This is where the problem for the "unmasker" arises. The unmasker attributes to us beliefs about color that are not just sometimes false, but always false. These are precisely the beliefs he is hoping to unmask. But Stroud's Davidsonian argument entails that the very attribution of these beliefs to us requires regarding most of them as true. Hence the unmasker is taking an incoherent positon. The unmasking project implicitly presupposes exactly what it claims to deny, namely that colors as commonsense understands them exist in physical objects in a mind-independent way.

Now, Stroud himself (2000, pp. 192-93) stops short of taking this to be a demonstration that naïve realism about color is true. The reason is that he thinks that both naïve realism and the "unmasking" project alike presuppose that we can get outside our system of perceptual experiences and beliefs and then compare them with physical reality – with the unmasker concluding that they don't match up, and the naïve realist concluding that they do. Since in fact we can't step outside this system and carry out such a comparison, Stroud concludes that we can't answer one way or the other the question whether physical objects really have color in the way common sense supposes. The most we can say is that even the unmasker has to *think about* objects as if they have color, but it doesn't follow that they really do have it.

It seems to me, however, that Stroud is wrong to weaken his conclusion in this way. Framing the issue in terms of an epistemological gap between the physical world on the one hand and our system of beliefs about it on the other seems to presuppose a Cartesian representationalist conception of knowledge. Now, I argued against this conception in chapter 2, and I would submit that Stroud's argument in fact provides further ammunition against it. In particular, I would propose that Stroud's argument is best interpreted as a retorsion argument, and as I also argued in chapter 2, a retorsion argument is best understood as a kind of *reductio ad absurdum* argument. Hence Stroud's argument, if otherwise unproblematic, can be taken to show that the unmasker's position entails a contradiction. But to show that a position entails a contradiction is to *refute* it. What Stroud should say is, not that the unmasker is in no better position than the naïve realist is in comparing the two sides of the epistemic gap, but rather that there is no gap in the first place. Since the unmasking view requires that there be a gap and the naïve realist view does not, the two views are not on a par. Stroud's argument really is an argument for naïve realism, and not merely a difficulty for the unmasking position.

Despite being a defender of naïve realism, Allen suggests (2016, pp. 168-69) that Stroud's argument fails, on the grounds that Stroud is attempting to "unmask" the unmasker and can be hoist with his own petard. Just as the unmasker claims that there really are no colors in physical reality even though there seem to be, so too does Stroud claim (according to Allen) that the unmasker doesn't really disbelieve in colors, even though he seems to disbelieve in them. And just as Stroud criticizes the unmasker for seeking an impossible external perspective from which he can compare physical reality and our system of beliefs about physical reality, so too (Allen claims) is Stroud open to the objection that he is seeking an impossible external perspective from which he can compare what the unmasker claims to believe which what he actually believes.

But this objection rests on a misunderstanding of Stroud's argument. Stroud is not attempting (or at least need not be interpreted as attempting) to "unmask" the unmasker, and in particular he is not claiming that there is a distinction between what the unmasker appears to believe and what he really does believe that is parallel to the distinction between the way physical objects appear to us and the way they really are. He is not claiming that the unmasker does not really believe that there are no colors. Rather, he is acknowledging that the unmasker *does* believe this, but arguing that this belief conflicts with other assumptions to which the unmasker is implicitly committed. In other words, Stroud is simply arguing that the unmasker is unwittingly committed to a contradictory set of beliefs – a pretty standard argumentative move in philosophy, and one that has nothing to do with seeking some external perspective by which what someone actually believes and what he thinks he believes might be compared.

There is a further problem with the "unmasking" project. It might seem that the unmasker removes color from the picture of the physical world presented to us in perception, but leaves the rest of the picture intact – like a page from a coloring book that has not yet been filled in. But as Berkeley argued, our perception of the other visible qualities is inextricably tied up with color, so that to unmask the latter is to unmask the former too. Mark Johnston writes:

> [U]nless the external world is colored it is invisible. For if the external world is not colored then we do not see the colors of external things. They are not visible... [I]f colors are not visible then no surface of a material object is visible. But if no surface of a material object is visible, then no material object is visible. Such is the consequence of denying that nothing corresponds to external color, the proper sensible of sight. Unless the external world is colored we do not see it and that means we do not see, period. (1997, p. 168)

Keep in mind that the "unmasker" assimilates color perceptions to sensations such as pain. According to the Lockean view, if you hold a pin between your fingers, the straightness and solidity you feel reflects qualities that are really there in the pin itself; whereas the pain you experience on being pricked by the pin does not reflect anything in the pin. You are really feeling *the pin itself* in the first case, but not in the second case. In the second case, you are feeling something entirely subjective that is merely caused by the pin. But if color is like pain, and all the other visible qualities are inextricably tied to color, then they are like pain as well. Like the experience of pain, a visual experience is not really an experience of a physical object, but only of something entirely subjective that is caused by the object. Visual experience is not really *seeing* at all, but a kind of feeling. (Cf. Decaen 2001, p. 210-12.) This is not only bizarre, but implies a representationalist conception of perceptual knowledge which, again, I argued against in chapter 2.

This only scratches the surface of the topic of color perception, and I have not discussed at all the other so-called secondary qualities. But a complete treatment of these matters would take us well beyond the philosophy of nature into issues in epistemology and the philosophy of mind. Suffice it for present purposes to note that, once again, a component of the traditional Aristotelian view of nature often dismissed as an anachronism is not only still defensible today, but is actually being defended today even by non-Aristotelian philosophers.

5.5 Is computation intrinsic to physics?

Talk of information, algorithms, software, and other computational notions is commonplace in the work of contemporary philosophers, cognitive scientists, biologists, and physicists. These notions are regarded as essential to an adequate scientific description and explanation of physical, biological, and psychological phenomena. Computation is thus routinely treated as something intrinsic to the material world, from the most complex cognitive processes down to the most elementary inorganic physical processes. Yet a powerful objection has been raised by John Searle, who argues that computational features are all essentially observer-relative rather than intrinsic to nature. If Searle is right, then computation is not a natural kind but rather a kind of human artifact, and is therefore unavailable for purposes of scientific explanation.

As I will argue, Searle's objection has not been, and cannot be, successfully rebutted by his naturalist critics. I will also argue, however, that computational descriptions do indeed track what Daniel Dennett calls "real patterns" in nature. The way to resolve this *aporia* is to see that the computational notions are essentially a recapitulation of the Aristotelian notions of formal and final causality, purportedly banished from science by the "mechanical philosophy" of the early moderns. Given a "mechanical" conception of nature, Searle's critique is unanswerable. If there is truth in computational approaches, then this can be made sense of, and Searle's objection rebutted, only if we return to a broadly Aristotelian philosophy of nature.

5.5.1 The computational paradigm

Fundamental to the notion that natural processes are computational is the idea of *information*. The term "information" has become something of a buzzword in contemporary pop science writing, and unfortunately it is not always used with precision. It is generally acknowledged, however, that the sense of the term operative in computer science, and thus in arguments to the effect that computational processes literally exist in nature, is not the everyday sense of the term but rather a technical sense.

The technical sense in question is essentially the one associated with mathematician Claude Shannon's celebrated theory of information (Shannon and Weaver 1949). Shannon was concerned with information in a *syntactic* rather than *semantic* sense. Consider the *bit*, the basic unit of information, which has one of two possible values, usually represented as either 0 or 1. To consider a bit or string of bits (e.g. "11010001") in terms of some interpretation or meaning we have attributed to it would be to consider it semantically. Semantic information is the sort of thing we have in mind when we speak of "information" in the ordinary sense. To consider the properties a bit or string of bits has merely as an uninterpreted symbol or string of symbols is to consider it syntactically. This is "information" in the technical sense. When instantiated physically, a bit corresponds to one of two physical states, such as either of two positions of a switch, two distinct voltage levels, or what have you.

As David Chalmers (1996, p. 281) points out, when physically instantiated, information in this technical sense essentially involves a causal correlation between a physical state of the sort in question and some effect at the end of a causal pathway leading from that state. Think, for example, of the correlation between a switch's being either up or down and the light to which it is connected being either on or off. The position of the switch carries a single bit of information, and any physical state that has the same effect down the causal pathway will carry the same information. Several switches (or, again, several distinct voltage levels or whatever) taken together will, naturally, carry more information. Following Chalmers, we can describe a combination of possible physical states (such as a combination of possible sets of positions of a number of switches) as an "information space." The structure of any information space will correspond to the structure of the set of possible effects down the causal pathway from the physical states that make up the information space.

Since it will be useful later on in our discussion, let me quote at length from Chalmers' example of the information carried by a compact disc. He writes:

> A disk [sic] has an infinite number of possible physical states, but when its effects on a compact-disk player are considered, it realizes only a finite number of possible information states. Many changes in the disk – a microscopic alteration below the level of resolution of the optical reading device, or a small scratch on the disk, or a large mark on the reverse side – make no difference to the functioning of the system. The only differences relevant to the disk's information state are those that are reflected in the output of the optical reading device. These are the differences in the presence of pits and lands on the disk, which correspond to what we think of as "bits"... The physical states of different pressings of the same recording will be associated with the same information state, if all goes well. Pressings of different recordings, or indeed imperfect pressings of the same recording, will be associated with different information states, due to their different effects...
>
> Each "bit" on the compact disk has an independent effect on the compact disk player, so that each location on the disk can be seen to realize a two-state subspace of its own. Putting all these independent effects together, we find a combinatorial structure in the space of total effects of a compact disk, and so we can find the same combinatorial structure in the information space that the compact disk realizes. (1996, p. 282)

As this example indicates, the amount of "information" in the sense in question that might be transmitted along a causal pathway is quantifiable, and that is what Shannon's theory of information is concerned with. That what comes out of the compact disc player when the disc is played counts as *music*, that the lyrics have a certain *meaning*, and so forth, is completely irrelevant to how much information is transmitted. Again, "information" is being used here in a *syntactic* rather than *semantic* sense. What is at issue is what effects a bit or string of bits has considered merely as an uninterpreted symbol or string of symbols and entirely apart from what meaning or interpretation we assign to them.

Now, computers are said to process information. This is what happens when (to stick with Chalmers' compact disc example) you place a CD-ROM into your computer and the text file you have saved on it appears onscreen as a document written in English. Of course, you won't find anything that looks like English words on the CD-ROM. What happens is that the electrical states of the computer serve as a causal pathway by which the information state embodied in the CD-ROM generates the images on the screen. The information on the CD-ROM is the *input*, the images on the screen are the *output*, and the computer moves from the former to the latter because it is running an appropriate *algorithm* or set of instructions. But of course, you also won't find anything in the computer that looks like a set of instructions. The algorithm is *itself* embodied as information in the relevant sense – that is to say, as a certain configuration of electrical states. As biologist John Mayfield writes, "a computer can be seen as a device in which one state (the input) interacts with another state (the current machine configuration) to produce a final state (the output)" (2013, p. 45). *Computation* just is this transition from states which can be characterized as embodying an informational input, via states which can be characterized as the embodiment of an algorithm, to states which can be characterized as the output of the algorithm.

A key property of computations is that you will not get more information out of them than went into them. As Mayfield puts it:

> Algorithmic information shares with Shannon information the property that it cannot be created during a deterministic computation. The information content of the output can be less than that of the input, but not greater. Thus, algorithmic information conforms with our intuitive notion that information cannot be created out of thin air. (2013, p. 50)

Now, many contemporary philosophers and scientists hold that computation can be found not only in the machines we design for that purpose, but also in the natural world. In particular, the notions of *information, algorithms*, and the like have been claimed to have application to the understanding of phenomena studied in physics, biology, and neuroscience. (Cf. Davies and Gregersen 2010.) Consider physicist John Wheeler's (1998, Chapter 15) famous "It from Bit" thesis. The idea here is that rather than physical states being metaphysically fundamental and information derivative, it is information (the "bit") that is metaphysically fundamental and the physical universe (the "it") that derives from information. Physicist Seth Lloyd (2006 and 2010) and others have developed the theme into the suggestion that the universe just *is* a gigantic computer. What exactly does all this mean, and why would anyone think it true?

Chalmers and physicist Paul Davies suggest illuminating interpretations. Following Russell, Chalmers notes that physics does not tell us the intrinsic nature of the fundamental entities it posits. "Physics tells us nothing about what mass *is*, or what charge *is*: it simply tells us the range of different values that these features can take on, and it tells us their effects on other features" (1996, p. 302). Having mass or charge, like carrying syntactic information, is simply a matter of being in one of several states in a space of different possible states that might generate various outcomes at the end of causal pathways leading from those states. Now if the fundamental entities of physics are essentially characterized in terms of their effects, and to be information in the syntactical sense is just to have certain characteristic effects, then what physics gives us (Chalmers proposes) is essentially an informational conception of its fundamental entities.

Davies (2010, p. 82), noting that the idea of "laws of nature" is metaphysically problematic when removed from the theological context in terms of which Descartes and Newton understood it, proposes grounding laws instead in information considered as the "ontological basement" level of physical reality. Physicist Rolf Landauer had put forward the thesis that the laws of physics are the algorithms according to which the universe computes. (Cf. Davies 1992, pp. 146-47.) Expounding Landauer's position, Davies notes that it opens up the possibility of seeing "the laws of physics [as] inherent in and emergent with the universe, not transcendent of it" (2010, p. 83).

When combined, Chalmers' and Davies' views suggest that the notion of the universe as a kind of computer provides a way of bringing the laws of physics "down to earth," as it were, and unifying them with the entities they govern. As we saw above, syntactic information is embodied in physical states correlated with some effect at the end of a causal pathway, and the algorithms by which this information is processed are themselves embodied as information, and thus embodied in such physical states. If the universe is a kind of computer, then, it is governed by the laws of nature in the same way a computer runs an algorithm, and the laws relate to the entities they govern the same way an algorithm is related to the physical states of a computer whose causal relations it describes.

The notions of information, algorithms, and the like have if anything played an even bigger role in biology. That genes carry syntactic or Shannon information about phenotypes is fairly uncontroversial, since this simply involves causal correlations between genetic factors and aspects of a phenotype. More controversial is whether there is *semantic* information to be found in biological phenomena – information with something comparable to the *meaning* or *intentional content* characteristic of thoughts and linguistic representations. Certainly biologists often describe the phenomena they study in ways that imply that there is such information. As philosopher of biology Alex Rosenberg notes:

> Molecular biology is... riddled with intentional expressions: we attribute properties such as being a *messenger* ("second messenger") or a *recognition* site; we ascribe *proofreading* and *editing* capabilities; and we say that enzymes can *discriminate* among substrates... Even more tellingly... molecular developmental biology describes cells as having "positional information," meaning that they *know* where they are relative to other cells and gradients. The naturalness of the intentional idiom in molecular biology presents a problem. All these expressions and ascriptions involve the representation, in one thing, of the way things are in another thing... The naturalness of this idiom in molecular biology is so compelling that merely writing it off as a metaphor seems implausible. Be that as it may, when it comes to information in the genome, the claim manifestly cannot be merely metaphorical, not, at any rate, if the special role of the gene is to turn on its information content. But to have a real informational role, the genome must have intentional states. (2006, pp. 99-100)

Now whether intentionality or semantic content can be given a materialist explanation is itself controversial. Like other critics of materialism, I think it cannot be. Rosenberg also thinks it cannot be, but since he is a materialist his solution is to take the eliminativist line according to which intentionality and semantic content are illusions. Accordingly, he denies that there really is intentionality or semantic information to be found in biological phenomena. However, he also holds that "the crucial question is not intentionality but programming" (2006, p. 108). In particular, in Rosenberg's view a genome still "programs the embryo" and runs "software" even if its doing so does not involve processing information of a semantic sort (pp. 107-8). (Rosenberg (pp. 103-5) happens to agree with the upshot of Searle's famous "Chinese Room" argument (Searle 1980), according to which running a program is not sufficient for intentionality or semantics.)

Others would go farther. For example, philosopher of biology Peter Godfrey-Smith thinks that "genes 'code for' the amino acid sequence of protein molecules" in a sense that is appropriately regarded as semantic, though he adds that this "does not vindicate the idea that genes code for whole-organism phenotypes, let alone provide a basis for the wholesale use of informational or semantic language in biology" (2007, pp. 109-10). Insofar as genes carry information vis-à-vis phenotypes, Godfrey-Smith thinks it only information of the syntactic or Shannon sort. Like Rosenberg, he also thinks there is at least a limited role for talk of "programs," in particular when describing the operation of gene regulation networks (pp. 111-12).

Biologist Richard Dawkins is particularly eloquent on the subject of programs in nature. In *The Blind Watchmaker* he writes:

> It is raining DNA outside. On the bank of the Oxford canal at the bottom of my garden is a large willow tree, and it is pumping downy seeds into the air... The cotton wool is mostly made of cellulose, and it dwarfs the tiny capsule that contains the DNA, the genetic information. The DNA content must be a small proportion of the total, so why did I say that it was raining DNA rather than cellulose? The answer is that it is the DNA that matters... DNA whose coded characters spell out specific instructions for building willow trees... It is raining instructions out there; it's raining programs; it's raining tree-growing, fluff spreading, algorithms. That is not a metaphor, it is the plain truth. It couldn't be any plainer if it were raining floppy discs. (1987, p. 111)

Others would go even farther in applying the notion of *semantic* information within biology. (Cf. Deacon 2010; Hoffmeyer 2010; Küppers 2010; Smith 2010.) For present purposes, however, we can simply note that at least the core computational notions of *syntactic* information and algorithms are widely applied within biology. Natural selection itself has been claimed by Dennett (1995) and Mayfield (2013) to amount to a kind of algorithm, and the evolutionary process to constitute a kind of computation or information processing.

The area in which computational notions are most famously thought to have application is, of course, the study of the mind. The best-known instance of this approach is the idea that the mind is a kind of software and the brain a kind of computer hardware which runs the software. The former thesis, that the mind is a kind of software, is one that Searle labels "Strong Artificial Intelligence" (or "Strong AI"), and it also goes by the name "Turing machine functionalism." The latter thesis, that the brain is a kind of digital computer, is one that Searle labels "cognitivism." Obviously the theses are related, but they are distinct, and Searle has presented distinct arguments against each.

His famous "Chinese Room" argument is directed against the first, "Strong AI" thesis. I will have nothing to say about that argument here. To be sure, I think Searle is simply and without qualification correct to hold that the mind is not a kind of computer program or software, though my reasons go beyond (even if they include) the ones he gives in that argument (Feser 2013a). But that is neither here nor there for present purposes. Here I want to focus instead on the "cognitivist" claim that the brain is a kind of computer, so that computation must be at least part of the story in a scientific account of human cognition, even if it is not the whole story.

In the work of philosophers like Paul Churchland, you will often find claims like the following:

> [T]he brain represents the world by means of very high-dimensional *activation vectors*, that is, by a pattern of activation levels across a very large population of neurons. And the brain performs computations on those representations by effecting various complex *vector-to-vector transformations* from one neural population to another. This happens when an activation vector from one neural population is projected through a large matrix of synaptic connections to produce a new activation vector across a second population of nonlinear neurons. (1998, p. 41)

This approach to studying the brain is developed in great detail in works of computational neuroscience. (Cf. Churchland and Sejnowski 1992.)

Now, if by "representations" such writers had in mind something like thoughts with conceptual content, then I think these sorts of claims would be false. In my view, the conceptual content of our thoughts cannot be explained in causal terms, or in any other terms acceptable to the materialist (Feser 2011b and 2013a). However, if what is meant is merely that there is information in the brain of the syntactic, Shannon sort, then the computationalist approach is certainly no less plausible here than it is in the case of physics or biology. Indeed, there can hardly be any doubt that the neural properties and processes described in such detail in books of computational neuroscience are real and important.

But is the specifically *computationalist* conceptual apparatus, here or in the other contexts we've considered, necessary to a correct description of the phenomena? Or is it just a dispensable and indeed misleading set of metaphors? This brings us to Searle's argument against cognitivism.

5.5.2 Searle's critique

Again, the argument in question is not to be confused with Searle's (1980, 1984) famous "Chinese Room" argument. In that argument, Searle's claim was that running a program does not entail having intentional content or meaning; as he famously summed it up, "syntax is not sufficient for semantics." Even if the brain could be said to process information in the syntactic sense Shannon was interested in, the "Chinese Room" argument entails that that would never by itself amount to the having of semantic information of the sort characteristic of thought. But the argument leaves open the question whether the brain really does process information in at least the syntactic sense.

Searle's (1992, Chapter 9; 2008) later argument against what he calls cognitivism is intended to show that it does not. It aims to show that computation is not only not the *whole* story about what the brain does, it is not even *part* of the story. The basic idea of the argument is very simple. Whatever else computation in the sense we're discussing might involve, at the very least it involves the physical instantiation of *symbols* or strings of symbols, whether 0s and 1s or some other kind of symbols. If left uninterpreted the symbols will not carry *semantic* information. They will still constitute *syntactic* information, but only insofar as we do think of them as *symbols*, even if uninterpreted ones. The syntactical rules that make up the algorithm according to which the inputted symbols generate a certain output are rules that govern physical states precisely *qua symbols*. For example, they will be rules according to which the computer will give a 0 as output when it gets a 1 as input, or whatever. And that the computer instantiates a certain algorithm will, as we have seen, itself amount to there being certain further physical states which count as instances of certain symbols or bits of syntactic information. So, computation boils down to the instantiation of symbols.

The problem is this. The status of being a "symbol," Searle argues, is simply not an objective or intrinsic feature of the physical world. It is purely conventional or observer-relative. And thus the status of being something that is running an "algorithm," or "processing information," or "computing," is also conventional or observer-relative rather than an intrinsic and objective feature of any physical system. This is obviously true where the computers of everyday experience are concerned. What they do constitutes the "processing" of "symbols" or "bits" of "information" according to an "algorithm" only because human designers and users of the machine *count* the electrical states as symbols, the transitions between states as the implementation of an algorithm, etc. But the same thing is true of anything else we might think of as a computer – a brain, a genome, or the universe as a whole. Its status as a "computer" would be observer-relative because a computer is simply not a "natural kind" but rather a sort of artifact. Searle draws an analogy:

> [W]e might discover in nature objects which had the same sort of shape as chairs and which could therefore be used as chairs; but we could not discover objects in nature which were functioning as chairs, except relative to some agents who regarded them or used them as chairs.

Similarly, he says:

> We could no doubt discover a pattern of events in my brain that was isomorphic to the implementation of the vi program on this computer. But to say that something is *functioning as* a computational process is to say something more than that a pattern of physical events is occurring. It requires the assignment of a computational interpretation by some agent. (2008, p. 95)

So, if a brain is a kind of computer, that can in Searle's view be true only in the trivial sense that we can *interpret* various brain states as symbols and various neural processes as computations if we like. But in that sense all sorts of other things are "computers" too. Searle writes:

> For any program there is some sufficiently complex object such that there is some description of the object under which it is implementing the program. Thus for example the wall behind my back is right now implementing the Wordstar program, because there is some pattern of molecule movements which is isomorphic with the formal structure of Wordstar. But if the wall is im-

> plementing Wordstar then if it is a big enough wall it is implementing any program, including any program implemented in the brain. (2008, p. 93)

But if the brain or any other natural system (such as the genome, or the universe as a whole) is computing only in the trivial and uninteresting sense in which a wall is "computing," then it is not computing in any sense that might be explanatorily useful in science or philosophy.

In short, Searle says, "computational states are not *discovered within* the physics, they are *assigned to* the physics" (1992, p. 210). They are no more a part of the furniture of the natural order of things than chairs are. Hence, just as no physicist, biologist, or neuroscientist would dream of making use of the concept of a chair in explaining the natural phenomena with which they deal, neither should they make use of the notion of computation.

Now an objection frequently raised against Searle is that more is required of something if it is to count as a computer than merely that we could interpret some isolated set of its states as a computation. It also has to have the right kind of causal organization (Block 2002, pp. 76-78; Chalmers 1996, pp. 219-20; Endicott 1996, pp. 103-7; Moural 2003, pp. 234-5; Rey 2002, pp. 215-17). It is not enough, for example, for a system plausibly to count as implementing the computation "1 + 2 = 3" that it has states corresponding to "1" and "+" and "2" and "=" which are followed by a state corresponding to "3." For what it does genuinely to count as addition, it must also be true that had the input been states corresponding to "2" and "+" and "3" and "=," the output would have been a state corresponding to "5"; that had the input been states corresponding to "3" and "-" and "2" and "=," the output would have been a state corresponding to "1"; and so on for other counterfactual inputs and outputs. We need an isomorphism not just between this or that particular computation and this or that particular state of the system, but between, on the one hand, the structure of a program as a whole and on the other, the causal structure of the entire physical system over time. And this will rule out cases like Searle's example of his wall implementing Wordstar.

But this, it seems to me, is not a serious objection to Searle. Searle acknowledges that a system's having an appropriate causal structure is a *necessary* condition for its implementing a program (1992, p. 209). His point is that it is not a *sufficient* condition. To recall his parallel example, having an appropriate causal structure is also a necessary condition for something's being a chair. Wood and steel have such a structure, but shaving cream, cigarette smoke, and liquid water do not, since they lack the solidity and stability to hold someone up. But whether some wooden or steel object counts as a chair is still observer relative, a matter of convention. Similarly, even though a system has to have the requisite causal structure in order to count as a computer, Searle's point is that it still will not count as one unless some observer assigns a syntactical interpretation to its physical states.

That a physical system's having the appropriate causal structure can only ever be a necessary and not a sufficient condition for its implementing a program is given further support by an anti-computationalist argument from Saul Kripke that is related to but distinct from Searle's. (Cf. Kripke 1982, pp. 35-37; Buechner 2011.) Consider Kripke's example of the "quus" function, which he defines as follows:

x quus y = x + y, if x, y < 57; = 5 otherwise.

The primary use Kripke makes of this odd example is, of course, to generate his famous skeptical paradox about meaning. A person's linguistic utterances and other behavior, and the words and images he calls before his mind, might all seem to show that he is adding when he says "1 + 2 = 3" and the like. But Kripke imagines a bizarre skeptic suggesting that for all we know, the person might really be carrying out "quaddition" rather than addition. If the person has never computed numbers higher than 57, then although we expect that when he computes "68 + 57" his answer will be "125," it may be that he is quadding rather than adding, so that the answer will actually be "5." Nor would it matter if he *had* computed numbers higher than 57. For there is always some number, even if an extremely large one, equal to or higher than which he has never calculated, and the skeptic can always run the argument using that number instead. Nor would it matter if the person in question said "I am adding and not quadding!", because just as *we* might be misinterpreting his use of words like "plus" and "adding," so too might he be misinterpreting his *own* use of those terms.

Now, one reason Kripke's paradox is philosophically interesting is that it might be claimed to show that there is no fact of the matter about what we mean by our utterances. I don't think it really does show this, for reasons I have explained elsewhere (Feser 2013a). But Kripke thinks it also has application as an argument against computationalism, and this seems to me correct. For whatever we say about what *we* mean when we use terms like "plus," "addition," etc., there are no physical features of a *computer* that can determine whether *it* is carrying out addition or quaddition, no matter how far we extend its outputs. No matter what the past behavior of a machine has been, we can always suppose that its next output – "5," say, when calculating numbers larger than any it has calculated before – might show that it is carrying out something like quaddition rather than addition. Of course, it might be said in response that if this happens, that would just show that the machine was malfunctioning rather than performing quaddition. But Kripke points out that whether some output counts as a malfunction itself depends on what program the machine is running, and whether the machine is running the program for addition rather than quaddition is precisely what is in question.

Obviously, Kripke's argument raises questions of its own. (See Buchner 2011 for a detailed exposition and defense.) Suffice it for present purposes to note how it bolsters Searle's point. Even if a physical system's having a certain causal organization is a necessary condition for its implementing a program for addition, it cannot be a sufficient condition, because that causal organization will also be consistent with its implementing quaddition rather than addition. And the point is completely general. There will be parallel quaddition-like counterexamples for *any* claim to the effect that a physical system's having a certain causal structure is sufficient for its implementing some specific program.

An objection raised against Searle by John Haugeland (2002, p. 391) is that the claim that syntactical features are observer-relative is falsified by the fact that there are empirical tests, based on stringent specifications, for whether something possesses such features. A related objection is raised by Jeff Coulter and Wes Sharrock (2002, p. 196) when they write that it is odd for Searle to suggest that "computation is merely an 'observer-relative' feature of a computer." The point of these objections seems to be that given how syntax and computation are defined, it is just a straightforward factual matter whether something has syntactical features or is a computer. But this just misses Searle's point. He isn't denying that there might be rigorously specifiable empirical criteria for whether something has syntactical features or is a computer. He is saying that whether there are or not, that something fitting those criteria counts as a computer is ultimately a matter of convention rather than observer-independent facts. Even if we had rigorously articulated empirical criteria for whether something is a chair, that anything counts as a chair in the first place would still be a matter of convention and thus observer-relative. Searle is saying that the same thing is true of whether something possesses syntactical or computational features. (Cf. Buechner 2008, pp. 158-59.)

Ronald Endicott (1996, p. 111) objects that the claim that the symbols posited by the computationalist are observer-relative rests on a false analogy with the symbols of natural languages. True, the symbols of English, German, and the like have the meanings they do only because they are assigned meanings as a matter of convention. But the symbols posited by the computationalist, says Endicott, are not like that. The computationalist takes them to get their meaning instead in a way described by some naturalistic theory of meaning, such as a causal covariation theory. And the biologist's talk of DNA codes and the like shows, in Endicott's view, that it is possible for there to be symbols in nature apart from interpreters.

But this line of objection both begs the question and misses the point. It begs the question in two ways. First, Searle has argued that causal covariation and other naturalistic theories of meaning all fail, so a critic can hardly take such a theory for granted when criticizing Searle. (Cf. Searle 1992, pp. 49-52. I think Searle is correct to reject such theories, for reasons set out in Feser 2011b and 2013a.) Second, whether DNA and related biological phenomena *literally* can be said to have computational features is, at least implicitly, precisely part of what is at issue between Searle and his critics. That computational notions are useful to the biologist would presumably be regarded by Searle as comparable to the fact that there are naturally occurring objects that we find it comfortable to sit on. The latter fact does not entail that it is not a matter of convention whether something is a chair, and the former fact (Searle would presumably say) does not entail that it is not a matter of convention whether DNA, the brain, or anything else counts as a computer.

Endicott misses the point when he speaks as if the issue has to do with whether the symbols posited by the computationalist get their meaning in something like the way the symbols of natural languages do, or instead the way naturalistic theories of meaning say they do. Semantics was the topic of Searle's Chinese Room argument, but not of his argument against what he calls cognitivism. Here the issue is, not how the symbols posited by the computationalist get their meaning, but rather whether it even makes sense in the first place to speak of symbols – even *uninterpreted* symbols – existing apart from human convention and apart from any observer.

Jeff Buechner (2008, chapter 5) raises several further objections against Searle. First of all, Buechner notes that computations, like the objects of mathematical discourse, are abstract objects. Now since conventionalist theories of mathematics are highly problematic, it is no less problematic to treat computation as if it were merely conventional or observer-relative (2008, pp. 160-65). It seems to me the obvious retort to this is that it just misunderstands what Searle is saying. Searle is not saying that computations considered as abstract objects are conventional or observer-relative; he is saying that whether such-and-such a physical system *implements* a computation is conventional or observer-relative. Buechner considers this possible reply (pp. 166-68 and 324, note 11). He concedes that in a physical system built by an engineer, there is a sense in which the fact that it implements a computation is observer-relative. But he suggests that such a system could instead be "assembled through evolutionary pressures," and implies that if Searle insisted that an intelligent designer would be necessary for such a system to count as implementing a computation, he would be committing himself to an untestable "Intelligent Design" theory.

But the problem with Buechner's response can be seen by once again considering Searle's parallel example of a chair. Something that just happened to be the sort of thing we would find it comfortable to sit on could in principle come about by evolutionary processes. All the same, it wouldn't count as a *chair* unless some observer decided so to count it, because chairs are not natural kinds but products of convention. Similarly, Searle need in no way deny that something as complicated as the computers human engineers construct could come about via evolutionary processes. He would deny only that this would, apart from an observer who assigns a computational interpretation to it, count as a computer. This no more commits him to "Intelligent Design" theory than denying that a chair-like object that arose via evolutionary processes would in the strict sense really be a chair commits one to "Intelligent Design" theory. And if Buechner digs in his heels and insists that such a product of evolution *must really be* a computer and not merely something to which we might assign a computational interpretation – that is to say, if he says that it meets conditions *sufficient*, and not merely *necessary*, for being a computer – then he is just begging the question against Searle.

Buechner also concedes (2008, pp. 169-72) that there is a degree of convention or observer-relativity in symbols and syntax, just as there is in the system of numerals we use to do mathematics, but he thinks this does not suffice to establish Searle's position. With any system of numerals, "the laws of arithmetic must be respected and human limitations must be respected" (p. 170). What Buechner has in mind by the first constraint, it seems, is that if a system of numerals allowed us to count as true a statement like "2 + 2 = 5" (for example), then it would obviously be deficient for the purposes of doing arithmetic. What he has in mind by the second constraint is that certain symbols would not be useful for us for the purposes of doing mathematics given e.g. our perceptual limitations. For instance, we just find it harder to read "||||||||" than "8" (and so on for other numbers), so that it would be practically impossible for us to do arithmetic via a stroke notation instead of a decimal notation. Now since it is not within our power to change these constraints, the choice of which numerals to use is not entirely observer-relative even if it is to some degree. But the same point can be made about the choice of syntax and symbols.

The problem with this as a reply to Searle is that once again, all that has been shown is something Searle has already conceded, viz. that having a certain physical structure is a *necessary* condition for a system's carrying out a certain computation. Searle's point, though, is that it is nevertheless not a *sufficient* condition. To be sure, Buechner adds a further point, namely that whereas in the case of computers made by us, human designers choose which symbols to use within the constraints in question, in the case of a natural computer like a brain, it is evolution which "chooses" among the various possible symbols fulfilling the constraints in question. He adds that these symbols could still be there even if we do not recognize them as such. Once again, though, Buechner's appeal to evolution is either a *non sequitur* or begs the question. If evolution produced something that was chair-like, it wouldn't follow that it had produced a chair, and if evolution produced something symbol-like, it wouldn't follow that it had produced symbols. And if Buechner simply insists that this would follow, then he is assuming precisely what is at issue, since Searle's whole point is that no natural processes – including evolution – could of themselves produce something that was literally a computer.

5.5.3 Aristotle and computationalism

So, Searle's critics have failed successfully to rebut his argument against cognitivism. That is not to say, however, that they haven't a leg to stand on. For Searle's critique to be decisive, he needs not only to give an argument *against* the claim that computation is intrinsic to the natural world. He also needs to show that there are no good positive arguments *for* the claim that it is intrinsic to the natural world. Are there any good positive arguments for that claim?

It seems to me that there are. We've already seen why at least some physicists, biologists, and neuroscientists would characterize the phenomena they study in terms of notions like *information*, *algorithms*, and the like. Searle would have to say that these are at best merely useful fictions, and that everything that has been put in these computational terms could be said without recourse to them. But that does not seem to be the case. To see why, first consider once again Kripke's "quus" example. Kripke's skeptic claims that there is no fact of the matter about whether any of us is ever doing addition rather than "quaddition." The common sense view, of course, is that there is a fact of the matter, and that the fact is that we are doing addition and not "quaddition." Hence if we came across someone whose arithmetical behavior seemed perfectly normal except that when he calculated "68 + 57" his answer was "5" instead of "125," we wouldn't conclude that he really was "quadding" after all. We would conclude instead that he was doing addition but, in this case, doing it badly. We would regard his answer as the result of a typographical error or momentary confusion, or perhaps even delirium, temporary insanity, dementia, brain damage, or what have you. We would not think of him as a properly functioning system carrying out "quaddition," but as a malfunctioning system carrying out addition.

We needn't worry for present purposes about what would rationally *justify* our taking this view, since meaning skepticism isn't our topic. (Again, see Feser 2013a for further discussion of Kripke's paradox.) What I want to consider here is the following sort of parallel case. Recall some of the computationalist claims I cited earlier. Rosenberg says that the genome "programs the embryo." Churchland says that "the brain represents the world by means of very high-dimensional activation vectors" and "performs computations on those representations." Now, for the program Rosenberg says the genome is running, the program Churchland says the brain is running, and for any other program someone wants to attribute to a natural process, we can construct a "quaddition"-like paradox. For instance, we can imagine what we might call a "quembryo" program which, when the genome runs it, produces the same results that the embryo program does except that the embryo does not develop eyes. Now, consider a human embryo that never develops eyes. Should we say that the genome that built this embryo was running what Rosenberg would call the embryo program but that there was a malfunction in the system? Or should we say instead that the genome was actually running the "quembryo" program so that there was no malfunction at all and things were going perfectly smoothly?

A Kripke-like skeptic would, of course, say that there is no fact of the matter. But notice that Searle, since he holds that there are no programs at all really running here in the first place, would also have to say that there is no fact of the matter. But that simply doesn't seem plausible. If there really is such a thing as a difference between a properly functioning organism and a malfunctioning one then it seems to follow that Rosenberg's postulated embryo program captures something about the facts of the situation that our imagined "quembryo" program does not. Using computer jargon, our hypothetical "quembryo" skeptic might say of the embryo's lack of eyes: "Maybe that's not a bug, but a feature!" But he would be wrong. The lack of eyes *is* a bug, and not a feature. Unless we are skeptics about the very distinction between properly functioning and malfunctioning organisms – and it is hard to see how biology would be possible given such skepticism – then it seems we have to agree that there really is something to the claim that what we have here is a malfunctioning system running the embryo program, as opposed to a properly functioning system running the "quembryo" program.

If the computer scientist's distinction between "bugs" and "features" has application to natural phenomena, so too does the distinction between "software" and "hardware." For (to stick with the embryo example) the lack of eyes is as dysfunctional in one human embryos as it is in another. A natural way of putting this is that all human embryos are running the *same* program, that the very same software is, as it were, being implemented in different pieces of hardware. That is why what is a "bug" or "feature" for one embryo is also a "bug" or "feature" for the others. Searle's view seems to be that there is nothing true in the computationalist's description of a natural physical system that cannot be captured by a description of the causal processes taking place in the system. But that is not correct. For there is a distinction to be made between *normal* and *aberrant* causal processes, and there is a distinction to be made between a *general type* of normative causal process and specific *token instances* of that type. The computationalist's language captures these distinctions in a way that a mere description of which causal processes happen to be taking place does not.

To borrow some jargon from Dennett (1981), Searle supposes, in effect, that everything that is true of a natural object or process can be captured by taking the "physical stance" toward it, but in fact there are aspects that can be captured only by taking the "design stance," and these are precisely the ones captured by the computationalist description. The computationalist description captures what Dennett (1998) calls "real patterns" in nature, patterns irreducible to the purely causal description to which the "physical stance" confines itself. For it is only by taking the "design stance," which is defined by consideration of *proper function* – where regarding the genome as running embryo software rather than "quembryo" software is at least one way of doing this – that we can make sense of the fact that the lack of eyes is an aberration, and that this is true of human embryos as such, not just of this or that particular embryo.

Now if all of this is correct, then we seem to have what Aristotle calls an *aporia*, a puzzle arising from the existence of apparently equally strong arguments for two or more inconsistent claims – in this case, equally strong arguments both against and for the claim that there is computation in nature. And the way to resolve it, I suggest, is to see that while Searle's position is unavoidable if we take for granted the essentially "mechanistic" conception of nature to which he and his naturalist critics are both committed, the computationalist approach can be made sense of if we adopt instead a broadly Aristotelian conception of nature. In their use of computational notions, contemporary naturalists have unwittingly recapitulated the formal and final causality that they, like their early modern "mechanical philosophy" forebears, thought had been banished for good.

Recall that "information" in the technical, syntactic sense essentially involves a causal correlation between a physical state and some effect at the end of a causal pathway leading from that state. Now, any physical state has any number of effects along a causal pathway. For instance, the physical states of the compact disc in the example from Chalmers cited earlier have, among their effects, the sounds that come out of the CD player when the disc is played. But those physical states have many other effects as well. For example, there is the electrical activity that occurs in the circuitry of the CD player which in turn causes the sounds to emerge from the speakers; and there is the shaking of the nearby walls that might take place if the volume is turned up too loud. Now, we say that the physical states of the compact disc carry "information" about the *sounds* they cause, specifically, rather than about the electrical activity in the CD player or the shaking of the walls. The reason for that, of course, is that the designers of compact discs made them for the *purpose* of allowing us to play back the sounds in question, rather than for the purpose of generating electrical activity or causing walls to shake. It is the existence of that purpose that allows us to identify the sounds as the specific effect down the causal pathway about which the physical states of the compact disc carry information.

But no such observer-relative purposes can be appealed to in the case of the information the computationalist attributes to physical states occurring in nature. That is, of course, why Searle says there is no information to be found in such states. But if we suppose that Aristotelian teleology or final causality is a real feature of nature after all, then we can make sense of such naturally occurring information. In particular, if we suppose that a physical state of type *S* inherently "points to" or is "directed at" some particular type of effect *E* down the causal pathway – rather than to some earlier effect *D* or some later effect *F* – then we have a way of making it intelligible how *S* carries information about *E* rather than about *D* or *F*. Without such teleology, though, it is hard to see why there would be anything special about *E* by virtue of which it would be the effect about which *S* carries information. (Cf. Feser 2011b.)

Consider also that when we speak of a pocket calculator running a program or algorithm for addition rather than "quaddition," this is easy to make sense of given that the designers of the calculator designed it for the purpose of doing addition rather than quaddition. But how do we make sense of the genome running the embryo program rather than the "quembryo" program, given that there is no human observer who assigns this purpose to it? If we suppose that there are such things as Aristotelian substantial forms after all, then we have a way of making this intelligible. For there to be a fact of the matter that the genome is running the embryo program rather than the "quembryo" program is for the genome to have the sort of *intrinsic* tendency toward certain characteristic operations that distinguishes a substantial form from a merely accidental form. (Cf. Ross 1990 and 2008, Chapter 7.) Moreover, as Mayfield notes, "an important requirement for an algorithm is that it must have an outcome" (2013, p. 44), and "instructions" of the sort represented by an algorithm are "goal oriented" (p. 13). The genome algorithm, for example, has as its "goals" or "outcomes" the sort characteristic of the embryo program rather than the "quembryo" program. It is hard to make sense of this except as an instance of Aristotelian immanent teleology.

Other aspects of the computationalist conception of nature also echo the Aristotelian conception. For instance, when Mayfield notes (as we saw earlier that he does) that "the information content of the output [of a computation] can be less than that of the input, but not greater," he is essentially recapitulating the principle of proportionate causality, according to which whatever is in an effect must in some way or other be in its total cause.

Some of the moves made by Searle's critics also at least gesture, inadvertently, in a broadly Aristotelian direction. For example, in response to Searle's claim that his wall is running Wordstar, Endicott objects that before we can plausibly attribute a program to some physical system, we have to consider "*non-gerrymandered* physical units" and "a *physical system whose parts have the disposition to causally interact in the way specified by the program*" (1996, p. 105). But a "non-gerrymandered" physical unit occurring in nature arguably suggests one marked off from others by virtue of having a substantial form, and a part having a "disposition" causally to act in certain specific ways arguably suggests one that is "directed toward" a certain kind of manifestation as toward a final cause.

Other writers have explicitly noted the Aristotelian implications of computational descriptions of natural phenomena. The neuroscientist Valentino Braitenberg has said that "the concept of information... is Aristotle *redivivus*, the concept of matter and form united in every object of this world" (quoted in Floridi 2008, p. 16). Philosopher of science John Wilkins (2014) calls information "the new Aristotelianism" and the "New Hylomorphism" (though unlike Braitenberg he does so disapprovingly, considering the notions in question to entail a regress to an outmoded conception of nature).

I imagine that Searle would share Wilkins' attitude, perhaps allowing that Aristotelian and computationalist arguments mutually reinforce one another, but concluding that they should simply all be thrown out together. Indeed, Searle explicitly maintains that not only computation, but function and teleology more generally are all observer-relative. Of biological phenomena, he writes:

> Darwin's account shows that the apparent teleology of biological processes is an illusion.
>
> It is a simple extension of this insight to point out that notions such as "purpose" are never intrinsic to biological organisms...

> And even notions like "biological function" are always made relative to an observer who assigns a normative value to the causal processes...
>
> In short, the Darwinian mechanisms and even biological functions themselves are entirely devoid of purpose or teleology. All of the teleological features are entirely in the mind of the observer. (1992, pp. 51-52)

Searle would also deny that there is any level of physical reality that can only accurately be described from the functional point of view represented by Dennett's "design stance," as opposed to the purely causal level represented by the "physical stance." Searle writes:

> Where functional explanations are concerned, the metaphor of levels is somewhat misleading, because it suggests that there is a separate functional level different from the causal levels. That is not true. The so called "functional level" is not a separate level at all, but simply one of the causal levels *described in terms of our interests...* When we speak of... functions, we are talking about those... causal relations to which we attach some *normative* importance... [But] the normative component... [is in] the eye of the beholder of the mechanism. (1992, pp. 237-38)

Now there are several problems with this. For one thing, there are different respects in which biological phenomena might seem to exhibit teleology. The adaptation of an organism to its environment is one apparent instance of biological teleology; developmental processes, and in particular the fact that some growth patterns are normal and others aberrant, are another. As several writers have pointed out, while Darwinism might explain away the first sort of example, it doesn't follow (contra Searle) that it explains away the second (Ariew 2002 and 2007; Grene 1974; Turner 2007). For another thing, the Aristotelian would argue that it is a confusion to suppose that one can entirely replace teleological explanations with causal ones, because even the simplest causal regularity will itself *presuppose* teleology. As we saw in chapter 1, for the Aristotelian, if *A* regularly generates *B* rather than *C* or *D* or no effect at all, that can only be because generating *B* is the outcome toward which *A* is inherently directed as toward a final cause. If we don't recognize such rudimentary teleology we will be stuck with Humean skepticism about causality.

A third problem with Searle's position is that if we say that there is no teleology inherent in mind-independent reality, we are pretty clearly left with two options. We could, on the one hand, say that there is no teleology *at all*, *anywhere*, not even in the mind. That would be an eliminativist position, and it would be difficult at best to make such a position coherent. For if there is no teleology or "directedness" of any sort, then there would be no "directedness" of the kind associated with the intentionality of thought. And it is notoriously difficult coherently to deny the existence of intentionality, since the very denial is itself a manifestation of intentionality.

Certainly Searle is no eliminativist about intentionality (1992, p. 6), nor, it seems, about teleology or "directedness" in general. His view would seem to be the second alternative, according to which there *is* teleology *in the mind* even if there is none in mind-independent reality. But this would seem to entail either Cartesian dualism or property dualism, with all their associated problems – the interaction problem, epiphenomenalism, and so forth. To be sure, Searle claims not to be a dualist (2008b), but like other critics I find it hard to see how his view differs from dualism except verbally. Consider what Searle says about the sort of "directedness" associated with the intentionality of the mental:

> [I]ntentional notions are inherently normative. They set standards of truth, rationality, consistency, etc., and there is no way that these standards can be intrinsic to a system consisting entirely of brute, blind, nonintentional causal relations. There is no normative component to billiard ball causation. (1992, p. 51)

This is said in the context of his remarks about the absence of teleology from biological phenomena. The clear implication is that the human body and brain consist "entirely of brute, blind, nonintentional causal relations" of the "billiard ball" type, whereas the mind is the seat of the intentionality, rationality, normativity, etc. which cannot be "intrinsic" to the body and brain thus understood. That sounds pretty close to the Cartesian dichotomy between matter conceived of as pure mechanism devoid of thought and mind conceived of as pure thought irreducible to mechanism.

It is true that Searle regards the mental as *caused by* the physical, but then Cartesian substance dualists and property dualists also often affirm a causal relation between the physical and the mental. And while these dualists have famously had difficulty in explaining exactly *how* this causal relation works, Searle too admits that:

> [W]e don't have anything like a clear idea of how brain processes, which are publicly observable, objective phenomena, could cause anything as peculiar as inner, qualitative states of awareness or sentience, states which are in some sense "private" to the possessor of the state. (1997, p. 8)

I discuss Searle's relationship to property dualism at greater length elsewhere (2015b), and have in any case criticized Cartesianism in earlier chapters. Suffice it for present purposes to say that from the Aristotelian point of view, the intractability of the debate between materialism and Cartesian forms of dualism is a consequence of what they have in common, namely the "mechanistic" conception of nature that supplanted the Aristotelian conception – a conception which leaves no place for the teleological, and thus no place for the intentional. Like the Aristotelian, Searle has been critical of both materialism and Cartesianism, but from the Aristotelian point of view Searle's own position is unstable, threatening to collapse back into one or the other of these alternatives, precisely because he is also committed to the same "mechanistic" picture they are. The key is to reject that picture and return to the one it supplanted. Contemporary computationalism, for all the flaws in it rightly identified by Searle, has the merit of gesturing precisely in the direction of such a return, however inadvertently.

6. Animate nature

6.1 Against biological reductionism

6.1.1 What is life?

A living thing, according to the traditional Aristotelian view, is a self-moving thing. But given the ambiguity of the term "moving," what this amounts to is better expressed by the Scholastic language of *immanent causation.* (Cf. Donceel 1961, pp. 26-28; Gardeil 1956, Chapter 2; Kluber-tanz 1953, pp. 47-50; Koren 1955, Chapter 1; Oderberg 2007, pp. 177-83; Oderberg 2013.) A causal process is immanent when it originates within the agent and terminates within it in a way that tends toward the agent's own self-perfection or completion. This is to be contrasted with *transeunt* (or *transient*) *causation,* which terminates outside the agent. A snake's digestion of the mouse it has eaten would be an example of an immanent causal process. Digestion begins when the meal is eaten and ends when its nutrients have all been absorbed into the bloodstream, and the result of the process is that the animal is enabled to survive, grow, and reproduce. A boulder's rolling down a hill during an earthquake and bumping into another boulder, which in turn bumps into a third, would be an example of a transeunt causal process. The source of the boulders' motion is entirely outside them, and does not terminate in anything like completion or self-perfection in any of them.

Living things no less than non-living things exhibit transeunt causation. Like a boulder, an animal might roll down a hill and bump into another animal. An immanent causal process might also have transeunt effects as a byproduct, such as the waste that an animal defecates after digesting a meal, which might contribute to polluting a nearby body of water. But living things are living because they alone exhibit immanent *as well as* transeunt causation. It is in this sense that they are self-movers. They move or change themselves in the sense that they carry out activities that contribute to their own completion or perfection.

As the talk of completion and self-perfection indicates, immanent causation is *teleological.* Digestion results in the nourishment of *the snake*, not of the mouse and not of some hybrid of snake and mouse. (Cf.

Pasnau and Shields 2004, p. 34.) Indeed, the mouse has altogether disappeared by the end of the process, whereas the snake continues. Hence the process *points* or *aims toward* the realization of the ends of the snake, specifically. Now, it might seem that non-living things exhibit a similar kind of teleology. For example, we say that a coffee machine can turn itself on in the morning, that a computer can run a self-diagnostic routine, and so on. But these are examples of artifacts, which have merely accidental forms and derivative teleology. Living things, by contrast, are true substances, with substantial forms and intrinsic teleology. The parts of a coffee machine or computer have no *built in* tendency to pursue the ends distinctive of those kinds of devices. They have to be made to do so by human designers. A snake *does* have a built in tendency to pursue ends such as digestion. Machines seem life-like only if we ignore this crucial distinction between substantial and accidental form.

It is important to add at once, however, that it is not intrinsic teleology *per se* that is definitive of immanent causation or of life. Recall that, for the Aristotelian, there is teleology wherever there is even the simplest inorganic causal regularity. The phosphorus in the head of a match *aims* or *points toward* the outcome of generating flame and heat; the brittleness of a glass *aims* or *points toward* a manifestation such as shattering; and so forth. These are examples of transeunt causation despite their teleological character, because they do not involve anything like the *perfection* or *completion* of an agent in the way that digestion involves the perfection or completion of the snake. It is directedness toward that particular *sort* of end – again, the perfection or completion of the causal agent itself – that is definitive of immanent causation and thus definitive of life.

These examples of inorganic teleology also show that it is not a good objection to the Aristotelian account of life to allege that, precisely *because* that account is teleological, it does not sit well with modern physics' non-teleological conception of nature. For one thing, as I have emphasized many times now, the absence of some feature from physics' *representation* of nature simply does not entail that that feature is absent from nature *itself*. That physics eschews teleological explanation merely reflects its mathematically oriented methodology, and by itself has no metaphysical implications. For another thing, there are powerful arguments for attributing teleological features to nature in general, and not just in the biological context. For example, there is the argument from the indispensability of computational notions when describing even inorganic phenomena, which I defended in the previous chapter. There is also the argument, influential within contemporary analytic metaphysics and philosophy of science, that we need the notion of causal powers to make sense of what physics and chemistry tell us, and that we need the notion of teleology (or "physical intentionality," to use the jargon favored by some contemporary powers theorists) in order to make sense of causal powers. I have expounded and defended this line of argument at length elsewhere (Feser 2014b, pp. 88-105). Of course, the critic of Aristotelianism will reject these arguments, but the point is that it is no good to reject the Aristotelian account of life *merely* on the grounds that it is teleological, since the Aristotelian has independent arguments which purport to show that we need to affirm teleology anyway, whatever we say about the nature of life.

Another possible objection would be that we don't need the Aristotelian account of life, because there are better alternatives available. Mark Bedau (1996) suggests that there are four main accounts of the nature of life in play in contemporary biology and philosophy. The first is the view that to be alive is to have a metabolism. The second holds that life is to be defined in terms of a longer list of characteristics, which may include metabolism but must refer to other characteristics as well – for example, reproduction, purposeful behavior, the possession of parts with functions, and/or capacity for evolution. The third holds that there is no one characteristic or set of characteristics that all and only living things have in common, but at best a "family resemblance" between living things. On this view, any living thing will have *some* of the characteristics found on lists like the ones adherents of the second view would draw up, but there won't be a common core that every single living thing possesses. The fourth view, which is the one Bedau favors, holds that life is to be defined in terms of "supple adaptation" to changes in the environment, of the kind seen in evolution.

One problem with appealing to such accounts in criticizing the Aristotelian view is that these alternatives are all themselves problematic. As Bedau notes, one problem with accounts like the second and third is that they raise the question *why* the characteristics they cite tend to be clustered together in living things. Whatever the answer to that question is would seem to be a more plausible candidate for revealing the nature of life than the cluster of characteristics themselves is. A problem with the appeal to metabolism, Bedau notes, is that there are, arguably, metabolizing entities that are not living, such as a candle flame. One problem with the "supple adaptation" appealed to by Bedau is that it seems to apply not only to living things, but also to cultures and economic markets, which adapt to their environments but would not ordinarily be regarded as alive. Bedau seems willing to bite the bullet and regard these things as alive, but as Margaret Boden (1996, pp. 23-24) points out, even if we were to go along with this implausible proposal, there are other problems. First, Bedau's account implies that it is *populations* of organisms that are alive in the most fundamental sense – since they are what adapt in the relevant respect – with individual organisms alive only in a secondary sense. But this gets things the wrong way around. Second, his account would seem to imply, implausibly, that a population that ceased evolving would no longer count as living. Third, and equally implausibly, it seems to imply that an organism or population that arose in a way other than evolution (such as direct creation by God) would not count as living.

But as David Oderberg (2013, pp. 214-16) points out, another and deeper problem with these accounts of the nature of life is that on analysis they tend to *presuppose* rather than replace immanent causation. For example, metabolism "is probably the paradigmatic example of immanence: the organism takes in matter/energy, uses it for its sustenance, growth, and development, and expels what is noxious or surplus" (p. 214). Moreover, understanding metabolism in terms of immanent causation explains why fire doesn't count as metabolizing and thus is not alive. For unlike agents engaged in immanent causation, fire is not a *substance*, but merely a modification of a substance or substances. Meanwhile, the "supple adaptation" to its environment of a population of organisms presupposes various kinds of activity in the individual organisms that make up the population, and this activity will involve immanent causation (metabolism, growth, etc.). A characteristic like reproduction also involves immanent causation insofar as it is an active process internal to the organism that is an ordinary function of mature or perfected members of its kind. This differentiates it from processes that might superficially appear similar to reproduction – such as the splitting of a rock, which is something that *happens to* the rock rather than an activity it carries out, which occurs due to causes entirely *external* to it, and which *damages* or *diminishes* rather than perfects or completes the rock (Oderberg 2007, pp. 179-80).

The point is best understood by recalling the Aristotelian distinction (introduced in chapter 1) between the *essence* of a thing and the *properties* or *proper accidents* that flow or follow from that essence. To take a stock example, the essence of a human being is to be a rational animal (if, for the sake of argument, the reader will go along with the traditional Aristotelian definition) and the capacity for humor is a property that flows from this essence. The capacity for humor is not itself part of the essence, but is rather a byproduct of it, a consequence of rationality. The manifestation of a property can be blocked, which is why some people can appear virtually humorless.

Now, the trouble with the alternative accounts of the nature of life under consideration is that they all focus on what are really at best *properties* of life rather than the *essence* of life (Oderberg 2007, pp. 177-78). By contrast, the Aristotelian account of life in terms of immanent causation captures precisely the essence of which the other characteristics are properties. This is, the Aristotelian proposes, the right way to understand Bedau's insight that to define the essence of life in terms of either a necessary and sufficient cluster of characteristics or a looser "family resemblance" cluster only raises the problem of explaining why the characteristics in question tend to be found together in living things. The Aristotelian answer is that the items on the right list of characteristics (whatever that turns out to be) are not parts of the essence of life but rather properties that flow from the essence, which is immanent causation. And it is because the manifestation of a property can be frustrated that some items on these proposed lists don't always appear in every single organism, even if they appear in most of them. For example, the reason some individual organisms and species of organism do not reproduce is not because reproduction is not a true property of living things, but rather because in some living things the manifestation of this property has been blocked (e.g. by chromosomal abnormalities). (Cf. Oderberg 2007, pp. 178-79.)

What should we say of borderline cases, such as viruses? Oderberg argues (2007, pp. 191-92), quite plausibly, that viruses lack immanent causal activity and thus are not truly alive. They do not take in or process nutrients, as truly metabolizing entities do; they do not grow; and even their capacity to replicate is arguably not true reproduction, insofar as it does not involve an active internal process:

> Virus replication, to [speak] metaphorically, is more like a genetic version of photocopying than genuine biological reproduction. The paper does not put itself into the copier – outside forces do that. And it is not the paper that expends energy in being copied – it is the copier that expends the energy. (Oderberg 2007, p. 284, n. 20)

But whether viruses are alive is, of course, a matter of controversy. Whatever the right answer, the Aristotelian does not claim that the matter can be settled from the armchair. What he does claim is that the *way* empirical considerations can settle the matter is by telling us whether or not viruses exhibit immanent causation.

In the previous chapter, I argued against reductionism in physics and chemistry. Does reductionism hold true in biology? Are living substances as the Aristotelian conceives of them reducible to non-living ones? The answer, naturally, hinges on whether immanent causation is reducible to transeunt causation. The answer to *that* question is that it is not reducible. (Cf. Koren 1955, pp. 18-19; Oderberg 2007, pp. 193-200; Oderberg 2013, pp. 216-23.)

The first problem for the reductionist here is an "apples and oranges" problem. Immanent and transeunt causation are simply different in kind, and not merely in degree. The difference between causal activity that perfects the agent and causal activity that does not is like the difference between a circle and a polygon. You can add as many sides to a polygon as you like, and you will never get a circle. Of course, you might get something that *looks like* a circle to sense organs that are incapable of making sufficiently fine discriminations, but that is not the same as getting an actual circle. Similarly, you can add to a transeunt causal process all the further transeunt causal processes you like, but you will never get immanent causation out of it. The most you will get is something that might *look like* immanent causation, just as a polygon with sufficiently many sides might look like a circle. That is precisely what we have in the case of computers, robots, and other complex machines that might appear superficially to be alive. As I have already said, in fact these are not truly alive, because they are artifacts with mere accidental forms rather than genuine substances with substantial forms, and a living thing is a kind of substance. Naturally, the critic of Aristotelianism would reject the distinction between substantial and accidental forms, but *merely* to reject it as a way of rebutting the argument I'm developing here would be to beg the question.

As Henry Koren notes (1955, pp. 18-19), another difference between immanent and transeunt causation is that the latter always involves the actualization of a potential, whereas the former, at least in principle, need not. What Koren has in mind is the case of a living agent which is *purely actual* and devoid of potentiality. Insofar as such an agent is, as it were, always already actualized by its very nature, it can be seen as a kind of limit case of immanent or self-perfective activity. Transeunt causation, by contrast, always involves an agent bringing about an effect in something external to it, which entails actualizing some potential in that thing. (The notion of a purely actual agent is, of course, at the core of the Aristotelian conception of God. But one not need affirm God's existence to see the force of Koren's point. That such an agent is at least in principle *possible* is all that his argument requires.)

A further consideration is that, as I emphasized when discussing reductionism in chemistry, any attempted reductionist account of an entity is going to fail whenever it has properties and causal powers that are unanalyzable in terms of the sum of the properties and causal powers of its parts. But we clearly do have that in the case of living things. For example, the self-perfecting or completing nature of an immanent activity like a snake's digestion of a mouse cannot be captured except by reference to the snake considered as a whole. It is true that the parts of the snake – eyes, skin, and so on – are also thereby nourished, but since these parts themselves exist only for the sake of the whole, that they are nourished is not the primary end of the immanent activity, but a secondary end, subordinate to the end of nourishing the snake. If irreducible higher-level causal powers of a non-immanent kind already pose a problem for reductionism in chemistry, higher-level features are hardly going to be *less* of a problem for reductionism in biology, where we have immanent causation added to the mix.

To forestall some irrelevant objections, note that nothing that has been said has anything at all to do with vitalism or any other commitment to some non-physical principle. On the contrary, for the Aristotelian, most kinds of living things – plants and non-human animals, for example – are entirely physical or corporeal. To deny that living things are irreducible is not to hold that they are non-physical, but entails merely that there are irreducibly different kinds of physical things. If someone points out to you that circles are irreducible to polygons, it would be quite ridiculous to accuse him of holding that circles must possess some mysterious non-physical principle that makes them different from polygons, or that circles are not geometrical figures the way polygons are. It is equally silly to accuse someone who argues that living things are irreducible to non-living things of being committed thereby to the existence of *élan vital* or otherwise denying that they are entirely physical.

Note also that the irreducibility of immanent causation to transeunt causation has nothing at all to do with complexity. An extremely simple thing will exhibit immanent causation and therefore life as long as its activity is self-perfective. A thing will fail to be alive as long as it exhibits only transeunt causation, no matter how many and how complex are the chains of transeunt causation to be found within it. The difference between a circle and a polygon has nothing to do with the former having greater complexity; indeed, there is an obvious sense in which it is simpler. Similarly, the difference between immanent and transeunt causation, and thus between living and non-living things, has nothing essentially to do with complexity.

For that reason, nothing in what has been said hinges on what one thinks of "design arguments" of the kind associated with William Paley and contemporary "Intelligent Design" theorists. If someone pointed out to you that circles differ in kind from polygons, it would be absurd to accuse him of insinuating thereby that circles must have been specially created by a divine designer. It is similarly absurd to accuse someone who regards immanent causation as irreducible to transeunt causation as insinuating thereby that living things are too complex to have arisen except by intelligent design. Whether circles are a kind of polygon, and where circles come from, are entirely separate questions. Whether immanent causation is reducible to transeunt causation, and where entities exhibiting immanent causation come from, are also entirely separate questions.

Now, a critic might give up reductionism without embracing the Aristotelian position. He might opt instead for eliminativism. That is to say, he might concede both that immanent causation is definitive of life and that immanent causation is irreducible to transeunt causation, but then deny that there really is any such thing as immanent causation. He will thereby be committed to denying that there is any such thing as *life*, but he may be willing to bite that bullet. He may say that, *strictly* speaking, there really are no living things, but only things that *seem* to be alive. (Cf. Jabr 2014.) This would be analogous to admitting that circles are irreducible to polygons but at the same time insisting that only polygons actually exist and that what we think are circles are really all just polygons that have so many sides that they seem to be circles.

But this cannot be right. Note first of all that the consistent eliminativist will have to deny that *he is himself* alive. More to the point, he will have to deny that he himself really carries out any immanent causal activity. But that is manifestly false, for thinking – including thinking about the nature of life, eliminativism, and so on – is itself an immanent causal activity. Gathering evidence, reasoning through the steps of an argument, and so forth all have as their end the perfection of the thinker as a rational creature. The thinker goes from ignorance to knowledge, is thereby perfected qua thinker, and this outcome remains the same whether or not it goes on to have any transeunt causal byproducts, such as the relating of his knowledge to other people. Thus, the eliminativist has to carry out immanent causal activity in the very act of denying that there is such a thing as immanent causal activity. His position is simply incoherent. To get around this problem, the eliminativist might of course decide to deny the existence of thought just as he denies the existence of life, but I already argued in an earlier chapter that this position is incoherent too.

Notice that I am *not* giving a simplistic argument to the effect the eliminativist breathes, moves his arms and legs, and does other things that common sense regards as the hallmarks of living things. Obviously the eliminativist would respond that he regards these activities as differing only in degree and not in kind from the activities that machines and other non-living things carry out, so that this commonsense retort begs the question. I am appealing, not to untutored common sense, but rather to the notion of immanent causation; and I am saying that, whatever one wants to say about breathing, moving one's limbs, and the like, *thinking* is an activity that cannot coherently be analyzed in terms of transeunt causation alone.

So, there is no way to be a *consistent, across-the-board* eliminativist. The would-be eliminativist will have to admit that he and other rational creatures are alive, at least in the form of Cartesian thinking substances – though I have also already argued against reducing the thinking subject to such a substance. In any event, denying that plants and animals are alive is not, at the end of the day, much more plausible than denying that we are. As with other forms of anti-realism, the eliminativist view of life is subject to the "no miracles" objection defended in an earlier chapter. The things that common sense takes to be alive certainly behave *as if* they were, and the simplest explanation of this is that they are in fact alive. There is no motivation for denying this other than to avoid having to admit that the biological realm cannot after all be assimilated to the mechanical world picture. But as I have been arguing throughout this book, there is ample independent reason to reject that picture.

6.1.2 Genetic reductionism

Kim Sterelny and Paul E. Griffiths (1999, p. 137) speak of an "antireductionist consensus" in contemporary philosophy of biology, and the problems with genetic reductionism in particular illustrate the sorts of arguments that underlie this consensus. Two conceptions of the gene are sometimes distinguished. There is, first, the gene in the Mendelian sense of a hypothetical cause for a pattern of inheritance of traits (e.g. eye color). Second, there is the gene in the biochemical sense of a sequence of nucleotides. It is often claimed that genes in the first sense have been reduced to genes in the second sense. But as John Dupré (1993, chapter 6; 2012, chapter 8) points out, that is not the case. Typically, there are many genes in the biochemical sense involved in the production of any one trait, and any one gene in the biochemical sense is involved in the production of many different traits. Context determines how a gene will operate. Hence there simply is no smooth one-to-one mapping of genes in the Mendelian sense onto genes in the biochemical sense. This is known as the "many-many problem" (Dupré 1993, p. 123). (Cf. Hull 1974, p. 39, and Rosenberg 1994, pp. 19-20.)

The problem becomes particularly acute, in Dupré's view (1993, pp. 128-31), when we consider that ontogeny (which concerns the development of a particular organism) and phylogeny (which concerns the evolutionary history of a kind of organism) have very different classificatory needs. "Evolution selects functions rather than structures," he notes, so that "wherever there are functionally homogeneous but structurally diverse gene products, the classifications of genes relevant for evolutionary investigations will be structurally heterogeneous" (p. 129). In evolutionary contexts, genes are conceived of by the biologist as *genes for* various morphological or behavioral traits; and, especially with complex traits, different possible alternative DNA segments might play this role. Moreover, the gene so conceived of is not taken to be *sufficient* causally to determine that the trait will be manifested. It may just raise the probability of the trait's occurrence. So, the relationship between the gene conceived of in functional terms and the gene conceived of in molecular terms is even messier when evolutionary explanation is factored in than it is where the development of the individual organism is concerned.

As Dupré (2012, pp. 134-36) also points out, the common assertion that the genome contains all the information required in order to build an organism is problematic. The word "information" is ambiguous.

If the term is used in the ordinary semantic sense, then the claim in question is manifestly false. There is no "information" in the genome in the same sense in which there is information in a book or a lecture. If the term is used in the technical information-theoretic sense (which I discussed in the previous chapter), then the idea is that there is a reliable causal link between a gene in the biochemical sense and a certain trait. But in that case the claim that genes carry information is trivial, because there are lots of other factors involved in the production of traits that carry "information" in that sense. There is nothing special about genes in that case.

Dupré (2012, p. 133) distinguishes between *what something does* and *how something does what it does*, where the former has to do with biological function and the latter with the mechanisms by which it performs that function. What biochemistry reveals, Dupré points out, is the latter but not the former. (Cf. Dupré 1993, pp. 124-26.) Dupré's distinction essentially recapitulates the distinction between *formal-cum-final causes* on the one hand, and *material-cum-efficient causes* on the other. He and other anti-reductionists in the philosophy of biology, like anti-reductionists in philosophy of chemistry and philosophy of science in general, are essentially rediscovering the Aristotelian anti-atomist point that all four causes are needed for a complete description of natural phenomena.

With biological phenomena no less than with the quantum and chemical phenomena described in the previous chapter, the lower-level features of a system cannot properly be understood in abstraction from the higher-level features that the reductionist purports to reduce to (or eliminate in favor of) the lower-level ones. Structures at the molecular level are classified as genes in the first place only because they correspond to the description first independently formulated at the Mendelian level. They cannot be so classified apart from that. (Cf. Sterelny and Griffiths 1999, p. 138.) Moreover, the higher-level features of a cell in part determine how the lower-level features at the molecular level will operate. Dupré writes:

> Downward causation seems a very natural way to think of much of what I have been saying about molecular biology. What causes the human genome to behave in the particular ways it does – for example, various sequences being transcribed or not at varying rate, changes in conformation and spatial relation of chromo-

> somes, and so on – is a variety of features dispersed over the surrounding parts of the cell. The behaviour of the part is to be explained by appeal to features of the whole. (2012, p. 139)

Sterelny and Griffiths (1999, p. 141) note that a possible reductionist reply would be to propose that the gene as understood at the Mendelian level can be reduced to a DNA sequence *plus* the larger context in which it operates. But as they also note, there are several problems with this proposal. For one thing, correlating genes understood at the Mendelian level with genes understood at the molecular level would still require specifying "unmanageably large chunks of the molecular context" and even then the relationship "would still be one-to-many," thus failing to solve the many-many problem (1999, p. 141). For another thing, reductionism as traditionally understood takes the causal powers of the micro-level parts to be independent of the macro-level features that are to be reduced. The classical reductionist aim is to *decompose* the whole into an aggregate of these micro-level parts. The proposal in question would effectively abandon this ambition.

The reductionist may at this point respond that the trouble is with too narrow a construal of reductionism, and not with reductionism itself (Sterelny and Griffiths 1999, p. 143). Sterelny and Griffiths suggest that on at least one version of reductionism, all that is required of a reductionist explanation of some phenomenon is that it identify a *mechanism* by which that phenomenon occurs, rather than leaving its operation "spooky" or "occult" (p. 120). They give as an example the way continental drift theory was not endorsed by most geologists until a plausible mechanism (plate tectonics) was discovered by which it could intelligibly operate. Molecular genetics can be said to give a reductionist account of Mendelian genetics insofar as it identifies the mechanism by which the latter operates.

The problem with this suggestion, though, is that it is hard to see what serious *anti*reductionist position is incompatible with it. In particular, "reductionism" in this thin sense poses no challenge at all to Aristotelian philosophy of nature. Again, the Aristotelian holds that a complete explanation of a natural phenomenon must make reference to both its formal-cum-final causes *and* its material-cum-efficient causes. The latter is essentially what Sterelny and Griffiths have in mind when they speak of identifying mechanisms.

6.1.3 Function and teleology

Since the presence of irreducible properties and causal powers is the mark of a substantial form, the failure of "decompositional" reductionism in biology (as Sterelny and Griffiths characterize it) is a vindication of the Aristotelian notion of formal cause. The vindication of final cause is evident from the failure of attempted reductionist analyses of biological function and other teleological notions.

One such account is the "systemic" or "causal role" analysis of function associated with Robert Cummins (1999). The basic idea here is that the function of some part of a system is to be identified with the causal role it plays in that system. For example, the heart can be said to have the function of pumping blood throughout the circularity system, because this is the causal role it plays within that system. Since the causal role in question is an *efficient* causal role, if such an analysis succeeds, then the notion of a function will have been reduced to the notion of efficient causation. But a well-known problem with this analysis is that it entails too broad a conception of function. The heart also plays the role of causing a thumping sound, in which case Cummins's analysis would have to take *that* to be a function of the heart as well. But no biologist would say that making a thumping sound is a biological function of the heart. Rather, it is a byproduct of its carrying out its function. As Elliott Sober writes, "the trouble is that the distinction between function and mere effect seems to get lost in Cummins's theory" (1993b, p. 86). (Another problem is that, as we have seen, the Aristotelian argues that efficient causality itself presupposes at least a thin kind of final causality, viz. the bare directedness of a cause towards its characteristic effect or range of effects. Hence even if Cummins's analysis succeeded, it would at most reduce a certain complex *kind* of teleology, but not all teleology as such.)

Another reductive account is the "etiological" approach of Larry Wright (1999). On Wright's analysis, a thing's function is to be analyzed in terms of what it does and how this led to its coming to be. For example, the headlights on a car illuminate the roadway in front of the car, and this is the reason why the designers of the car put them there. Hence, illuminating the roadway is the function of the headlights. The heart pumps blood, and the fact that it pumps blood is the reason why it was favored by natural selection. Hence pumping blood is the function of the heart. But Wright's analysis too is open to counterexamples. For instance, a man's obesity might cause him not to exercise, and the fact that it causes this is also among the causes of his obesity, insofar as the lack of exercise only reinforces obesity (Boorse 1976). So, on Wright's analysis, it would seem that we'd have to say that his obesity has the function of preventing exercise. But that is absurd.

Currently, however, the most popular approach is the Darwinian analysis of function associated with Ruth Millikan (1984), which is essentially a modification of Wright's account. On Millikan's analysis, the function of a biological trait is to be understood in terms of the causal factors that led to its being favored by natural selection. For example, the heart causes blood to be circulated throughout the body. It also causes a thumping sound, but the fact that it circulates the blood, rather than the fact that it makes a thumping sound, is the reason natural selection favored it. Hence circulating blood, rather than making a thumping sound, is the function of the heart. The sort of counterexample that afflicts Cummins' theory is thereby avoided. But the emphasis on natural selection also allows Millikan to avoid counterexamples of the kind that afflict Wright's theory. Obesity causes one to refrain from exercise, but obesity was not favored by natural selection, so that we are not led by Millikan's account to attribute a biological function to it.

But there are nevertheless several problems with Millikan's analysis. For one thing, it implies that you cannot know a thing's function without knowing its evolutionary history. But that is not the case. We can know what hearts, eyes, ears, feet, and the like are for whether or not we know anything about evolution (Sober 1993b, p. 85; Fodor 1998, p. 210). Nor is the point merely epistemological. A biological trait could surely have a function whether or not it arose via natural selection. For instance, if organisms with eyes arose either by spontaneous generation or by way of special divine creation rather than by natural selection, their eyes would still have the function of allowing the organisms to see (Fodor 2000, p. 85). It is hard to see how the Darwinian analysis of function can deny this without begging the question.

There is also the objection that something could have been favored by natural selection and yet not have a biological function, such as "a parasite disguising itself within a host's body" (Godfrey-Smith 2014, p. 64), which need not serve a function relative to the host. As Peter Godfrey-Smith notes, one could respond by modifying the analysis so that in order to serve a function, the trait favored by natural selection must be one that has a "beneficial effect" on the larger system in which it is embedded (ibid.). But to be "beneficial" to an organism is to serve some end that it has. Hence this response would bring teleology back in, when the whole point was to analyze it away.

Then there is the *indeterminacy* objection often raised against the Darwinian analysis of function – the problem that a trait's evolutionary history does not in fact determine a *unique* functional description of the trait (Enç 2002; Fodor 1990, chapter 3; Perlman 2002; Putnam 1992, chapter 2; Walsh 2002). To cite a popular example, a frog will snap its tongue toward bugs that flit by it, though also at pellets that are thrown its way. So should we say that the biological function of the underlying neural mechanism is *to catch bugs* or *to catch small moving things*? Millikan would say that the first option is the right answer, because the fact that this behavioral tendency allowed frogs to catch bugs is the reason it was favored by natural selection. But the problem is that this is not the only way to tell the evolutionary story. We could equally well say that natural selection favored a tendency to catch small moving things, because most of those small moving things happened to be bugs. There is nothing in natural selection itself that favors the one description over the other. Hence there is nothing in natural selection itself that determines which functional description is the correct one.

In short, being favored by natural selection is neither necessary nor sufficient for a trait's having a biological function. Now, it might seem that the reductionist could bite the bullet and concede that neither Millikan's analysis nor any other reductionist account succeeds in capturing the notion of function as traditionally understood, but then argue that in reality there simply is nothing more *to* biological function than whatever can be captured by such reductive analyses. The idea would be to eliminate the more robust commonsense notion of function and replace it with a deflationist theoretical conception – to replace teleology with "teleonomy," as it is sometimes called (Godfrey-Smith 2014, p. 64). Hence if biological function is indeterminate on a reductionist analysis, then what we should conclude according to this proposal is not that such an analysis is wrong, but rather that there is no objective fact of the matter about which of the various possible functions we might attribute to a trait is really the correct one.

But there are two problems with this proposal. First, it faces the same objection that plagues other anti-realist moves in philosophy of science. The commonsense notion of function has enormous utility. It would be practically impossible to do biology without it. How could this be if it does not correspond to anything in objective reality? (This is an application of the "no miracles" argument for scientific realism, which I defended in chapter 3.)

Second, this deflationist strategy cannot coherently be carried out in a consistent way. To follow it out consistently, we'd have to say that there is no objective fact of the matter about what human perceptual and cognitive faculties are for. If we say *that*, however, then we will also have to say that there is no objective fact of the matter about whether those faculties are malfunctioning in any particular case (in hallucinations, fallacious reasoning, mental illness, etc.). Objectively speaking, there will be nothing to make any output of these faculties better or worse than any other. The distinction between genuine empirical evidence and illusion, and between good and bad reasoning, will collapse – taking all arguments, including arguments for reductionism, down with it.

It is worth adding that it is in any case tendentious to argue as if an analysis of function in terms of natural selection would necessarily be deflationist or entail reducing the teleological to the non-teleological. As Ernst Mayr notes, "after Darwin established the principle of natural selection, this process was widely interpreted to be teleological, both by supporters and by opponents" (2004, p. 62). Indeed, Darwin himself sometimes spoke of evolution in teleological terms (Gilson 1984, pp. 80-89; Lennox 1993). Of course, the reductionist would argue that this was a mistake and that a consistent working out of Darwin's principles should lead to a thoroughgoing rejection of teleology. The point, though, is that this is a thesis that *would* have to be *argued* for. It will not do simply to take it for granted that natural selection is a non-teleological process, so that successfully to analyze function in terms of natural selection would *ipso facto* be to reduce the teleological to the non-teleological. Indeed, as we will see below, in fact there are serious problems with the supposition that natural selection is non-teleological.

Finally, as André Ariew points out (2007, p. 177), function is not the only biological phenomenon that Aristotle thought required a teleological analysis. Another is the regularity of the growth patterns common to members of a biological kind, and about this, Ariew says, "Darwin's theory is silent" (p. 180). The problem here is how to distinguish growth patterns that are normal from those that are aberrant, and the Aristotelian answer is that there is no way to do so without reference to the end state toward which the process naturally points. Biologist J. Scott Turner (2007, p. 147) makes a similar point about the "self-regulation and self-correction" of an organism, which he argues cannot be understood without attributing a kind of "intentionality" or purposiveness to the processes in question. He writes:

> Though Darwin and Wallace delivered the death blow to the purported intentionality of what organisms *are*, they did not invalidate the very different kind of intentionality that underpins what organisms *do*. (2007, p. 147)

As we saw in the previous chapter, Alex Rosenberg would respond that such phenomena can be explained in terms of the "program" or "software" by which the genome constructs and regulates an organism. But as we also saw, this is no *alternative* to teleology at all, but on the contrary presupposes teleology for its intelligibility.

6.1.4 The hierarchy of life forms

As I noted in chapter 1, the traditional Aristotelian view is that there are at least three irreducibly different forms of living substance – the vegetative, the sensory, and the rational. (Cf. Gardeil 1956; Koren 1955; Oderberg 2007, pp. 183-93; Phillips 1950.) Is this division still defensible today?

The first thing to say is that there *are* indeed genuine substances of *some* sort at the level where these three forms are to be found – that is to say, the level of familiar everyday living things such as trees, dogs, and people, as opposed to the micro-level parts of these organisms or the larger populations of which they are members. There are various reasons the theorist might lose sight of this fact, obvious though it is to common sense. For example, one might be influenced by general reductionist and eliminativist arguments in metaphysics. Just as some philosophers hold that there are no stones but only particles arranged stone-wise, so one might suggest that there are no dogs but only particles arranged dog-wise. Or one might be moved by borderline cases such as siphonophore jellyfish or dictyostelid slime molds, with respect to which it is difficult to tell whether one is dealing with a single organism or a population of organisms. (Cf. Boulter 2013, chapter 4; Godfrey-Smith 2014, chapter 5.) Such examples might tempt one to think that there is no fact of the matter concerning *any* living thing about whether it is really a single individual or a population. Or one might be led by arguments influential in the "units of selection" controversy to conclude that it is either genes or populations of organisms, rather than individual organisms of the familiar sort, that are operated upon by natural selection. (Cf. Rosenberg and McShea 2008, chapter 6; Sober 1993b, chapter 4.)

The general metaphysical reductionist and eliminativist arguments in question make no appeal to specifically biological features but are intended to apply to all physical objects. I have answered those arguments at length elsewhere (Feser 2014b, chapter 3) and have also already addressed the issue in earlier chapters of this book. As to the borderline cases, it would simply be fallacious to infer, from the existence of a handful of examples that are difficult to interpret, the conclusion that in the vast majority of cases where we clearly and distinctly appear to have individual organisms, this appearance is illusory. Hard cases make bad law, and they make bad metaphysics too. In the unproblematic cases (trees, dogs, people, etc.), the general principle that irreducible properties and causal powers entail the presence of a substantial form and thus a true substance shows that these everyday living things are indeed genuine substances. We should interpret the problem cases in light of these unproblematic ones, rather than the other way around. And as Stephen Boulter argues (2013, chapter 4), an Aristotelian interpretation of cases like the ones in question is available. In the case of siphonophores, he takes the parts of a siphonophoric colony to be true individual substances insofar as they are capable of carrying out the operations distinctive of living things independently of the colony. In the case of dictyostelid slime molds, he argues that spores start out as true biological individuals but lose this status on incorporation into the mold. (As Boulter emphasizes, in interpreting such cases it is crucial to keep in mind the Aristotelian principle that a term such as "individual" can have an *extended, analogical* sense. See Feser 2014b, pp. 256-63 for discussion of the analogical use of terms.)

As to the "units of selection" problem, it is clear that however it is to be resolved, it casts no doubt on the reality of trees, dogs, people, and other familiar living things as genuine substances (as opposed to mere aggregates of their parts, or mere parts of a population). Boulter (2013, p. 86) argues that evolutionary biology in fact presupposes the reality of such individuals, on the basis of considerations like the following: First, the individual organism is commonly treated by biologists as the *basic* unit of selection, even if there are others. Second, it is the births and deaths of these individual organisms that are counted by population biology. Third, the concept of "fitness" applies *primarily* to such individual organisms. Fourth, adaptations too are primarily features of these individual organisms. Fifth, the comparative method in biology is essentially a matter of comparing individual organisms with other individual organisms (rather than with parts of organisms or with populations). Sixth, the major transitions in evolution mark the appearance of new kinds of individuals.

What these considerations together with those adduced earlier in this chapter show is that there are genuine substances of at least the *vegetative* sort, in the technical Aristotelian sense of that term. That is to say, there are substances that metabolize, or take in nutrients and eliminate wastes so as to sustain themselves; that grow in the sense of increasing in size from within rather than merely by accretion; and that reproduce in the sense of generating new, distinct individuals of the same kind as themselves. For example, a tree does these things, and insofar as its carrying out of these operations is irreducible to the aggregate of the activities of its parts, it counts as a true substance. Any living substance that carries out some variation of these activities, but nothing further that differs in kind from them, is a purely vegetative substance. (Of course, things other than plants carry out such activities, such as fungi. But again, "vegetative" is being used here in a technical sense, and does not correspond in meaning to terms like "vegetable" or "plant" in the familiar senses of those terms.)

Now, the question with which I started out this section is whether there really are other kinds of living substance that are irreducible to this vegetative kind, and the traditional Aristotelian position is that there are. For example, there are *sensory* living substances, which possess the basic vegetative capacities but, in addition, three irreducibly different capacities of their own. The first and most fundamental is sentience, which is the capacity for conscious awareness of stimuli. The second is appetite, which is the capacity either to seek or to avoid the stimuli one is aware of. The third is locomotion, which is the capacity to move oneself either toward or away from the object of appetite. Obviously, these three form a kind of package. The final cause of locomotion is to allow the organism to react to sensed stimuli, with appetite being the bridge between locomotion and sentience.

The empirical evidence bears this out insofar as these three capacities all tend to be found together in organisms that have any one of them. But as Aristotelian philosophers argue (Klubertanz 1953, p. 58; Koren 1955, p. 139; Oderberg 2007, pp. 185-86), the connection is metaphysically stronger than a mere empirical correlation. Locomotion would be positively harmful in organisms that lacked sentience, since they would move about without being able to know whether what they are moving themselves toward was beneficial or dangerous. Appetite would result in nothing but frustration in a creature that lacked locomotion. Sentience would be pointless, at least with respect to stimuli that it was beneficial for the organism either to acquire or to avoid, in the absence of locomotion – though perhaps one might argue that there could, at least in theory, be creatures that were sentient but existed in a world of stimuli that could neither benefit nor harm them and which they merely contemplated (cf. Strawson 1994, chapter 9).

The key issue for the traditional Aristotelian view of the hierarchy of life forms is whether the most basic of these capacities, sentience, is irreducible to anything that merely vegetative forms of life are capable of. It might seem that it is *not* irreducible insofar as plants grow toward the light and sink roots in the direction of water, a Venus fly trap will react to the presence of an insect and the *Mimosa pudica* will respond to touch, and so forth. Isn't this evidence that something like sentience exists in plants?

No, it is not. The Aristotelian does not deny that merely vegetative forms of life can in certain ways be sensitive to external stimuli. The claim is rather that they lack *conscious awareness* of these stimuli – that, as contemporary philosophers of mind would put it, they lack *qualia*. And sensitivity to external stimuli of the kind plants exhibit doesn't entail the presence of qualia. For example, that the roots of a plant grow in the direction of water doesn't entail that it feels thirst, and that the Venus fly trap reacts to the presence of an insect doesn't entail that it feels hunger – any more than a smoke alarm experiences the smell of smoke, or the motion detector in an outdoor security lamp has a visual experience of someone crossing in front of it.

Moreover, plants lack crucial features that lead us to attribute conscious awareness to animals (Koren 1955, pp. 72-73; Tye 2017). In animals there are specialized sense organs associated with their various forms of awareness – eyes with visual awareness, ears with hearing, and so on. Plants lack such organs. Furthermore, sensation in animals is associated with a variability of response that is not present in plants. Unless it is in some way damaged, a plant will simply grow toward the light or sink its roots downward in response to the relevant stimuli. A properly functioning animal, by contrast, may respond in a number of ways to stimuli presented to it. For example, it might immediately leap toward the prey it sees, or sneak up toward it slowly, or refrain from acting at all if it sees another, stronger predator in the vicinity or some barrier it is afraid to cross. A conscious experience functions as a kind of *intermediary* between external stimuli and different possible behavioral responses, an intermediary that makes this variability of response possible. That plants lack such variability is thus a reason to think they lack anything like such intermediary conscious experiences. Furthermore, as Oderberg notes (2007, pp. 186-88), a sentient plant could not move itself *as a whole* either toward or away from anything it sensed in its environment that was either beneficial or dangerous. As noted already, in the absence of such locomotion, sentience would be pointless or even harmful. Hence the movements of which plants are capable are best thought of as merely mechanical rather than on the model of animal locomotion.

What for any contemporary philosopher must be the most telling consideration in favor of the irreducibility of sentience, however, is the notorious intractability of what has come to be known as "the qualia problem" or the "hard problem of consciousness" – the problem of explaining qualitative conscious experience in scientific or naturalistic terms. Even prominent naturalist philosophers like Thomas Nagel (2012), John Searle (1992), Galen Strawson (2008), and David Chalmers (1996) have been highly critical of existing materialist solutions. Jerry Fodor once famously wrote:

> Nobody has the slightest idea how anything material could be conscious. Nobody even knows what it would be like to have the slightest idea about how anything material could be conscious. So much for the philosophy of consciousness. (1992, p. 5).

Twenty years later, Alva Nöe saw no reason to revise this judgement:

> Science... lacks even a back-of-the-envelop [sic] concept explaining the emergence of consciousness from the behavior of mere matter. We have an elaborate understanding of the ways in which experience depends on neurobiology. But how consciousness arises out of the action of neurons, or how low-level chemical or atomic processes might explain why we are conscious – we haven't a clue. We aren't even really sure what questions we should be asking. (2012)

Colin McGinn concludes that "I think the time has come to admit candidly that we cannot resolve the mystery" (1991, p. 1).

Now, the debate is often presented as a conflict between materialism and dualism, and the problem formulated as a question about whether qualia are physical or non-physical. That is not how the Aristotelian understands the issue. For the Aristotelian tradition, dogs, horses, birds, and other non-human sentient creatures are as purely physical as plants and stones are. The Aristotelian claim is that sentient creatures, though entirely physical, are an irreducibly different *kind* of physical thing from merely vegetative creatures, just as vegetative creatures are an irreducibly different kind of physical thing from stones. The Aristotelian objects, not to the claim that conscious experience is physical, but rather to the desiccated, reductionist conception of the physical to which most contemporary naturalists are committed. As noted in earlier chapters, the conception in question is that of the mechanical world picture with which Descartes, Hobbes, Locke, Newton, and the other founding fathers of modern philosophy and modern science replaced the Aristotelian philosophy of nature. Accordingly, an Aristotelian interpretation of the contemporary debate over the "hard problem" would emphasize that the difficulty consciousness poses for materialism indicates the need for a revision to this standard modern conception of matter. And that is exactly what some naturalist philosophers (such as Nagel, Strawson, and Chalmers) have concluded, even if they wouldn't all sympathize with a specifically Aristotelian way of carrying out such a revision.

This brings us naturally to what the Aristotelian tradition regards as a third irreducible kind of living substance, namely *rational* animals or human beings. For the state of debate in contemporary philosophy lends support to this aspect of the Aristotelian position as well. In contemporary philosophy of mind, the question of whether rationality can be given a reductionist explanation is framed in terms of questions such as whether a naturalistic account can be given of the "propositional attitudes" (believing, desiring, and so forth) and of their "intentional content." An example of a naturalistic account of the propositional attitudes would be the functionalist view that to have a belief, a desire, or the like is to be in a state that plays the causal role of mediating between sensory input and behavioral output in a certain way. The functionalist may then go on to analyze the way these causal roles relate to brain activity in terms of the relationship between computer software and the hardware in which it is implemented. An example of a naturalistic account of the intentional content of the propositional attitudes would be the causal theory of meaning, according to which a certain belief will have the content that *the cat is on the mat* if it is caused in the right sort of way by the presence of a cat on a mat.

As the popularity of the notion of "artificial intelligence" and the associated computer model of the mind indicate, there are many philosophers and scientists who find such theories plausible. But as with materialist theories of consciousness, materialist accounts of rationality have also been subjected to vigorous criticism from prominent mainstream philosophers. Searle (1992), Hubert Dreyfus (1992), and Hilary Putnam (1988, 1992) have leveled powerful objections to computationalism (Putnam having once been a supporter). Naturalistic theories of intentional content face notorious difficulties, such as indeterminacy problems that parallel those afflicting naturalistic analysis of biological function. (Cf. Crane 2016 for a useful survey.) McGinn (1993, chapter 4) speculates that the intentional content or meaning of our thoughts may turn out to be as impenetrable to naturalistic explanation as he takes consciousness to be. Nagel judges that:

> In light of the remarkable character of reason, it is hard to imagine what a naturalistic explanation of it, either constitutive or historical, could look like...
>
> A reductive account of reason, entirely in terms of the properties of the elementary constituents of which organisms are made, is even more difficult to imagine than a reductive account of consciousness. (2012, pp. 86-87)

And here too Nagel thinks that the solution will require a retreat from the desiccated conception of matter that naturalists have inherited from the anti-Aristotelian revolution of the early modern thinkers:

> [I]f one asks, "*Why* is the natural order such as to make the appearance of rational beings likely?" it is very difficult to imagine any answer to the question that is not teleological. (1997, p. 138. Cf. Nagel 2012, pp. 88-93)

In the case of human rationality, the traditional Aristotelian view is that we *do* have a phenomenon that is not merely irreducible to lower forms of life, but is incorporeal. (Cf. Feser 2013a and 2018, and Oderberg 2007, chapter 10.) Even revisionist naturalists like Nagel are unwilling to go that far. For present purposes, however, this further immaterialist thesis is irrelevant. What matters is that, as with reductionist accounts of consciousness, reductionist accounts of rationality are highly controversial at best within contemporary philosophy, and have come in for attack even from within the naturalist camp itself.

Needless to say, what matters at the end of the day is not the fact that some striking quotes can be mined from the writings of contemporary naturalist philosophers, but rather the quality of the arguments that have been given both for and against reductionist accounts of consciousness and rationality. Now, evaluating those arguments would require a survey of the vast literature in contemporary philosophy of mind, and that is obviously beyond the scope of a general book on the philosophy of nature, such as this one. But I have, in any event, had much to say elsewhere in defense of an antireductionist positon with respect to both consciousness (Feser 2006, chapters 4 and 5; 2015b) and rationality (Feser 2006, chapters 6 and 7; 2011b; and 2013a). For present purposes it is sufficient to note that a neo-Aristotelian defense of the irreducibility of sentient and rational forms of life can call upon a large and impressive body of contemporary antireductionist argumentation that was developed almost entirely independently of any Aristotelian influence or motivation.

It is important to add that the skepticism among some naturalists about reductionist accounts of consciousness and rationality finds echoes even where the nature of life is concerned. We have already seen that there is no agreement among contemporary biologists and philosophers of biology even about how to *define* life. A reductionist *explanation* of life is no less elusive. What we have are at best a variety of highly controversial and sketchy speculations, but no actual worked-out theory. As Nöe sums up the situation:

> Science has produced no standard account of the origins of life. We have a superb understanding of how we get biological variety from simple, living starting points. We can thank Darwin for that. And we know that life in its simplest forms is built up out of inorganic stuff. But we don't have any account of how life springs forth from the supposed primordial soup. This is an explanatory gap we have no idea how to bridge. (2012)

Similarly, Nagel writes:

> Indeed, when we go back far enough, to the origin of life – of self-replicating systems capable of supporting evolution by natural

selection – those actually engaged in research in the subject recognize that they are very far from even formulating a viable explanatory hypothesis of the traditional materialist kind...

Although scientists continue to seek a purely chemical explanation of the origin of life, there are also card-carrying scientific naturalists like Francis Crick who say that it seems almost a miracle. (2012, pp. 89 and 123-4)

I will return to the topic of the origin of life below, but for the moment let it be noted that these three phenomena that even many prominent naturalists take to be mysterious and perhaps inexplicable in reductionist terms – life, consciousness, and rationality – correspond exactly to the three irreducibly different forms of life posited by the Aristotelian, viz. vegetative, sentient, and rational. The traditional Aristotelian view is that even the most fundamental form of life, the vegetative, is irreducible to any purely inorganic phenomenon, and the difficulties that have faced materialist attempts to explain life are just what we should expect if this is true. The Aristotelian holds that sentient life is irreducible to merely vegetative life, much less to anything inorganic, and that consciousness has persisted as a "hard problem" for materialism is precisely what we should expect if *that* is true. And the Aristotelian holds that rational life is irreducible to mere sentience, much less to vegetative or inorganic phenomena, which should lead us to expect that propositional attitudes and their intentional content would be as difficult to explain reductively as contemporary philosophers have indeed found them to be.

In short, the state of play in contemporary philosophy and science if anything reinforces rather than undermines the traditional Aristotelian doctrine of three fundamental forms of life. When we combine considerations like those just summarized with the fact that a general Aristotelian philosophy of nature is (as I have been arguing throughout this book) perfectly defensible today, there can be no doubt that this more specific Aristotelian doctrine is also still perfectly defensible.

But how, it might be asked, does this doctrine square with the sort of taxonomy of life forms that a modern biologist would give? There is, for example, the three-domain system of classification into *Bacteria*, *Archaea*, and *Eukarya*, with their various sub-classifications. There is the older classification into the five kingdoms *Monera*, *Protista*, *Plantae*, *Fungi*, and *Animalia*, with their sub-classifications. And there are other proposed systems of classification. But none of them correspond to the Aristotelian distinction between vegetative, sensory, and rational forms of life. So does the latter conflict with the findings of modern science?

No, it does not, for as Oderberg points out (2007, pp. 183-93), the newer systems of classification and the traditional Aristotelian division are essentially addressing different questions. At least where natural substances (as opposed to artifacts) are concerned, the Aristotelian draws a sharp distinction between (a) the *essence* of a thing or *what it is*, and (b) the *origin* of a thing or *where it came from*. The Aristotelian doctrine of the hierarchy of forms of life is concerned to address the first issue. Modern systems of biological taxonomy, by contrast, are concerned with the second. In particular, they are concerned to classify organisms in terms of inferred evolutionary descent. They have, as Oderberg puts it, thereby been "all but evacuated of metaphysical content" (2007, p. 184), collapsing the question of what an organism is into the question of how it got here.

6.2 Aristotle and evolution

6.2.1 Species essentialism

However, recent years have seen a revival of interest in an essentially Aristotelian notion of the essence of a biological species (Walsh 2006; Oderberg 2007, chapter 9; Devitt 2008; Elder 2008; Dumsday 2012; Boulter 2013; and Austin 2017 and 2018). As Michael Devitt (2008, p. 353) has noted, biologists and philosophers of biology sometimes distinguish between *functional biology* and *evolutionary biology*. (Cf. Mayr 1961 and Kitcher 1984.) Functional biology is concerned with the structure of an organism, the functions its parts serve, the developmental processes by which those parts form, their genetic basis, and so forth, all considered in a more or less ahistorical manner. Evolutionary biology is concerned with the historical origins of organisms and their traits. These two forms of inquiry can be carried out more or less independently of one another.

As Devitt (2008, p. 354) argues, an excessive focus on the historical or evolutionary approach can make it seem as if anything we might want to identify as the essence of a group of organisms captures only their relations to other organisms rather than any intrinsic essence. But if instead we look at things from the approach of functional biology, the need to affirm an essence intrinsic to organisms of the same kind is clear. Simply as a prerequisite to carrying out such inquiry, biologists group organisms according to common traits and treat these *as if* they reflected some common intrinsic nature (Devitt 2008, pp. 351-55). They do this whatever views they explicitly hold on the question of essentialism, and this approach is fruitful. We find that organisms within these groupings really do reliably tend to manifest certain common properties, to exhibit certain common characteristic behaviors, and so on. We need an explanation of why this is so, and the best explanation is that there *really is* an essence intrinsic to organisms of the same kind – that it is not just a useful fiction. Moreover, this metaphysical judgment seems to be confirmed empirically, by the results of genome research. Though robust essentialism of the traditional Aristotelian kind is commonly supposed to be dead, Devitt says, "many claims that biologists make day in and day out about the living world require species to have natures that they do not have according to this consensus" (p. 380).

It is often claimed that evolution is incompatible with essentialism. For example, it is sometimes said that essentialism requires, whereas evolution rejects, the fixity of species. But the objection rests on confusion. (Cf. Sober 1993b, pp. 146-7; Oderberg 2007, pp. 204-7; Boulter 2013, p. 106.) For one thing, there is nothing in essentialism that requires that exactly the same species exist now as have existed in the past. The essentialist need not deny that some species have gone extinct and that new ones have come into being. He claims only that every species, including the extinct ones and the newer ones, has an essence – that there is an objective fact of the matter about what it is, and that this fact is constant. What makes a *Tyrannosaurus rex* a *Tyrannosaurus rex* is the same now as it was millions of years ago, even though there are no longer any instances of this particular kind of organism. Neither is there anything in the thesis that things have unchanging essences that entails that one species cannot give rise to another. As Sober writes:

> [A]n atom smasher can transform (samples of) lead into (samples of) gold. However, this does not undermine the idea that the chemical elements have immutable essences. Likewise, the fact that a population belonging to one species can give rise to a population belonging to another does not refute essentialism about species. Essentialists regard species as perennial categories that individual organisms occupy; evolution just means that an ancestor and its descendants sometimes fall into different categories. (1993, pp. 146-47)

David Stamos (2003, p. 122) objects that it is one thing for an *individual sample* of a kind of element to give rise to a sample of another kind, but another thing altogether for one *species* of thing to give rise to another. But this is not a difference that makes a difference to the point at issue. For one species to give rise to another is just for the individual members of one species to give rise to individual members of another kind, so that Sober's point stands (Oderberg 2007, p. 205). (Whether considerations *other* than essentialism entail that one species could not give rise to another is a question to which I will turn below.)

A second evolution-based objection is that essentialism is inconsistent with the various notions of species that have become common in biology since Darwin. There is, for example, the "biological concept," which takes a species to be an isolated and interbreeding population of organisms; and the "phylogenetic-cladistic concept," which defines a species in terms of its evolutionary history and reflects the modern taxonomic practices referred to earlier. On these accounts, what species an organism belongs to is to be determined by reference to its relations to other organisms, rather than anything intrinsic to it. Yet a thing's essence, as the Aristotelian understands it, is supposed to be something intrinsic to it.

But Devitt (2008, pp. 356-58) notes that we need to distinguish the question of what makes it the case that an organism belongs to a certain group from the question of what makes some group a species as opposed to a subspecies, a genus, or whatever. Essentialism is an answer to the first question, whereas the various species concepts in question are answers to the second. Hence, whatever one thinks of those concepts, they are not in competition with essentialism.

But those concepts are problematic in any case. The biological species concept faces several objections, such as that it cannot account for organisms that reproduce asexually, and that it puts the cart before the horse insofar as reproductive isolation is to be explained in terms of a difference in species rather than the other way around. (Cf. Sterelny and Griffiths 1999, pp. 187-90; Oderberg 2007, p. 214; Godfrey-Smith 2014, pp. 102-3.) A well-known problem with the phylogenetic-cladistic concept is that considered by itself it is circular. For the basic idea of the phylogenetic-cladistic concept is that a species comes into being when an existing lineage splits in two and goes extinct when either it in turn splits or its members all die out. But we cannot judge that a lineage really has split into two species without a prior understanding of what a species is. (Cf. Okasha 2002, p. 201; Oderberg 2007, pp. 216-17; Devitt 2008, p. 356.) Hence its defenders acknowledge that the phylogenetic-cladistic concept needs to be supplemented with some other species concept. Yet if some other species concept is doing the job (such as the biological species concept, if it were otherwise acceptable), then the phylogenetic-cladistic concept drops out as otiose (Oderberg 2007, p. 217).

The phylogenetic-cladistic concept also has some absurd implications. For example, as Devitt notes, it implies that "no matter how dramatically a lineage changes it will not form a new species unless it splits" (2008, p. 369). Hence it would follow, absurdly, that even if human beings had evolved from Protista with no splits, there would have been no change in species. There is also the absurd implication that a species goes out of existence every time there is a split, no matter how similar some of the organisms after the split are to those that existed before the split (Devitt 2008, pp. 369-70). The phylogenetic-cladistic concept would also entail, no less absurdly, that two organisms would not be of the same species, no matter how similar they were morphologically and even genetically, if they did not belong to the same evolutionary lineage (Oderberg 2007, pp. 218-19). For example, it would follow that an organism that was particle-for-particle identical to a dog, and not only acted like the dogs we know but could mate with those dogs, was not really a dog if it did not descend from the dogs we know (e.g. if it was engineered by extraterrestrials, or specially created by God, or came about by spontaneous generation). By the same token, it would follow from the phylogenetic-cladistic concept that the very first organisms belonged to no species, since they did not descend from any organisms (Oderberg 2007, pp. 219-20).

Another objection sometimes raised against essentialism is that there is no set of traits that is common to every member of any species and which could play the role that an essence is said to play. One problem with this sort of objection is that it typically ignores the Aristotelian distinction between the essence of a thing, the proper attributes (or "properties" in the technical Aristotelian sense) that flow from the essence, and the merely contingent attributes that do not flow from the essence. The proper attributes of a kind of organism will be manifest in a mature and healthy specimen, but as I noted in chapter 1, they may fail to manifest if the specimen is injured or otherwise defective (as in a dog that is missing a leg). Contingent attributes are even more likely to be absent in some cases, though it is possible that such an attribute might be present in every instance and even give the false appearance of being a proper attribute. To borrow an example from Oderberg (2007, pp. 209-11), occasionally a tiger appears that lacks well-defined stripes. Because such tigers also usually exhibit poor health, we have reason to think that stripes are a proper accident of tigers and that a tiger lacking them would be defective. Alternatively, it could turn out that stripeless but healthy tigers are possible and that the having of stripes is after all a contingent rather than proper attribute of tigers. But either way, this would not show that tigers lack an essence, because the essence of a thing is distinct from whatever proper or contingent attributes it may or may not have.

In any case, as even one critic of essentialism concedes, "there are important genetic similarities between members of a single species... [and] species taxa are distinguished by clusters of covarying traits" (Okasha 2002, p. 197). Devitt (2008, p. 371) suggests that such clusters are precisely what essences can be identified with. Other essentialists argue that an organism's essence should be sought, not in either its genotype or phenotype *per se*, but in the "developmental program" that maps the one onto the other. (Cf. Boulter 2013, pp. 111-13; Austin 2017.) But as Oderberg (2007, pp. 201-3 and 234-40) emphasizes, it is a mistake, in biology no less than in chemistry, to suppose that the essence of a thing can be sought *only* at the micro-level or in features otherwise below the level of morphology. We saw in the previous chapter that we cannot so much as identify the micro-level features of an inorganic substance that the reductionist is interested in without reference to the macro-level features they underlie, and the same is true with organisms. An organism's genetic features, its morphology (understood broadly to include not just body plan, but also mode of reproduction, behavioral habits, etc.), and the developmental processes that lead from the one to the other should *all* be taken account of in determining its essence.

To be sure, and as Christopher Austin (2017) notes, the modern biologist is bound to object that morphology is not entirely a consequence of genetic or other factors intrinsic to an organism. Environment plays a role as well. This is sometimes raised as an objection against intrinsic essences. However, to affirm the reality of intrinsic essences is by no means to deny the role of environment. As Austin argues, what the Aristotelian essentialist is committed to is the reality of *dispositions* that follow from an organism's essence. As theorists of dispositions and causal powers often emphasize, a disposition will typically manifest in tandem with the operation of other powers, and the *way* it manifests can vary depending on variations in these concomitants. (Cf. Feser 2014b, pp. 53-68.) These concomitant powers can include those operating in the environment external to the agent. So, there is no reason why it cannot be true that morphological features are the consequence both of the organism's essence and its environment, operating together. Indeed, as Austin (2017) argues, evolutionary developmental biology (or "evo-devo") is "a framework in which morphological variation is derived from invariant, functional causal mechanisms which serve as highly conserved 'deep homologies', underwriting a vast array of organismal diversity."

Then there is the objection that essentialism requires sharp boundaries between species, whereas evolution requires that the boundaries be vague (Ereshefsky 1992, pp. 188-89; Hull 1992). One possible reply to this is proposed by Sober:

> [E]ssentialism is a doctrine that is compatible with certain sorts of vagueness. The essentialist holds that the essence of gold is its atomic number. Essentialism would not be thrown into doubt if there were stages in the process of transmuting lead into gold in which it is indeterminate whether the sample undergoing the process belongs to one element or to the other. (1993b, p. 148. Cf. Devitt 2008, p. 371)

To be sure, the notion of a vague essence is problematic (Oderberg 2007, pp. 226-27). At least for the Aristotelian essentialist, there is always a fact of the matter about whether a natural substance is of one type or the other, whether or not we know what type that is. But the point is that vagueness would not by *itself* suffice to refute essentialism. The anti-essentialist would have to show that there being *some* cases where there is no fact of the matter about what a thing is entails that there are *no* cases where there is a fact of the matter about what a thing is. But this *can't* be shown, because in fact the very idea that there are indeterminate cases presupposes that there are at least some determinate ones, by contrast with which the indeterminacy is measured (Oderberg 2007, pp. 227-28).

In any event, evolution simply doesn't require vagueness in the first place. Lead and gold have certain properties in common (they are both metals, after all), but it doesn't follow that there is no sharp boundary between them. Similarly, that a species *S1* and its descendent species *S2* will have certain traits in common doesn't entail that there is no sharp difference between their essences. Even if, among the intermediary groups of organisms in between *S1* and *S2*, it is hard to determine where one ends and the other begins, it doesn't follow that these intermediary groups lack sharp essences. The vagueness might be merely epistemic rather than ontological, and if we have independent reason to believe in essences (which we do) then we have good reason to conclude that that is all that is going on.

As Boulter (2013, pp. 107-11) argues, evolution if anything actually *presupposes* essentialism insofar as it holds that new species come into being and go extinct as a matter of objective fact, and not merely as a matter of changes in our classificatory practices. That entails that there must be objective facts that distinguish one species from another, which (as the Aristotelian argues) in turn requires essences. Other writers (Walsh 2006; Oderberg 2007, pp. 212-13) point out that evolution presupposes essentialism in another respect. An organism exhibits a certain malleability in the very process of maintaining itself, insofar as it exhibits homeostasis and otherwise adapts to changes in its environment. This malleability is the source of the variation that leads to evolutionary change. But it is itself grounded in the nature of the organism. Evolution could not occur unless there were a fact of the matter about what an organism is that determines what sorts of mutations and adaptations it is capable of.

6.2.2 Natural selection is teleological

Jerry Fodor and Massimo Piattelli-Palmarini (2011) make a similar point when they emphasize that the environmental circumstances that evolutionary theory tells us shape the phenotypes of organisms do so only *through* factors endogenous to the organism. Naturally, this includes the genome, but it involves more than that. To take an example Fodor appeals to elsewhere (2007), in order to explain why pigs lack wings, it will not do to suggest that wings would have been maladaptive in the environmental circumstances in which pigs evolved. Rather, one has to note that the anatomy and physiology of pigs simply rule out their having wings. In order to have wings, a pig would have to have a radically different musculature, metabolism, weight, and so forth. It would have to be very *unlike a pig*. Hence, explanations of traits in terms of natural selection, which appeal to factors extrinsic to the organism that determine the fitness of traits, cannot be the whole story. You might say that they piggyback on factors that flow from the intrinsic essence of the organism. As Fodor says:

> [T]heories... seeking to co-opt natural selection... attempt to explain why we are so-and-so by reference to what being so-and-so buys for us, or what it would have bought for our ancestors... But, in point of logic, this sort of explanation has to stop somewhere. Not all of our traits can be explained instrumentally; there must

> be some that we have simply because that's the sort of creature we are. (2007)

If Fodor and Piattelli-Palmarini inadvertently give aid and comfort to Aristotelian essentialism, they do the same for teleology, albeit only implicitly. To see how takes some spelling out. Their book is best known for its critique of Darwin's theory of natural selection. They do not deny that evolution has occurred, but only that natural selection provides a good explanation of *how* it occurred. The argument is complex and often misinterpreted, and thus needs to be set out in some detail if it is to be understood correctly. So I will devote the next few pages to exposition, and return later to the question of the teleological implications of the argument.

The argument begins by drawing an analogy between Darwinian evolutionary theory and B. F. Skinner's behaviorist psychology (Fodor and Piattelli-Palmarini 2011, chapter 1). In Skinner's theory of operant conditioning, the subject is treated as a "black box," the inputs to which are a set of behavioral traits at a certain time together with reinforcements, and the output from which is a set of behavioral traits at a later time. Darwin's theory of natural selection can also be taken to describe a kind of black box, the inputs to which are a distribution of phenotypic attributes within a population of organisms at a certain time together with their environmental circumstances, and the output from which is a distribution of phenotypic attributes within the population at a later time.

Fodor and Piattelli-Palmarini describe a number of further parallels between the theories. The first is *iterativity*. The outputs of the process Skinner describes are subject to further operant conditioning, and the outputs of the process Darwin describes are subject to further evolution. The second is *environmentalism*. The tendency of Skinner's theory is to emphasize the subject's environment rather than endogenous factors as what is crucial to psychology, and the tendency of Darwin's theory is to emphasize the organism's environment rather than endogenous factors as what is crucial to evolution. Cognitive science helped correct the behaviorist overemphasis on the environment, and evolutionary developmental biology has helped correct the Darwinian overemphasis on the environment. A third parallel is *gradualism*. Operant conditioning results in dramatic transformations in behavioral traits only by way of the accumulation of smaller transformations, and natural selection results in dramatic transformations in phenotypic attributes only by way of the accumulation of smaller transformations. The fourth parallel concerns *monotonicity*. The idea here is that if a reinforcement increases the strength of a habit in one case it will do so in the next one, and if selection increases fitness in a certain ecological context in one case it will do so in the next.

The fifth parallel is *locality*. Only what actually happens in proximity to a subject can influence the conditioning of behavior, and only what actually happens in proximity to organisms can influence natural selection. What happened in the past (apart from whatever traces it has left in the present), what will happen in the future, what happens in distant locations, and other causally distant factors can have no effect on what is selected here and now. The sixth parallel is *mindlessness*. Human minds can be affected by causally isolated factors insofar as they can *mentally represent* them. For example, we can think about *counterfactual* situations – what *would have* happened *if* such-and-such conditions had obtained. But behaviorist psychology rules out any role for mental representations, and the whole point of natural selection is to eliminate the need for appeal to anything like a designing mind. Finally, the two theories are parallel in that they posit similar *mechanisms*. In Skinner's psychology, the subject is, in abstraction from the effects of prior learning, a random generator of stimulus-response dispositions. Such randomly generated traits are then filtered via reinforcement from the environment. Similarly, in Darwinian evolution, phenotypic variations arise via random mutations and then are filtered by natural selection.

Now, there is an insuperable problem with the behaviorist account of learning (Fodor and Piattelli-Palmarini 2011, pp. 101-6). Suppose we say that an animal has learned to respond in a certain way to certain stimulus. What exactly is the right way to characterize the stimulus and the response? To borrow an example from Fodor and Piattelli-Palmarini, if the animal learns to choose a card with a yellow triangle on it over a card with an X on it, should we say that the animal has learned to choose yellow triangles over Xs, or yellow objects over Xs, or triangles over Xs, or closed figures over Xs, or should we opt instead for one of the many other possible characterizations of the stimulus? Should we say that the animal has learned to walk toward the stimulus, or to move toward it in general (since it might instead swim toward it if we filled the path with water), or to turn right, or to turn east, or should we opt instead for one of the many other possible characterizations of the response?

The problem is one of *indeterminacy* in the sense operative in thought experiments like the one from Quine discussed in chapter 2 and the one from Kripke discussed in chapter 5. Given the constraints of behaviorist theory together with the behavior the animal actually exhibits, there can be *no fact of the matter* about which of the characterizations is correct, since any of them is consistent with the behavior. To be sure, there could be further tests which elicit further behavior indicating that it is (say) triangles rather than yellow objects that the animal has learned to respond to. But this new repertoire of behavior would itself be susceptible of various alternative interpretations, so that the problem of indeterminacy would just reappear.

Now, as Fodor and Piattelli-Palmarini note, what the behaviorist has to appeal to in order to characterize the animal's behavior one way or the other are counterfactuals. For example, in order to judge that what the animal is responding to are triangles, specifically, one has to say that if the animal had encountered a yellow square and a green triangle, then it would have chosen the latter. But counterfactuals can have an influence only insofar as they are mentally represented, and the behaviorist eschews mental representation. This is one reason behaviorism fails as a theory of human psychology. There is a fact of the matter for a human subject about whether he is choosing yellow triangles, or yellow objects, or triangles, or whatever, but it is precisely the mental representations the behaviorist shuns that determine what the right answer is.

However, as Fodor and Piattelli-Palmarini note (2011, pp. 106-9), similar indeterminacy problems notoriously afflict any naturalist theory of the mind even when it is willing to affirm the existence of mental representations. Recall the frog example from earlier in this chapter. The standard naturalistic account of the representational content of the frog's perceptual states is going to make reference to the causal relations between those states and what they represent. For example, the naturalist might hold (as Millikan does) that a certain perceptual state will represent bugs because those among the frog's ancestors who had such perceptual states were better able to catch bugs, and this is the reason they were favored by natural selection. But why should we say that the perceptual state in question represents *bugs* as opposed to *small moving things*? For suppose that in the environment in which the frog's ancestors evolved, the only small moving things that it ever encountered happened to be bugs. Then, even though the expression "bugs" and the expression "small moving things" don't have the same sense, they would in this context be co-extensive, i.e. they would *refer* to the same things. Hence to describe the situation as one in which *natural selection favored frogs which snapped their tongues at small moving things* would be no less correct than describing it as one in which *natural selection favored frogs which snapped their tongues at bugs*. So the causal factors in question do not suffice to determine that the frog's perceptual state represents bugs rather than small moving things.

At this point the naturalist might appeal to counterfactuals to solve the problem. If the frog's ancestors had been snapping their tongues at small moving things other than flies, then they would not have been favored by natural selection. This, the naturalist might suggest, shows that it really is after all bugs rather than small moving things that the frog's perceptual state represents. But the trouble with this response, as Fodor and Piattelli-Palmarini point out, is that while *we* know this counterfactual to be true, *natural selection* does not, because natural selection doesn't know anything at all. It is, as noted above, both mindless and sensitive only to the actual local causal situation. What *would have* happened in some counterfactual situation can have no effect on it. So counterfactuals cannot solve the indeterminacy problem.

This brings us at last to the problem that Fodor and Piattelli-Palmarini think all of these parallels pose for the theory of natural selection as an account of adaptation. To take a further example from Fodor (2007), consider polar bears. The theory of natural selection purports to explain how they are adapted to their environment by saying that they were selected for having white fur. But should we say that they were selected for *being white*, or for *matching their environment*? The expressions "being white" and "matching their environment" don't have the same sense, but given that the environment in which polar bears evolved was white, these expressions are co-extensive in that context. Hence, Fodor and Piattelli-Palmarini argue, the features of the causal situation that would seem to justify the claim that polar bears were selected for being white would equally well justify the claim that they were selected for matching their environment. But in that case, there can be no fact of the matter about which of these features natural selection selected for.

Of course, the Darwinian biologist might at this point appeal to the counterfactual claim that if the environment in which polar bears had evolved had been green, then they would not have been favored by natural selection. Hence, the argument might go, it must be *being white* that natural selection selected for, and not *matching their environment*. But again, natural selection is mindless, and sensitive only to actual local causal circumstances. So it cannot be affected by what would have been the case in some counterfactual situation. Hence, even though *we* can know the counterfactual to be true, its truth does not contribute anything to the causal factors that actually influence natural selection itself.

The parallel with the other cases, then, is this. If learning worked the way behaviorist psychology says it does, then there could be no fact of the matter about what exactly it is that a human subject learns. Since there is a fact of the matter, behaviorism fails as an explanation of human learning. If there was nothing more to the representational content of a mental state than what naturalistic theories of meaning say there is, then there could be no fact of the matter about what exactly is the content of any of our mental states. Since there is a fact of the matter, naturalistic theories fail as explanations of representational content. Similarly, if evolution worked the way the theory of natural selection says it does, then there could be no fact of the matter about what features are selected for. Since the theory says that there is a fact of the matter and purports to explain it, the theory fails by its own standard. The theory's notion of "selection for" is as indeterminate as the behaviorist's stimulus-response pairings are, and as the contents of mental representations are on a naturalistic theory of meaning.

Fodor and Piattelli-Palmarini consider various possible responses to this argument (2011, pp. 111-38). The first goes as follows. What the theory of natural selection is trying to explain are those traits that are correlated with fitness. So, suppose there is a trait T that contributes to fitness and that this trait is linked in a law-like way to another trait T' that does not contribute to fitness but is, as it were, a kind of free-rider. For example, T could be the heart's activity of pumping blood and T' could be its tendency to make a thumping sound. Now, it might seem that this entails the sort of indeterminacy problem that Fodor and Piattelli-Palmarini are calling attention to. Both T and T' are correlated with fitness, because T is correlated with fitness and T' is correlated with T. So, it might seem that there is no fact of the matter about whether natural selection selects for T or selects for T'. But we can easily solve this problem by saying that what natural selection selects for are traits that are *directly* correlated with fitness. Now, T is correlated with fitness directly, but T' is correlated with it only indirectly. Hence we have good reason to say that what natural selection selects for is T rather than T', and the indeterminacy problem is solved.

But as Fodor and Piattelli-Palmarini argue, this purported solution is illusory. For we need to know *what makes it the case* that T is tied to fitness directly, and T' only indirectly. And the answer is going to be framed in terms of counterfactuals. The claim is going to be that if T had existed in the absence of T', natural selection would still have favored it, whereas if T' had existed in the absence of T, natural selection would *not* have favored *it*. Hence it is T rather than T' that is directly linked to fitness. But again, even if *we* can know that this counterfactual is true, *natural selection* cannot be influenced in any way by the situation the counterfactual describes, since natural selection is mindless and affected only by actual local causal circumstances. If the theory of natural selection cannot explain fitness, then, again, it fails by its own standards.

There is a connection here to the "spandrel" problem famously raised by Stephen Jay Gould and Richard Lewontin (1979) in criticism of the tendency of some Darwinians to suppose that almost any trait of an organism can be explained as an adaptation. A spandrel is the triangular space created by the meeting of two arches, and in churches is often filled with an illustration of some sort. It might seem that spandrels are intentionally put into a church in order to provide a surface on which to paint such illustrations, but in fact they are an unintended byproduct of the construction of arches. The illustrations are just a way to make use of a space that would otherwise serve no purpose. Similarly, Gould and Lewontin argue, some traits are not really adaptations, but rather a by-product of traits that are adaptations. You might say that the point that Fodor and Piattelli-Palmarini are making (a point that is far more radical than the one Gould and Lewontin were making) is that natural selection cannot determine of two traits which one is actually an adaptation and which is merely a spandrel.

Darwin was, in Fodor and Piattelli-Palmarini's view (2011, pp. 115-16), misled by the analogy he drew between natural selection and selective breeding. Since a breeder has a mind, there can be a fact of the matter about which trait he is breeding for. Since natural selection is in some respects analogous to selective breeding, it seems that there can also be a fact of the matter about what *it* is selecting for. But natural selection is mindless, and when we delete from the situation the mental representations that are present in the case of selective breeding, we delete along with it anything that can make it the case that there is a fact of the matter about what is selected for. The spandrel analogy can be similarly misleading. Architects intend to build arches and do not necessarily intend to build spandrels, which are, again, merely a byproduct of what is intended. Comparing natural selection to what architects do can thus make it seem that there is a fact of the matter that it is one trait rather than another that is selected for. But this is an illusion insofar as natural selection, unlike an architect, is mindless and thus lacks anything analogous to intentions.

This sort of error underlies another possible reply to their argument considered by Fodor and Piattelli-Palmarini (2011, pp. 119-22). The critic might suppose that we should understand the outcomes of natural selection in terms of what Mother Nature has intended (Dennett 1995), or how "selfish genes" manipulate us (Dawkins 1989), or what the "blind watchmaker" of evolution has designed (Dawkins 1987). By attributing mental properties to nature in this way, we can make sense of the idea that some traits are selected for and others are not. But of course, the trouble with these anthropomorphic descriptions is that they are mere metaphors and not literally true. Nature is not literally a mother who intends anything, genes are not literally selfish or manipulative, and evolution is not literally a watchmaker or any other kind of designer. These may or may not be *useful* fictions, but they are *fictions* all the same. As Fodor and Piattelli-Palmarini write, "fictions can't select things, however hard they try. Nothing cramps one's causal powers like not existing" (2011, p. 121). (Cf. Stove 2006.)

Another reply they consider is the suggestion that there are laws of nature that determine what is selected for (2011, pp. 122-27). The problem with this proposal is that laws of nature are supposed to hold universally, whereas fitness depends on context. Fodor and Piattelli-Palmarini emphasize that the problem is *not* that laws of selection would be of the *ceteris paribus* type. They have no objection to such laws. But with *ceteris paribus* laws, there are at least idealized circumstances in which the law would be strictly true. By contrast, what makes for fitness is so *thoroughly* dependent on contingent circumstances that the idealization a *ceteris paribus* law requires is not possible. For example, there are no plausible idealized circumstances which would justify choosing between the purported law that *being big is better for fitness than being small* and the purported law that *being small is better for fitness than being big*. In the actual world, big and small organisms are both so common that neither generalization would plausibly count even as a *ceteris paribus* law.

A further reply considered by Fodor and Piattelli-Palmarini (2011, pp. 127-30) appeals to an example due to Sober (1993a, pp. 98-100). Consider a sieve with holes of such a size that they allow smaller marbles to fall to the bottom while preventing larger marbles from doing so. Suppose the smaller marbles are all red and the larger ones are of different colors. Since only small red marbles make it to the bottom, it might seem that we have an indeterminacy problem of the kind Fodor and Piattelli-Palmarini claim faces natural selection. That is to say, it might seem that there is no fact of the matter about whether the sieve is selecting for small marbles or selecting for red marbles. But in fact the sieve is clearly selecting for small marbles insofar as it is the size of the holes, and not anything to do with the color of the marbles, that determines which marbles reach the bottom. Now, natural selection, the critic might say, can avoid indeterminacy in a similar way as long as there are features in an environmental context that are analogous to the size of the holes.

But Fodor and Piattelli-Palmarini respond that one problem with this example is that it makes reference to *endogenous* features of the sieve, whereas the theory of natural selection emphasizes *exogenous* features, namely the circumstances of an organism's environment. To be sure, many Darwinians would simply incorporate endogenous features of organisms into their account of natural selection. But there is in any case a deeper problem with Sober's example, which is that it *is* in fact plagued by just the sort of indeterminacy it claims to avoid. For why should we suppose in the first place that the marbles that reach the bottom are the ones selected for? Why not suppose instead that the sieve is selecting for the marbles that stay on top? The answer can only be that the designer of the sieve, or Sober himself, intended for the sieve to be conceived of as selecting for the marbles that reach the bottom. But then selection for in this case reflects the intentions of some mind, whereas natural selection is mindless. So the cases are not really parallel after all.

The final possible reply that Fodor and Piattelli-Palmarini consider (2011, pp. 131-38) is the proposal that explanation in terms of natural selection shouldn't be understood as a predictive theory that accounts for adaptations by subsuming particular cases under general laws. Rather, it is merely an explanatory schema that tells us that for any particular adaptive phenotypic trait, it was selected for enabling the organism to deal with some aspect of its environment. One problem with this, Fodor and Piattelli-Palmarini argue, is that the schema is in danger of becoming a mere truism rather than a genuine empirical claim. Stripped of predictive bite, the thesis that *adaptive traits are always selected for dealing with some environmental factor* ends up like the thesis that *bachelors always turn out to be unmarried*.

Another problem is that the proposal essentially turns evolutionary explanations into historical narratives about various particular contingent causal sequences, comparable to a historical narrative describing the many different ways various people happened to became rich. Unlike nomological explanations, such historical narratives make no reference to any necessary connections between properties, and thus support no counterfactual claims. But counterfactual claims are what is needed in order to solve the indeterminacy problem facing the notion of "selection for." Hence conceiving of evolutionary explanations as historical narratives cannot solve the indeterminacy problem. Fodor and Piattelli-Palmarini conclude:

> [I]f there are no nomologically necessary generalizations about the mechanisms of adaptation as such, then the theory of natural selection reduces to a banal truth: 'If a kind of creature flourishes in a kind of situation, then there must be something about such creatures (or about such situations, or about both) in virtue of which it does so.' Well, of course there must; even a creationist could agree with that. (2011, p. 137)

Several further objections were raised against Fodor and Piattelli-Palmarini following the publication of the first edition of their book. For example, some argued that if T rather than T' is the trait that is causing increased reproductive success, then that by itself suffices to show that T is being selected for in the sense that matters to the theory of natural selection (Godfrey-Smith 2010; Sober 2010). But as Fodor and Piattelli-Palmarini point out (2011, pp. 179-80), this essentially just stipulates as *true by definition* the thesis that the trait that causes increased fitness is the one selected for. In that case, that T causes increased fitness cannot provide an *explanation* of why it was selected for, any more than the fact that someone is a bachelor explains why he is unmarried. Yet the theory of natural selection claims to provide just such an explanation.

Another charge is that Fodor and Piattelli-Palmarini are committed to the implausible claim that there is no fact of the matter about which of two correlated traits T and T' is causally responsible for reproductive success (Block and Kitcher 2010). But that is not the claim Fodor and Piattelli-Palmarini are making. Indeed, like their critics, they consider that claim "preposterous" and agree that there *is* a fact of the matter and that we can know what it is (Fodor and Piattelli-Palmarini 2011, p. 181). What they are claiming is rather that *natural selection* cannot distinguish the trait that causes reproductive success from the trait that is merely correlated with what causes it.

An objection raised by Rosenberg (2013) is that Fodor and Piattelli-Palmarini's critique is irrelevant because the theory of natural selection doesn't need the notion of "selection for," but only the notion of "selection against." Fodor's response is worth quoting at length:

> Since it is tautological that there can't be selection for/against a neutral trait, it follows that, if there is selection at all, then it is selection for a trait iff it isn't selection against it. Still, let's assume, for the sake of argument that whiteness wasn't selected for in polar bears. What, in that case, was selected against? Being pink? Being green with orange stripes? Do Darwinists believe that there used to be green polar bears with orange stripes, but they all got eaten up by predators? If not, what does [Rosenberg] think is gained by rejecting selection for in favour of selection against? (Marshall 2014, p. 253)

I have set out Fodor and Piattelli-Palmarini's argument at length because it is complex and often misunderstood. But I want to make a different use of it than they do, because while I think their replies to their critics are correct, the argument doesn't actually show quite what they say it does. What they claim to show is that there is no way to solve the indeterminacy problem facing the notion of "selection for." But what they actually show is rather that there is no way to solve that problem *given the assumption of metaphysical naturalism.* In particular, they show that the problem cannot be solved *given a non-teleological conception of nature*. Since Fodor and Piattelli-Palmarini are, like their critics, committed to a naturalistic and non-teleological conception of nature, it is understandable that they would frame their conclusion the way they do. But if we look at the situation from a teleological and non-naturalistic point of view, it takes on a very different complexion.

Recall that in chapter 1, I drew a distinction between *extrinsic* and *intrinsic* teleology. A thing or process has extrinsic teleological features when those features are in no way intrinsic to it, but entirely imposed from outside it. An example would be the time-telling function of a watch, which is in no way intrinsic to the nature of the metal parts that make up the watch but exists only relative to the intentions of the designers and users of the watch. A thing or process has intrinsic teleological features when those features follow from its very nature or essence. An example would be an acorn's tendency toward the end of becoming an oak, which is built into an acorn simply by virtue of being an acorn.

As several philosophers have noted in recent years (Ariew 2002 and 2007; Shields 2007, pp. 68-90; Feser 2010), this distinction corresponds to a distinction between two ways of understanding the thesis that there is teleology in nature. The first might be labeled *Platonic teleological realism*, which holds that there really is teleology in natural objects and processes, and that it is in them in something like the way that the time-telling function is in a watch. That is to say, it is imposed from outside by a mind – such as the *nous* of Anaxagoras, the demiurge of Plato's *Timaeus* (hence the "Platonic" label), or the divine designer of William Paley's *Natural Theology*. The second view might be labeled *Aristotelian teleological realism*, which holds that there really is teleology in natural objects and processes, and that the acorn example rather than the watch example provides the correct way to understand it. That is to say, on this view teleology is *intrinsic* to natural objects and processes. It follows from their very natures, and thus would still be there whether or not there was some divine or other mind external to them.

As I have noted elsewhere (Feser 2010 and 2013b) there is also a third view, which can be labeled *Scholastic teleological realism* since it was held by Scholastic writers like Aquinas. It is a variation on the Aristotelian position, since it takes natural objects and processes to have teleological features intrinsically, by virtue of having the natures they do. But it also gives a nod to the Platonic teleological realist position in that it takes the divine intellect to be the ultimate cause of things having the natures they do. The position essentially conceives of *final* causality in a way that parallels Aquinas's concurrentist conception of *efficient* causality. Concurrentism is a middle ground position between the occasionalist view that God is the direct efficient cause of everything that happens, and the deist view that efficient causes in nature operate entirely independently of God. Concurrentism holds that natural objects really do have efficient causal power (contrary to occasionalism) but that this causal power cannot operate without continual divine cooperation or concurrence (contrary to deism). Scholastic teleological realism takes an analogous middle ground position with respect to final causality. It holds that the *proximate* ground of the teleological features of a natural object or process is its nature or essence, and thus is intrinsic to it (contrary to Platonic teleological realism), but also that these teleological features have the divine intellect as their *ultimate* ground (contrary to atheistic versions of Aristotelian teleological realism). (See the articles of mine cited above for more detailed discussion of the view.)

Now, Fodor and Piattelli-Palmarini do in several places briefly mention, if only immediately to dismiss, the idea that an appeal to God might solve the "selection for" problem (2011, pp. 120, 122, 141-2, and 155). That is understandable given that they are "fully signed-up atheists" (p. 240), indeed "outright, card-carrying, signed-up, dyed-in-the-wool, no-holds-barred atheists" (p. xv). But suppose one holds that there are compelling arguments for the existence of God (Feser 2017) and also finds the theory of natural selection to be plausible apart from Fodor and Piattelli-Palmarini's critique of it. Then one has the materials for a theistic solution to the "selection for" problem – either a Platonic teleological realist solution, or (if one is a theist who is also an Aristotelian) a Scholastic teleological realist solution. Fodor and Piattelli-Palmarini reject this solution, but only because they take atheism for granted. They don't give any actual argument against it.

Or suppose that, like Thomas Nagel (2012), one is an atheist but takes seriously the Aristotelian idea that there is teleology intrinsic to natural objects and processes, which can accordingly be known and studied even if one does not think it requires a divine cause. Then one has the ingredients for an atheistic variation on an Aristotelian teleological realist solution to the "selection for" problem. Fodor and Piattelli-Palmarini don't even mention this possibility, much less dismiss it. Perhaps they are not aware that there is such a thing as an Aristotelian teleological realist approach distinct from the Platonic teleological realist position. Or perhaps they take it for granted that modern science has shown that there is no teleology of any sort in nature. But as I have been arguing in this book, modern science has shown no such thing, and neither has it refuted the other main elements of the Aristotelian philosophy of nature. Hence for all Fodor and Piattelli-Palmarini have shown, the "selection for" problem might be solved, and the theory of natural selection therefore salvaged, if one adopts an atheistic brand of Aristotelian teleology.

In fact, this seems at least implicitly to have been the approach of Darwin himself. As James Lennox (1993) and others have noted, in several places Darwin affirmed the existence of a kind of natural teleology, and even connected it to the operation of natural selection. (Cf. Gilson 1984, pp. 80-89; Depew 2015; Rothman 2015, chapter 13.) But it was, of course, a teleology divorced from any notion of divine design (Lennox 1993, p. 418). Writes Lennox:

> Selection explanations are inherently teleological, in the sense that a value consequence (Darwin most often uses the term 'advantage') of a trait explains its increase, or presence, in a population...
>
> Darwin essentially re-invented teleology... The concept of selection permits the extension of the *teleology* of domestic breeding into the natural domain, without the need of conscious design. As in domestic selection, the good served by a variation continues to be causally relevant to its increasing frequency, or continued presence, in a population – but the causal mechanism, and the locus of goodness, shifts. (1993, pp. 410 and 417)

Similarly, when commenting on Darwin's position, David Depew proposes conceiving of "natural selection as properly final causality," and elaborates as follows:

> Proper final causality is causality that runs through a process whose constituent moments, to the extent that something does not interfere, emerge as they do *because* they have a good effect – as in the case of the eye... [I]n Aristotle's technical terms Darwinian adaptations do have properly final causes. They reliably have certain effects and they come to be precisely because they have these good effects. (2015, p. 126)

A natural way to read this would be as holding that a tendency to select for traits that are advantageous – for T as opposed to T', to put it in the terms used above – is in some way intrinsic to the very nature of the evolutionary process itself. The evolutionary process is inherently *directed toward* this end. This would solve the "selection for" problem in a non-theological way, though also in a non-naturalistic way insofar as it affirms the existence of Aristotelian intrinsic teleology. To be sure, as Depew emphasizes (pp. 131-2 and 135), the kind of teleology Darwin affirmed is very modest. It does not entail any recherché claims about cosmic progress of the kind associated with evolutionists like Pierre Teilhard de Chardin. But it does break with the contemporary naturalist's insistence that scientific explanation confine itself to efficient causes alone.

Of course, the naturalist might say that Darwin was simply mistaken, and that had he followed out consistently the implications of his own theory, he would have abandoned teleological notions altogether, as later Darwinians have. But then the naturalistic Darwinian will be stuck with the "selection for" problem, and without *some* kind of teleology there is no way to solve it. Alternatively, the naturalistic Darwinian might say that Darwin was right to affirm teleology, but that the kind of teleology the Darwinian needs can be given a reductive analysis via Millikan's account of biological function or the like. But as we have seen, such reductive analyses fail.

We are left with two ironies. First, while it is routinely asserted that Darwin banished teleology from biology, the truth (as Fodor and Piattelli-Palmarini have shown) is that his theory actually *presupposes* teleology. Second, this conclusion is actually truer to Darwin's own understanding of natural selection than the standard contemporary naturalistic interpretation of Darwinism is. *If you want to be a Darwinian evolutionist, you need to be an Aristotelian.*

6.2.3 Transformism

Of course, the converse doesn't hold. You can be an Aristotelian without being an evolutionist. After all, Aristotle himself was an Aristotelian without being an evolutionist. But do we need to say something stronger? Does Aristotelianism actually *rule out* evolution? In particular, does it rule out the possibility of one species giving rise to another?

No, it doesn't, though it *does* rule out a philosophical naturalist or mechanistic interpretation of how that would happen – that is to say, an interpretation that eliminates formal and final causes and the rest of the Aristotelian philosophical apparatus. Consider Rosenberg's proposed naturalistic account of how primitive organisms could arise out of inorganic chemical processes (2011, chapter 3). Rosenberg characterizes the problem as that of providing "an explanation of how, starting from zero adaptations, any adaptation at all ever comes about" (p. 50). He rightly emphasizes that a genuinely naturalistic explanation "can't cheat" by smuggling in adaptation from the get-go (p. 70), and can't make implicit use of any other teleological concepts either:

> The explanation we need can't start with even a tiny amount of adaptation already present. Furthermore, the explanation can't help itself to anything but physics. We can't even leave room for "stupid design," let alone "intelligent design," to creep in. If scientism needs a first slight adaptation, it surrenders to design. It gives up the claim that the physical facts (none of which is an adaptation) fix all the other facts. (p. 50)

Rosenberg further sets the stage for his account as follows:

> Natural selection requires three processes: reproduction, variation, and inheritance. It doesn't really care how any of these three things get done, just so long as each one goes on long enough to get some adaptations. Reproduction doesn't have to be sexual or even asexual or even easily recognized by us to be reproduction. Any kind of replication is enough. (p. 59)

He later says instead that in addition to "replication and variation... fitness differences [are] the last of the three requirements for evolution by natural selection" (pp. 64-65).

With these criteria in hand, Rosenberg devotes several pages to sketching out scenarios in which inorganic molecules can be said to replicate, vary, differ in their fitness, and thereby give rise to "adaptation." And he has no trouble doing so given how *broadly* he construes the key concepts: The formation of crystals counts as an example of "replication"; the chemical difference between sugar and Splenda counts as an example of "variation"; an inorganic molecule's being able to "persist or replicate or both" counts as "adaptation"; and so on.

Thus does Rosenberg purport to show how adaptation can arise from non-adaptation in a way that doesn't "cheat" by smuggling in adaptation at the beginning. But this is like proudly proclaiming that you didn't cheat on your exam, when you knew in advance that the professor would only ask you questions you had an answer for. It's true but uninteresting. For given how broadly Rosenberg construes the key notions, you might as well say that *pebbles* are "well-adapted" to their environment. After all, they "replicate" (when one pebble is broken in two); they "vary" (the new pebbles are smaller than the original, and differ from it and from each other in shape); they "inherit" features from their parents (the new pebble is solid and rough, just like Dad – a chip off the old block); and they differ in their "fitness" (the new pebbles are smaller and thus less easily broken than their ancestors). Descent with modification, in rock gardens no less than botanical gardens!

But what does any of this have to do with *organic* phenomena, with *biological* adaptation? Nothing at all. Certainly Rosenberg does nothing to justify the claim that it does, other than to make the obligatory hand-waving reference to the Miller-Urey experiment and hydrothermal vents, with a passing concession that "molecular biologists don't yet know all the details... or even many of them" about how organic processes might arise from inorganic ones (2011, p. 67). Since what is at issue is whether biological adaptation really can be explained in terms of the stuff about crystals, Splenda, etc., to leave out these "details" is just to fail to answer the question at all. Rosenberg is like someone who contracts to build you a house, clears the ground a little, and then takes off without doing anything else – dismissing your concerns about the absence of a foundation, framework, walls, electrical, plumbing, etc. as mere quibbling over "details." Needless to say, these are not mere details. They're the *house*.

What Rosenberg owes us is an account of how *biological* adaptation, specifically – and not mere "adaptation" in the loose sense that a resilient inorganic molecule or a pebble exhibits – can arise from physical processes that initially involve no *biological* adaptation at all. That means he owes us an account of what *life* is – an account that makes it evident exactly *how* the sort of "adaptations" he describes add up to the kind that a *living* thing exhibits. Yet the nature of life is a question which Rosenberg's account does not directly address. He just speaks of "adaptation" *sans phrase*, and insinuates, without argument, that having given an account of processes that might in some *extended* sense of the word be called "adaptation," he has thereby given an account of life.

Now, as I have argued, living things are substances which exhibit immanent causation as well as transeunt causation, where immanent causation is a species of teleology. Rosenberg, who is as staunchly reductionist and anti-teleological a naturalist as they come, in effect treats living things as aggregates rather than true substances and as governed by efficient causation alone. So, his approach is doomed from the start. The factors to which he confines himself could never give rise to life, no matter how much detail he adds to the story. Substantial form and final causality are in the *outcome* of any process that give rise to life, so they must in some way or other be there at the origins of the process, and Rosenberg (like other naturalists) rules them out from the get-go.

Of course, this presupposes the Aristotelian *principle of proportionate causality*, according to which whatever is in an effect must pre-exist in its total cause in some way or other, whether formally, virtually, or eminently. I discussed this principle in chapter 1 and have defended it at greater length elsewhere (Feser 2014b, pp. 154-59). But it is worth emphasizing that Rosenberg himself is implicitly committed to something like this principle, as are other Darwinian naturalists. This is precisely the reason why they refuse to affirm the existence of irreducible teleology at the level of living things. For they think both that there is no teleology of any kind at the level of inorganic physical and chemical phenomena, and that living things arose out of such phenomena. Hence, they reason, if teleology in no way exists in the one, then it cannot exist in the other. The principle is also presupposed by both sides in the debate over punctuated equilibrium (Sterelny 2007). One side argues that the saltations posited by punctuated equilibrium models cannot have arisen from the known mechanisms of evolution, so they must not have occurred. The other side argues that there is fossil evidence that such saltations *have* occurred, so that there must be more to the mechanisms of evolution than is usually supposed. Both sides are essentially agreeing with the Aristotelian principle that a cause must be proportional to its effect, and simply applying it in different ways.

Now, the principle of proportionate causality together with a robust anti-reductionism might seem to make evolution even *less* likely than it would be given Rosenberg's reductionism, not more likely. For it might appear that for living things to arise from inorganic precursors, or for one species to give rise to another, would be for an effect to have something that was not first in its cause. For this reason it is sometimes claimed that evolution and Aristotelian philosophy of nature cannot be reconciled. (Cf. Chaberek 2017.) However, since the middle of the twentieth century, the general tendency of Aristotelian-Thomistic philosophers has been to argue that they *can* be reconciled, so long as evolution is not construed in mechanistic or reductionist materialist terms. (Cf. Ashley 1972; Bittle 1945, chapter 22; Carroll 2000; Clarke 2001, chapter 15; De Koninck [1936] 2008; Dodds 2012, pp. 199-204; Donceel 1961, chapters 3-4; Grenier 1948, pp. 540-51; Hugon [1927] 2013, pp. 368-76; Klubertanz 1953, pp. 412-27; Koren 1955, chapters 22-23; Maritain 1977; McCormick 1940, pp. 201-13; O'Rourke 2004; Phillips 1950, chapters 17-18; Reith 1956, pp. 261-65; and Royce 1961, pp. 345-50. It is worth noting that not all of the older works cited are even sympathetic with evolution, but nevertheless allow for the possibility in principle of a reconciliation with Aristotelian philosophy of nature.)

This should not be entirely surprising given some of the things that even Aquinas held to be consistent with Aristotle's philosophy of nature. For example, Aquinas thought that new kinds of animals could arise from existing kinds, and even that there could be spontaneous generation of new organisms out of putrefying matter. He writes:

> Since the generation of one thing is the corruption of another, it was not incompatible with the first formation of things, that from the corruption of the less perfect the more perfect should be generated. Hence animals generated from the corruption of inanimate things, or of plants, may have been generated then. But those generated from corruption of animals could not have been produced then otherwise than potentially. (*Summa Theologiae* I.72.1, ad 5)

> Species, also, that are new, if any such appear, existed beforehand in various active powers; so that animals, and perhaps even new species of animals, are produced by putrefaction by the power which the stars and elements received at the beginning. Again, animals of new kinds arise occasionally from the connection of individuals belonging to different species, as the mule is the offspring of an ass and a mare; but even these existed previously in their causes, in the works of the six days. (*Summa Theologiae* I.73.1, ad 3)

Now, the reason Aquinas believed in spontaneous generation was that he thought there was good empirical evidence that it actually occurred. Of course, he was wrong about that. But the point is that he did not try to explain away this apparent empirical evidence on the grounds that spontaneous generation would be impossible given Aristotelian principles. He thought it *was* possible. It is also true that the example of the mule is not nearly as dramatic as the kinds of transformations posited by modern evolutionary biologists. But the point is that the example illustrates that Aquinas did not think it impossible on metaphysical grounds that one kind of animal could give rise to another. Moreover, the spontaneous generation of new organisms out of inanimate matter certainly *is* a transformation as dramatic as any posited by modern biologists. It is also true that Aquinas thought that putrefying matter *alone* was not sufficient to generate new organisms. Putrefaction provided suitable matter for such generation, but he thought that the form had to come from celestial bodies:

> [I]n the case of animals generated from putrefaction, the formative power of is the influence of the heavenly bodies. (*Summa Theologiae* I.71.1, ad 1)

> An effect... [can be] virtually contained in the cause; as the form of the effect is virtually contained in its cause: thus animals produced by putrefaction, and plants, and minerals are like the sun and stars, by whose power they are produced. (*Summa Theologiae* I.105.1, ad 1)

This too presupposes scientific errors, but what matters for present purposes is that Aquinas thought that it is metaphysically possible for a combination of natural causal factors to generate new kinds of organisms.

But the most important lesson to take from the passages quoted is their reminder that the principle of proportionate causality is more subtle than is sometimes supposed by those who think there is a conflict between Aristotelian philosophy of nature and evolution. Recall that the principle says that what is in an effect must pre-exist in its *total* cause in *some* way, whether *formally*, *virtually*, or *eminently*. The most immediate and obvious causal factor in a living thing's generation is not necessarily the only factor or total cause. And what is in the organism can be in its total cause virtually or eminently rather than formally, and thus in a more subtle way than, for example, the way that the main features of a dog are evident in its parents. Thus does Aquinas say that new organisms are "potentially" in putrefying matter; that new species "existed beforehand in various active powers" of the celestial bodies and the elements working in concert; and that the forms of these new organisms are "virtually contained" in their causes.

A second relevant theme from Aquinas is his understanding of human embryonic development. He supposed that the process begins with a vegetative form of life, which gives way to a sensory or animal form of life, which in turn gives way to a rational or human form of life. (Cf. *Summa Contra Gentiles* II.89.11.) Here too he was simply mistaken scientifically. (Cf. Haldane and Lee 2003.) What matters for present purposes, though, is that he did not think such a transition was ruled out on metaphysical grounds. To be sure, he was not addressing the issue of whether one *species* could arise from another, but rather describing the way he supposed an individual living thing can give rise to another. But since the living things in question are of dramatically different types (e.g. vegetative versus sensory) the possibility of the latter kind of transition would lend plausibility to the possibility of the former kind.

A third relevant theme from Aquinas is the potentiality (or "potency") of prime matter to realize successively higher levels of the hierarchy of nature (Donceel 1961, pp. 62-63). Aquinas writes:

> [S]ince a thing is perfect in so far as it is actualized, the intention of everything existing in potency must be to tend through motion toward actuality... Now, among the acts pertaining to forms, certain gradations are found. Thus, prime matter is in potency, first of all, to the form of an element. When it is existing under the form of an element it is in potency to the form of a mixed body; that is why the elements are matter for the mixed body. Considered under the form of a mixed body, it is in potency to a vegetative soul, for this sort of soul is the act of a body. In turn, the vegetative soul is in potency to a sensitive soul, and a sensitive one to an intellectual one. This the process of generation shows: at the start of generation there is the embryo living with plant life, later with animal life, and finally with human life. (*Summa Contra Gentiles* III.22.7)

Now, what Aquinas is describing here is primarily a sequence of ontological levels rather than a temporal sequence. Still, the reality of the one sequence lends plausibility to the possibility of the other, and Aquinas himself appeals to a temporal sequence (the stages he of embryonic development, as he supposed it worked) to illustrate the reality of the ontological sequence.

For twentieth-century Aristotelian-Thomistic philosophers, the nub of the question of whether evolution could be reconciled with an Aristotelian philosophy of nature concerned precisely the hierarchy of life forms alluded to here by Aquinas, and which I discussed earlier. (Aquinas refers to these life forms as types of "souls," but nothing rides on that traditional terminology. A "soul" as Aquinas uses the term is just the substantial form of a living physical substance and has nothing to do with ghosts, ectoplasm, *élan vital*, etc.) Here it is essential to make some terminological clarifications. Our topic is commonly described as the question of whether new species could arise by evolution. But what is a "species" and what is meant by "evolution"? In traditional logic, a species is just a class of things defined in terms of a genus and a differentia. For example, to define a triangle as a closed plane figure with three straight sides is to say that triangles form a species that falls under the genus *closed plane figure* and are differentiated from other species in that genus by having three straight sides. "Evolution" is often used as a synonym for change.

So, if we think of the question of whether new species can arise by evolution in *these* senses of the key terms, then the answer is obviously that they can. For example, a new breed of dog can be defined as a "species" in this sense, and it arises or "evolves" from an older breed. But of course, this kind of "evolution" is trivial and uncontroversial. It is not the sort of thing people have in mind when they debate whether new species could arise via evolution.

Now, modern biology uses the terms "species" and "genus" in narrower senses than the ones just described. In biology, these terms are applied only to living things. Furthermore, a species is not just *any* more specific category and a genus is not just any more general category. Rather, *species* and *genus* are to be distinguished from the higher taxonomic levels *family*, *order*, *class*, *phylum*, *kingdom*, and *domain*. But as we've seen, precisely how to define a species is nevertheless a matter of controversy in modern biology and philosophy of biology, with competing "species concepts" (the biological species concept, the phylogenetic-cladistic species concept, etc.) each having their defenders.

Twentieth-century Aristotelian-Thomistic philosophers clarified the key metaphysical issues by focusing on the four levels of physical reality that Aristotelian philosophy of nature has traditionally taken to mark the sharpest divides in nature: the inorganic realm; purely vegetative forms of life; sensory or animal forms of life; and the rational or human form of life. For it is these four kinds of substance which, as we have seen, the Aristotelian takes to be the most clearly irreducibly different. Following Henry Koren (1955, pp. 300-2), let us refer to these as *philosophical species* (to indicate that the point is to mark a metaphysical distinction between irreducibly distinct kinds of substance, rather than the sort of distinction the modern biologist is making). Within the first three of these philosophical species, there are various subclasses. For example, within the sensory or animal realm, there are reptiles, birds, mammals, etc., as well as further subdivisions such as the distinction within the mammal category between cats, rodents, whales, apes, and the like. Again following Koren, let us call these subclasses *philosophical subspecies*. Within these philosophical subspecies, there are in turn various further subclasses. Koren simply refers to these as "lesser differences," such as the distinction between varieties of cockroach. Needless to say, the *philosophical subspecies* and *lesser differences* categories are not terribly precise. Many further and more clear-cut distinctions could be drawn, as of course they are by the modern biologist. There is also the important question of exactly how the metaphysical distinctions being drawn here map onto the classifications familiar from modern chemistry and modern biology. (See Oderberg 2007, chapters 5, 8, and 9 for detailed discussion of this issue.) But for the specific purposes of the present discussion, these matters are not relevant and can be ignored.

The metaphysical question we are interested in is whether one kind of substance can give rise to an irreducibly different kind of substance. The thesis that this is possible was, in older works, labeled "transformism," and since the term "evolution" is somewhat vague, the older term is preferable. The question of whether new species can arise via evolution is thus better formulated as the question of whether it is possible for there to be transformation of one philosophical subspecies to another, or one philosophical species to another. The thesis that it is possible for the simplest kind of inorganic substance to give rise, by purely physical processes and through a series of intermediate transformations, to the rational or human form of life, is one that – again following Koren (1955, p. 298) – we can call *universal transformism*. *Mitigated transformism*, as Koren calls it, is the thesis that only some transformations between philosophical species or subspecies are naturally possible.

For the mitigated transformist, since some transitions are not naturally possible, they require special divine action. Now, the twentieth-century Aristotelian-Thomistic philosophers I have been citing all took the view that the human intellect is incorporeal. For that reason, they held that even if something like the human body could arise through purely physical evolutionary processes, the intellect could not have. These philosophers were also all theists, and held that the transition from animals that were physiologically like human beings to the truly rational or human form of life would therefore have to involve special divine action. In this way, they were all mitigated transformists and advocated a kind of *theistic evolution*, at least with respect to human origins. Since evaluating the arguments for these claims requires an extended treatment of issues in philosophical anthropology and philosophy of mind, not to mention natural theology, they are beyond the scope of a general work in the philosophy of nature. I will, accordingly, have little further to say here about the question of human origins. (But for a defense of the incorporeality of the human intellect, see Feser 2013a and 2018, and for a defense of theism, see Feser 2017.)

But what about transformations between the inanimate, vegetative, and sensory philosophical species, and between the various philosophical subspecies and lesser differences? Transformations between

Koren's "lesser differences" (for example, from one variety of cockroach to another) are uncontroversial, precisely because they don't really involve a transformation of one irreducibly different kind of substance to another. Koren argues (1955, p. 302) that a transformation between philosophical subspecies is also unproblematic given Aristotelian metaphysical principles, on the grounds that it is comparable to other kinds of substantial change. On this view, transformation *within* a philosophical species – for example, transformation between different kinds of reptile or even between reptiles and birds, all of which fall within the philosophical species *sensory or animal life* – is as unproblematic as the change from water to hydrogen and oxygen, or from uranium to lead. This view is common among the authors I have been citing.

More controversial is the question of transformations between the inanimate, vegetative, and sensory or animal philosophical species. Obviously, if such transformations are possible, then the less radical sort of transformation between philosophical *sub*species will also be possible. Here there are two basic positions the Aristotelian might take. The first would be to hold that even though transformations between philosophical *sub*species are naturally possible, transformations between philosophical *species* are not, and would require special divine action. On this view, purely natural transformations within the inanimate realm can give rise to a wide variety of types of inanimate substance. Diverse lines of causality within the inanimate realm might even naturally converge in such a way as to provide the material cause of a living substance. But for these inanimate precursors to give rise to a truly living substance would require special divine action to introduce the needed substantial form. Once this most simple vegetative form of life exists, then through purely natural means, a wide variety of vegetative forms might evolve. Diverse lines of causality within the vegetative realm might even natural converge in such a way as to provide the material cause of a sensory or animal substance. But once again, special divine action would be required to introduce into the process a distinctively animal sort of substantial form. Once the simplest forms of animal life exist, purely natural evolutionary processes could give rise to a wide variety of animal forms, and diverse lines of causality could naturally converge to provide the material cause of a rational or human form of life. But once again, special divine action would be required to introduce a distinctively human substantial form.

This view might be called *Aristotelian theistic evolutionism*. It is a kind of evolutionism insofar as it affirms at least a mitigated transformism, theistic insofar as it posits special divine action as the partial cause of the most significant evolutionary transitions, and Aristotelian insofar as it interprets the process in terms of a metaphysics of substantial form, teleology, proportionate causality, etc. rather than in terms of a mechanistic or reductive materialist metaphysics.

Alternatively, the Aristotelian could argue that even transformations between philosophical species are naturally possible and therefore would not require special divine action. On this view, even the most complex kinds of sensory or animal life are contained at least *virtually* in the simplest kind of vegetative life – and indeed, contained virtually even in the simplest inanimate substances. (Again, for present purposes I put to one side the question of human origins.) The idea here would be that the nature of the elementary kinds of inanimate matter is such that, when they exercise their causal powers in concert in the right sort of way, the eventual result will be simple kinds of vegetative organic substances; and that the nature of these simplest vegetative substances is such that, when they together with the inorganic substances that make up their environment all exercise *their* causal powers in concert in the right sort of way, the eventual result will be simple kinds of sensory or animal substances. The properties and causal powers of the simplest inorganic substances are on this view naturally sufficient to generate this outcome, just as purely natural processes can produce water out of hydrogen and oxygen and lead out of uranium.

This view might be called *Aristotelian natural evolutionism*. It is a kind of evolutionism insofar as affirms either a universal transformism or a near-universal transformism (if an exception is made in the case of human origins). It posits a natural kind of evolution insofar as it holds that the transitions even between (all or most) philosophical species can occur without special divine action, just by virtue of physical substances exercising the causal powers that follow from their natures. It is Aristotelian insofar as, even if it were to posit an unqualified universal transformism, it would interpret the evolutionary process in terms of the Aristotelian metaphysics of substantial form, teleology, proportionate causality, etc. rather than in terms of a mechanistic or reductive materialist conception of nature.

I describe this view as positing a "natural" rather than "naturalistic" kind of evolution, because the term "naturalism" has come to be associated with reductive materialism. I do not describe this view as atheistic or even non-theistic, because an adherent of Aristotelian natural evolutionism could perfectly well hold that God is the cause of there being a world of inanimate physical substances with causal powers that eventually give rise to a variety of vegetative and animal forms of life. Of course, an adherent of the view might also deny this. The point, though, is that Aristotelian natural evolutionism *as such* does not require either atheism or theism, any more than (say) affirming that water can come from hydrogen and oxygen requires either atheism or theism.

Again, the Aristotelian-Thomistic writers I have cited generally take the view that the transition from sensory or animal life to rational or human life requires special divine action. Many of them (such as Koren) also hold that the transition from inorganic substances to vegetative life also requires special divine action. Thus they tended to be Aristotelian theistic evolutionists. By contrast, Thomas Nagel (2012), who tentatively proposes an atheistic neo-Aristotelian form of teleology as a way to make intelligible the origin within the material world of life, consciousness, and cognition, appears to be flirting with something like an atheistic form of Aristotelian natural evolution.

Now, such a position raises a number of questions. For example, exactly *how* are the properties distinctive of sensory or animal forms of life "virtually" present in vegetative forms of life, and indeed in the inorganic realm? One possible answer would be to hold that even inorganic and vegetative substances possess something analogous to a very rudimentary sort of awareness. The idea would be that it is only with the evolution of sensory or animal substances that awareness comes to be associated with a physiology sufficiently complex to manifest the awareness in behavior, and that the awareness itself becomes more complex the more complex the physiological features with which it comes to be associated. This would amount to a kind of *panpsychism*, and Nagel at least considers it as one possible way to solve the problem.

Of course, an objection that might be raised against this proposal would be that it blurs the distinction between sensory and non-sensory substances to the point of making *everything* a kind of sensory or animal substance. Yet this is not the only possible way to interpret the claim that the properties distinctive of sensory life are virtually present within vegetative and inorganic substances. Indeed, the problem with panpsychism is precisely that it really seems to make a kind of conscious awareness *actually* present in vegetative and inorganic substances, rather than merely virtually present. By contrast, when the Aristotelian holds, for example, that the parts of a substance are only virtually present in the whole, he means precisely that they are *not actually* there, but rather may *potentially* be drawn out of it.

However such problems are dealt with, they will require recourse to concrete empirical scientific considerations no less than to abstract metaphysical considerations. By no means does the Aristotelian suppose that these questions can be settled entirely from the armchair. On the contrary, his point is in part precisely that they cannot be. It is one thing to claim that inorganic natural causes could in principle converge in such a way as to generate an organic substance, or that the causal powers inherent in vegetative forms of life could in principle give rise to sensory or animal forms of life. It is another thing actually to identify specific causal powers in the relevant inorganic and vegetative phenomena that could do the job. By the same token, the principle of proportionate causality is flexible enough that the Aristotelian needs to be cautious about making peremptory declarations about what sorts of change are or are not possible in principle. Aquinas got the scientific details wrong, but insofar he did take the best science of his day seriously, he provides an example for the modern Aristotelian to follow.

Determining *which* of the various possible stories the Aristotelian could tell about the origins and development of life is the correct one is, accordingly, a very large task, and it is not one that I am going to attempt to carry out here. But I don't need to carry it out in order to establish the two points I have been arguing for: that an Aristotelian philosophy of nature does not as such rule out evolution; and that, in any event, evolution itself requires rather than undermines Aristotelian essentialism and teleology.

6.2.4 Problems with some versions of "Intelligent Design" theory

What should the Aristotelian think of the criticisms of evolution raised by "Intelligent Design" (ID) theorists? It depends on the Aristotelian and on the criticism. As I have said, Aristotelian philosophy of nature *per se* does not *require* evolution, as should be obvious enough from the fact that Aristotle himself was not an evolutionist. Equally obviously, then, an Aristotelian *qua* Aristotelian could accept some argument against evolution, whether raised by an ID theorist or by anyone else for that matter. Furthermore, the expression "Intelligent Design" is sometimes used so loosely that just any old claim to the effect that an intelligence of some sort or other is in some way or other involved in the origin of species is counted as a version of "Intelligent Design" theory. What I have called Aristotelian theistic evolution would, in that case, count as a kind of ID theory. So, in these (rather trivial) ways, Aristotelian philosophy of nature is compatible with ID theory.

However, some of what passes under the ID label is *not* consistent with Aristotelianism. One problem is theological, insofar as some ID arguments presuppose a conception of God and of divine action which does not sit well with Aristotelian-Thomistic natural theology. But since this issue is not relevant to the philosophy of nature, and since I have addressed it elsewhere (Feser 2013b), I will say no more about it here.

Another problem is that the conception of nature that some ID theorists are working with is at least implicitly *mechanistic* in the sense described in chapter 1, a conception which I have been criticizing throughout the course of this book. To be sure, ID writers sometimes object to this characterization of their position, as prominent ID theorist William Dembski has (2004, pp. 25 and 151). No doubt some specific ID arguments do not presuppose mechanism. But there is also no doubt that some ID arguments *do* presuppose it, as some of Dembski's own remarks make clear.

For example, take Dembski's discussion of Aristotle in the very book where he objects to the characterization of ID as mechanistic (2004, pp. 132-33). Dembski here identifies "design" with what Aristotle called *techne* or "art." As Dembski correctly says:

> The essential idea behind these terms is that information is conferred on an object from outside the object and that the material constituting the object, apart from that outside information, does not have the power to assume the form it does. For instance, raw pieces of wood do not by themselves have the power to form a ship. (p. 132)

This contrasts with what Aristotle called "nature," which (to quote Dembski quoting Aristotle) "is a principle in the thing itself." For example (and again to quote Dembski's own exposition of Aristotle), "the acorn assumes the shape it does through powers internal to it: the acorn is a seed programmed to produce an oak tree" – in contrast to the way the "ship assumes the shape it does through powers external to it," via a "designing intelligence" which "imposes" this form on it from outside (p. 132).

Having made this distinction, Dembski goes on explicitly to acknowledge that just as "the art of shipbuilding is not in the wood that constitutes the ship" and "the art of making statues is not in the stone out of which statues are made," "so too, *the theory of intelligent design contends that the art of building life is not in the physical stuff that constitutes life but requires a designer*" (p. 133, emphasis added). In other words, according to Dembski, living things are for ID theory to be modeled on ships and statues, the products of *techne* or "art," whose characteristic "information" is not "internal" to them but must be "imposed" from "outside." But that entails that they have only accidental rather than substantial forms, and only extrinsic rather than intrinsic teleology. And as we saw in chapter 1, reconceiving of natural objects (living or otherwise) in these terms is precisely how the mechanistic world picture departed from the Aristotelian philosophy of nature.

Another example is Dembski's assertion (2004, p. 140) that "law-like regularit[ies] of nature" such as "water's propensity to freeze below a certain temperature" are "as readily deemed brute facts of nature as artifacts of design" (unlike the "specified complexity" that he takes to be a genuine mark of design). Now, as we saw in chapter 1, for Aristotelian philosophy of nature, such regularities are paradigm examples of final causality. That *A* is reliably an efficient cause of an effect or range of effects *B* is, according to the Aristotelian, unintelligible unless we suppose that generating *B* is the end toward which *A* is naturally directed. From an Aristotelian point of view, efficient causal regularities are *as such* and whether or not they exhibit complexity ("specified" or otherwise) the opposite of "brute facts." They could seem to be candidates for brute facts only given a conception of nature on which causes are not *intrinsically* aimed or pointed toward their characteristic effects – that is to say, on the mechanistic conception of efficient causality which, as we saw in chapter 1, supplanted the Aristotelian conception.

Indeed, in another place (2002, p. 5), Dembski bemoans early modern science's abandonment of final causes or purposes, but immediately goes on to say: "Now I do not want to give the impression that I am advocating a return to Aristotle's theory of causation. There are problems with Aristotle's theory, and it needed to be replaced." What Dembski objects to is rather that what replaced it was "a view of science that could only end up excluding design" (ibid.). The implication, then, is that Dembski rejects the Aristotelian notion that teleology is *intrinsic* to natural substances and processes in favor of the Platonic-Paleyan conception of teleology as *extrinsic* or imposed on natural substances from outside by a designer. Again, that is precisely a mechanistic conception of causality, even if it is a theistic rather than atheistic version of mechanism.

From an Aristotelian point of view, framing criticisms of reductive materialistic evolutionary explanations in terms of "probabilities" and "complexity," as Dembski does, also gets the discussion off on the wrong foot. For example, the problem with Rosenberg's account of the origins of life is not that it is *improbable* that life could arise the way he says it does. The problem is that it is *impossible in principle* that it could arise in that way, since living things have substantial forms and intrinsic teleology and Rosenberg rules out substantial forms and intrinsic teleology from the start. Nor does the problem have anything to do with complexity. Even the least complex form of life imaginable could not arise even in principle given Rosenberg's reductionist constraints. And even the most complex and improbable thing imaginable, even if it were designed, would not be a *living* thing if it had only an accidental form and extrinsic teleology. To frame the issues in terms of "probabilities" and "complexity" is therefore implicitly to get the basic metaphysical issues wrong before the discussion even gets started, or at least to distract attention from them. In effect, Dembski is trying to play the mechanistic game the way Descartes, Newton, Boyle, and Paley played it, before the atheists and materialists took it over. The Aristotelian, by contrast, refuses to play that game at all.

Defending ID theory from the charge of mechanism, Robert Koons and Logan Paul Gage point out that Darwinian biologists are typically committed to mechanism, "yet, one rarely sees the critics' ire directed toward Darwinism" (2012, p. 80). Of course, this doesn't show that ID theorists aren't committed to a mechanistic picture of nature, but only that their critics need to be more consistent. In any case, as the reader will have noticed, I have certainly been critical of the mechanistic approach of Darwinian naturalists no less than of ID theorists.

Koons and Gage (pp. 80-82) also cite the work of Stephen Meyer (2009) as an example of ID argumentation that does not presuppose mechanism, and they note that Dembski has said critical things about mechanism. But what *Meyer* says is hardly relevant to whether *Dembski's* position is mechanistic. Moreover, though Dembski does sometimes say critical things about mechanism (as I have noted myself), we have seen that he *also* says things that clearly imply a mechanistic position *in the sense of* an anti-Aristotelian position (which is the only sense of "mechanistic" that matters for present purposes). The passages cited by Koons and Gage show only either that Dembski's statements on this subject have not been consistent, or (perhaps) that what Dembski objects to are atheistic and deistic forms of mechanism rather than mechanism *per se*.

Indeed, Dembski has more or less acknowledged that he has said things that seem to imply a mechanistic position. In response to earlier criticisms of mine, he has written:

> [I]ntelligent design is compatible with a nonmechanistic conception of organisms. Nonetheless, in fairness to Feser... [his] criticisms are understandable because intelligent design advocates, myself included, haven't always been as clear as we might in our use of design terminology, not clearly distinguishing external design from intelligence or teleology more generally. (2014, pp. 58-59)

If Dembski now wants to distance ID from any commitment to an anti-Aristotelian conception of nature, that is a welcome development, and he deserves credit for acknowledging that the Aristotelian-Thomistic critics of ID have had cause for complaint.

Unfortunately, the waters are still muddy at best, because in the very same book in which he makes this remark, Dembski also continues to say things that clearly assert or imply a rejection of fundamental elements of the Aristotelian philosophy of nature. For example, he writes:

> [T]he question remains to what degree nature, in its material aspect, is able to account for the various things that happen in nature... [I]f one is forced to answer this question... one will need to know what nature in its material aspect can be expected to do and then determine to what extent nature does things outside that expectation. The beauty of matter is that it is supposed to exhibit an unbreakable normativity. In consequence, deviation from that normativity can be taken as evidence for the influence of teleological principles not reducible to material processes. (2014, pp. 50-51)

It is hard to imagine a more anti-Aristotelian understanding of natural teleology. As I have said, for the Aristotelian, it is precisely the *normal* operation of nature, nature doing what we would *expect* it to do, that is the most obvious mark of teleology. Yet Dembski says that it is "*deviation*" from the normal course of things, nature "do[ing] things *out-*

side that expectation," that is the mark of teleology. Of course, such deviations can be a mark of a *kind* of teleology, namely the kind operative in miraculous interruptions of the natural order. But Dembski is not talking about miraculous or supernatural events here. He is talking about ordinary natural objects and processes, such as living things. His remarks imply a conception of nature on which its normal operations require no teleology at all, but only efficient causes – a hallmark of the mechanical world picture, as we saw in chapter 1.

Dembski also explicitly criticizes Aristotelian hylemorphism, which he characterizes as the view that natural substances are combinations of "matter and information" (2014, p. 92). His objection is that hylemorphism holds that "information" is always embodied in matter, whereas in Dembski's view, "information... can run on information in the absence of matter" (p. 94). He offers the example of his once running the Eudora email program on a Windows XP simulation, which was in turn running on a Windows 7 machine (p. 93). He then argues that the components of this machine can themselves be analyzed in informational terms, so that "it's information all the way down" (p. 94). The idea of matter as a substratum for information "seems entirely dispensable" (p. 95), especially since our "only access to matter [is] informational" anyway (p. 96) and "we don't know what matter is in itself" (p. 94). For all we know, Dembski says, the universe may be a "computer simulation running not on an electronic machine composed of integrated circuits but on a purely mathematical device, such as a Turing machine" (ibid.).

But this argument is a mess. First, while I acknowledged in chapter 5 that modern computational notions to some extent recapitulate Aristotelian notions, the match is by no means perfect. Certainly, given the many connotations of the word "information" (and *especially* given Dembski's promiscuous use of that term), it is simply incorrect to identify the Aristotelian notion of form with the notion of information. Second, to suggest that since what we *know* of matter is (Dembski claims) only its informational properties, it follows that there must *be* nothing more to the material world than information, is to commit a non sequitur. It is to confuse epistemology with metaphysics.

Third, computer programs and "purely mathematical device[s], such as a Turning machine" are, considered by themselves and apart from matter, mere abstractions. Hence to identify the universe with such a thing is to identify it with an abstract object, which it is not. Dembski is just committing the same fallacy which, in chapter 3, we saw is committed by ontic structural realists, and his position faces the same problems theirs does. He is also simply overlooking, without answering, the Aristotelian point that one of the reasons we need to affirm the existence of matter in addition to form is precisely to explain how what would otherwise be purely abstract gets tied down to a particular concrete individual thing, time, and place. Of course, the Aristotelian also allows that a form can exist in an intellect rather than in matter. But if Dembski means to say that the universe, considered as a kind of form or "information" (as he prefers to put it), exists in an intellect – which, given Dembski's theism, would be the divine intellect – then he is essentially committed to a kind of pantheism.

Finally, whatever one thinks of Dembski's argument, the bottom line is that it explicitly rejects the Aristotelian position that physical objects are composites of form and matter – which means that Dembski's position is, after all, flatly incompatible with Aristotelian philosophy of nature.

Then there is Dembski's claim that "the Aristotelian distinction between nature and design" is "pernicious" (2014, p. 53) and "prone to a certain fuzziness" (p. 55). How so? The basic idea of the distinction, the reader will recall from chapter 1, is that products of art or design have only accidental forms and extrinsic teleology, whereas natural objects have substantial forms and intrinsic teleology. For example, the form and function of a watch are imposed on its components from outside by a designer, whereas the form of an acorn and its tendency to grow into an oak follow from its very nature. The problem with this, Dembski says, is that since acorns didn't always exist, something must have caused them, and this could have been a designer. But if that is the case, then intrinsic teleology really collapses into extrinsic teleology after all. On the other hand, a materialist would argue that human designers are purely material things, just as acorns are. And in that case, Dembski says, the extrinsic teleology supposedly associated with human beings collapses into intrinsic teleology of the kind illustrated by the acorn. So, Dembski concludes (pp. 55-56), the distinction between nature and design is less clear than the Aristotelian supposes.

But this is just another non sequitur. First, what Dembski is describing are at best just situations in which only extrinsic teleology exists and intrinsic teleology turns out to be illusory, or where intrinsic teleology exists and extrinsic teleology turns out to be illusory. To think that this shows that there is no clear distinction between extrinsic and intrinsic teleology is like thinking that, since we can imagine situations where only black objects exist and no white ones do and situations where only white objects exist and no black ones do, it follows that there is no clear distinction between black and white.

Second, the situations Dembski describes don't in fact involve what he thinks they do. Even if acorns are created by an intellect, it simply doesn't follow that they don't after all have intrinsic teleology. To think otherwise is like thinking that, since a certain Euclidean triangle drawn on a piece of paper was drawn by a student, its having angles that sum to 180 degrees is not after all a property intrinsic to it qua Euclidean triangle, but is rather an observer-relative feature deriving from the mind of the student. So, the first situation Dembski describes is not really one in which intrinsic teleology collapses into extrinsic teleology. (It is worth adding that Aristotelians like Aquinas hold that the *proximate* ground of a natural object's teleology is its own nature, whereas the *remote* ground is the divine intellect. (Cf. Feser 2013b.) Hence for Aquinas, a substance's having intrinsic teleology does not exclude there being a sense in which its teleology derives from a divine intellect – in which case, again, Dembski's example doesn't show what he thinks it does. Perhaps Dembski would reject Aquinas's position, but if so, he gives no non-question-begging reason for doing so.)

Furthermore, as Dembski realizes, the materialist doesn't really think that all teleology in nature is, after all, intrinsic teleology of the sort the Aristotelian would attribute to the acorn. Rather, the materialist holds that there is *no* real teleology in nature of *any* sort, whether extrinsic or intrinsic, at all. Hence the second, materialist situation that Dembski describes is not really one in which extrinsic teleology collapses into intrinsic teleology.

The bottom line here too, though, is that since Dembski is explicitly critical of the distinction between nature and art, it is clear that his position does not sit well with an Aristotelian philosophy of nature.

Now, Dembski also emphasizes (2014, pp. 51 and 60-1) that the mechanistic conception of nature he is working with is adopted only for the sake of argument, as a premise in a *reductio ad absurdum* against materialism that the ID theorist can dispense with once that argument is completed. But there are two problems with this claim. First, Dembski's arguments against hylemorphism and the distinction between nature and design are *not* presented merely for the sake of argument in the course of criticizing materialism. They are presented precisely as criticisms of *Aristotelianism*. But hylemorphism and the distinction between nature and design are hardly incidental features of Aristotelian philosophy of nature. They are at the core of the Aristotelian critique of mechanism. Hence, whatever other ID theorists might think, and notwithstanding Dembski's remarks about the compatibility of ID theory in general with a non-mechanistic conception of nature, there can be no doubt that *Dembski's* own position is incompatible with Aristotelianism.

Second, Dembski often describes ID theory as far more than merely a *reductio ad absurdum* of materialism. He describes it (2004) as nothing less than "a new kind of science" that entails a "revolution" in how biology is done, and has even co-authored a textbook presenting the main ideas of this purported science (Dembski and Wells 2008). Now, while you can base a *reductio ad absurdum* argument on a premise you take to be false, you can hardly base a *science* on such a premise.

Perhaps Dembski would respond that ID theory qua science extends beyond the particular *reductio* argument in question, and he certainly characterizes ID very broadly. But that brings us to another problem. Dembski tells us (2014, p. 58) that the "textbook definition" of ID is "the study of patterns in nature that are best explained as the product of intelligence." The average reader would naturally suppose, given that definition, that Dembski takes the patterns in question to derive from an intellect. But Dembski says that that would be a mistake:

> [I]ntelligence can be a general term for denoting causes that have teleological effects. Intelligence therefore need not merely refer to conscious personal intelligent agents like us, but can also refer to teleology quite generally. (p. 59)

If this sounds odd, Dembski asks us to consider that "computer algorithms" are said to exhibit artificial intelligence even though they are not "capable of consciousness or of exhibiting personhood" (ibid.). He also cites the example of a view he attributes to atheist astronomer Fred Hoyle, to the effect that the universe can be said to have a kind of intelligence even though it is "not in any straightforward sense conscious, personal, or agentive" (pp. 59-60). Dembski suggests that "it seems reasonable to regard intelligence as including among its meanings teleology in general" (p. 60).

Now, if *this* is all that being an ID theorist requires, then naturally, Aristotelians would count as "ID theorists," since they affirm teleology. But by the same token, even many *atheists* (like Hoyle) and *materialists* (like those who embrace artificial intelligence) would also count as "ID theorists"! Dembski makes the "intelligent design" label so elastic that it ceases to be informative or interesting. It gets worse. Dembski also tells us that "design" can include "pattern, arrangement, or form, and thus can be a synonym for information" (p. 59) and can refer to "any causal process that brings form to a thing, regardless of whether it is teleological or nonteleological" (p. 64). Indeed, "design needs also to be regarded as a generic term for signifying intelligence or teleology," and "design explanations" are "explanations that explain by appealing to intelligence or teleology" (ibid.).

The problem with all of this should be obvious. If "design" can refer to intelligence, then "intelligent design" can mean "intelligent intelligence." Since everyone believes that there is such a thing as intelligent intelligence, everyone therefore counts as an "intelligent design" theorist. If "intelligence" can refer to teleology and "design" can also refer to teleology, then "intelligent design" can also mean "designed design." Since everyone believes that design is designed, everyone, once again, counts as an "intelligent design" theorist. And if "design" explanations appeal to intelligence or teleology, but any pattern, form, arrangement, or information can count as "design," then everyone who affirms that there are patterns, forms, arrangements, or information counts as an "intelligent design" theorist or a teleologist.

Of course, Dembski would not want to draw such ridiculous conclusions. The point, though, is that his use of terms like "intelligence" and "design" is so extremely imprecise that it invites such parody. More to the point, it is this imprecision that gives the illusion that his position is somehow compatible with Aristotelianism. (And Dembski accuses *Aristotelians* of "fuzziness"!) Nor are these the only terms that Dembski uses in so sloppy a way. He uses the term "information" (2004, 2014, and elsewhere) in several different senses, freely sliding from one to another without always making it clear which one is supposed to be doing the work in a given argument. In some places he insists that the "designer" that ID posits could in theory be something within the natural order, such as an extraterrestrial, so that there is no truth to the charge that ID has an essentially theological agenda. But elsewhere he insists that "specified complexity" cannot be given a naturalistic explanation, and even allows that positing a designer who is part of the natural order would only initiate an explanatory regress – which would imply that a genuine explanation *does* require an appeal to the supernatural. His main arguments all have an unmistakably realist thrust, and yet in response to a particular objection he suggests (2004, p. 65) that ID theory is perfectly compatible with a non-realist philosophy of science (though it does not seem to occur to him that his Darwinian opponents could make exactly the same move in response to some of his criticisms of them). And so on.

In short, Dembski seems intent on sidestepping potential objections to ID by making its basic commitments as flexible as possible. So long as certain *words* are preserved (especially "intelligence" and "design") he is happy to allow almost any *meaning* to be attached to them. This is the opposite of the kind of rigor one would hope for in a serious candidate for a "new science." And while it might appear to make ID *verbally* compatible with a wide range of metaphysical commitments, imprecision and incoherence do not entail compatibility in *substance*. In any event, as we have seen, verbal sleight of hand notwithstanding, Dembski's commitment to what *is* in substance an anti-Aristotelian conception of nature stands out as a clear and consistent theme of his work.

The reader sympathetic with ID should take note that these criticisms have nothing to do with evolution. Other than arguing that evolution requires Aristotelianism but that Aristotelianism neither requires nor rules out evolution, I have nothing to say about that subject here. Someone drawn to some other idea or argument associated with "Intelligent Design" could consistently reject evolution and endorse the criticisms I've raised against Dembski. But given the enormous influence of Dembski's ideas within the ID movement, no one should be surprised that Aristotelian-Thomistic philosophers have often been very critical of that movement.

6.3 Against neurobabble

As I have said, the vast tangle of issues and arguments that arise within the philosophy of mind are mostly beyond the scope of a general work on the philosophy of nature. Still, I began the main arguments of this book with a consideration of the thinking, conscious, embodied subject, and it is fitting to end the book by coming full circle and returning to that topic. But while I approached it before from a phenomenological point of view, in this chapter on animate nature I will approach it from a biological point of view – specifically, the point of view of neuroscience.

Neuroscience, no less than chemistry and biology, is often claimed to have vindicated reductionism or even eliminativism. For example, it is sometimes claimed that neuroscience has shown that we are really nothing but our brains, that consciousness plays no role in causing our actions, that introspection is unreliable, that the self is an illusion, and that free will is an illusion. (Tyler Burge (2010) has labeled sensationalistic claims of this sort "neurobabble," and Raymond Tallis (2009, 2011) calls it "neurotrash" born of "neuromania.") But as in the case of those other sciences, the reductionist and eliminativist claims made in the name of neuroscience do not withstand scrutiny. I have already argued in chapter 2 that the reality of the thinking conscious subject cannot coherently be denied, especially not in the name of science. It follows that any argument that appeals to neuroscientific findings in order to cast doubt on the reality of the thinking conscious subject must be mistaken. But the various specific neuroscientific arguments for reductionist and eliminativist conclusions are clearly bad even apart from that consideration.

Let's begin by setting out two general problems facing neuroscientific reductionism and eliminativism, and then come back later to the problems with certain specific claims made about consciousness, free will, etc. The first of these two general problems is a variation on a problem that we saw afflict reductionism in chemistry and biology, viz. that the relevant micro-level phenomena uncovered by science cannot even be identified or understood without constant reference to the commonsense macro-level phenomena they underlie, so that the latter cannot coherently be reduced to, or eliminated in favor of, the former.

Hence, consider any claim to the effect that some mental phenomenon *M* (a certain thought, sensation, or what have you) is correlated with some brain process *B*, and ought to be reduced to *B*, or taken to supervene upon *B*, or eliminated from our ontology altogether and replaced by *B*, or whatever. (Whether *M* and *B* are taken to be *types* of mental and neural phenomena, or rather individual *tokens* of mental and neural types, doesn't matter for present purposes.) For any such argument even to get off the ground, we first have to be able to identify *B*, as opposed to some other neural process (or indeed as opposed to some other kind of physiological process altogether), as the relevant process. But how can we do that?

No description of the anatomy and physiology of the brain, however detailed, can *by itself* ever tell us. In the specific respects relevant to this particular problem, one neural process seems to observation more or less like another. The person whose brain the neuroscientist is studying cannot pick out *B*. Typically he will not even know that there is such a specific process as *B* until someone with expertise in neuroscience tells him. But even the neuroscientist cannot, from the anatomy and physiology of the brain *alone*, pick out *B* as the process that is correlated with *M*. The neuroscientist has to rely on the *introspective reports* of the person whose brain is being studied, or the introspective reports of other people whose brains have been studied. For example, he has to know that when *B* occurs in a certain person's brain, the person reports having *M* (or that when other people's brains have been studied, they would report having *M* when *B* was occurring). Only on the basis of such reports can neuroscience establish a correlation between *M* and *B* and thereby provide evidence of the sort to which reductionists and eliminativists appeal.

Now, to make use of such introspective reports, the neuroscientist has to make certain assumptions. He has to assume that the person whose brain is being studied is, at least in general, providing *accurate* descriptions of what is going on in his mind during the course of the neuroscientist's examination of his brain. Accordingly, the neuroscientist has to assume that the person can correctly grasp and relate what is going on within his mind at a particular moment, that he can remember what was happening at the preceding moments and correctly judge whether there has been any change in his conscious experience, that the person correctly understands the questions the neuroscientist is putting to him and can correctly make the relevant logical inferences, and so on. In short, the neuroscientist has to assume that he is dealing with a single conscious rational subject who persists over time and provides accurate information about the contents of his mind. If the neuroscientist is wrong about these assumptions, the entire evidential base of the correlations between mental phenomena and neural phenomena that he takes himself to have discovered will collapse.

The problem for reductionism and eliminativism should be obvious. If the reductionist or eliminativist claims that introspection is in general unreliable, or that consciousness has no effect on what we do (and thus no effect on what the person is saying to the neuroscientist examining him), or that consciousness is an illusion altogether, or that there is no self that persists from moment to moment, or any similar claim, then he will be implying that the introspective reports the neuroscientist is relying on are all false, so that the alleged evidence of a correlation between *M* and *B* is all worthless. He will, accordingly, be undermining any basis for thinking that there is anything special about *B* that makes it suitable either to reduce *M* to or to replace *M* with. If introspective reports are worthless, then we might as well say that what is going on in some other part of the brain – or indeed, what is going on in one's kneecaps, or earlobes, or fingernails, or anywhere else – is what is really responsible for the phenomena that common sense regards as mental. All such claims will have the same amount of evidential support – namely, none at all. (Cf. Olafson 2001, pp. 72-75.)

Note that it will not do to resort to a less extreme form of reductionism and suggest that while the mental phenomena in question are all *real*, they are really *nothing but* phenomena of the kind that can be described in terms of the anatomy and physiology of the brain, so that we can translate whatever the introspective reports say into the language of anatomy and physiology. Remember, the reason we had to resort to introspective reports in the first place was precisely because what can be expressed in terms of a description of the anatomy and physiology of the brain *is not adequate* to tell us which brain processes are the relevant ones. The commonsense mentalistic language of the introspective reports captures a level of reality that neuroscientific practice itself implicitly presupposes is both real and irreducible to what can be expressed in anatomical and physiological language.

So, that is the first general problem for neuroscientific reductionism and eliminativism. The second is that the picture of human nature that reductionism and eliminativism would put in place of the commonsense mentalistic picture is simply a non-starter. As Frederick Olafson notes (2001, pp. 67-71), these naturalistic approaches typically work with what he calls a "transmission" model of knowledge. That is to say, they model the brain on a device such as a television, radio, or computer which receives input comparable to electronic signals sent from some external source, and then produces a representation analogous to the image on a screen or a sound emitted by a speaker. They then try to identify processes in the brain that might plausibly be said to correspond to such representations – an "inner screen," as Olafson puts it, that is analogous to a television or computer screen insofar as it represents whatever external source lay at the beginning of the transmission.

There are two basic problems with this model. First, no plausible candidates for inner "representations" of the kind posited are forthcoming. For one thing, as I have said, identifying any brain process *B* as a plausible stand-in for a mental process *M* is going to presuppose the reality and irreducibility of *M*, when the whole point is to reduce or replace *M*. For another thing, even if we identify *B*, attributing a precise content to it faces notorious and insuperable indeterminacy problems analogous to those we have seen face naturalistic accounts of biological function and of the notion of "selection for." The physical properties of *B* will not by themselves be sufficient to determine that what *B* represents are (for example) *bugs* rather than *bugs or small moving things*, in a human being no less than in a frog. (Cf. Feser 2011b and 2013a for detailed discussion of this issue.)

Second, even if those problems could be solved, there is the further problem that we have to ask *for whom* these purported neural representations are representations. An image on a television screen is a representation *for* the person viewing it, who takes it to stand for whatever person, thing, or situation at the other end of the transmission is causing it. So if some neural process *B* is a representation comparable to what is on the screen, who is "viewing" *it?* Not the person in whose brain *B* is to be found, since that person will typically not even know about *B*. The temptation is to posit some further neural process that in some sense monitors or scans the first one. But then we are treating this part of the brain as if it were, like the whole person, itself a kind of perceiver. We are committing what M. R. Bennett and P. M. S. Hacker (2003, chapter 3) call the "mereological fallacy" (because it involves attributing to a part what is really true only of the whole) and what John Searle (1992, pp. 212-14) calls the "homunculus fallacy" (because it involves treating a part of the brain as if it were a homunculus or "little man" inside the head).

Part of the problem with doing this is that activities like perceiving are properly attributed to *persons as a whole*, and it is dubious at best to think they can intelligibly be attributed to *parts* of persons such as brain processes, any more they can be attributed to kneecaps or thyroids. Another part of the problem is that the homunculus move just pushes back a stage the problem it was supposed to be solving, and thus doesn't really solve it at all. The problem was to explain, for example, how human perceivers know external physical things. The proposed answer amounts to saying that the way this works is that there is inside a human perceiver a smaller physical thing (a brain process) and a smaller perceiver (a further brain process) that knows that smaller physical thing. That is, of course, no explanation at all, but just a relocation of the problem.

So, the second general difficulty with reductionist and eliminativist arguments that appeal to neuroscience is that they are bound to be explicitly or implicitly committed to a representationalist picture of human knowledge that is simply unworkable. (Recall that I have in earlier chapters raised further objections to this "representationalist" picture.)

Let us turn now to some specific reductionist or eliminativist claims often made in the name of neuroscience. In the phenomenon known as "blindsight," a subject's primary visual cortex has been damaged to the extent that he is no longer capable of having conscious visual experience in at least certain portions of his visual field. (Cf. Weiskrantz 2009.) But he is nevertheless able to identify distant objects in those portions of the field, by color, shape and the like (by pointing to or reaching for the objects, say, or by guessing). Though blind, the subject can "see" the objects in front of him in the sense that information about them is somehow getting to him through his eyes, even though it is not associated with conscious experiences of the sort that typically accompany vision.

What this tells us, Alex Rosenberg concludes, is that "introspection is highly unreliable as a source of knowledge about the way our minds work" (2011, p. 151). Indeed, Rosenberg claims that "science reveals that introspection – thinking about what is going on in consciousness – is completely untrustworthy as a source of information about the mind and how it works" (pp. 147-8). In particular, "the idea that to see things you have to be conscious of them" is "completely wrong" (p. 149). But there are three problems with these claims. First, the "blindsight" evidence cited by Rosenberg does not in fact show that introspection is unreliable *at all*, let alone "highly" or "completely" unreliable. Second, even if it is partially unreliable, it doesn't follow that to see things you needn't be conscious of them. Third, the blindsight cases in fact *presuppose* that introspection is at least partially reliable.

Take the last point first. The blindsight subject tells us that he has no visual experience at all of the objects he is looking at – that he cannot see their colors or shapes. How does he know this? By introspection, of course. The description of the phenomenon as "*blind*sight," and the argument Rosenberg wants to base on this phenomenon, presupposes that the subject is right about that much. If he's wrong about it, then that entails that he *really is* conscious of the colors, shapes, etc. – and such consciousness is, of course, precisely what Rosenberg wants to deny is necessary to vision. Moreover, the argument also presupposes that the subject can tell the difference between being blind and having conscious visual experience – something the subjects in question did have in the past, before suffering the neural damage that gave rise to the blindsight phenomena. Hence, their introspection of that earlier conscious experience must also be at least partially reliable.

So, the subject cannot be *completely* wrong if Rosenberg's argument is even to get off the ground. But isn't he at least partially wrong? Well, wrong about *what*, exactly? Rosenberg says that the example shows that introspection "is highly unreliable as a source of knowledge about the way our minds work," and he asks rhetorically:

> After all, what could have been more introspectively obvious than the notion that you need to have conscious experience of colors to see colors, conscious shape experiences to see shapes, and so on, for all the five senses? (2011, p. 151)

But this is sloppy. Strictly speaking, what we are supposed to know via introspection *by itself* are only our immediate conscious episodes – "I am now thinking about an elephant" or "I am now experiencing a headache" or the like. No one maintains that the claim that "You need to have conscious experience of colors to see colors, etc." is directly knowable via introspection, full stop. The most anyone would maintain is that introspection *together with other premises* might support such a claim. So, even if the claim turned out to be false, that would not show that introspection itself is unreliable. It could be instead that one of the other premises is false, or that the inference from the premises is fallacious.

Now, blindsight subjects also say that it feels like they are guessing, even though their judgments are more accurate than guesses. Doesn't this show that introspection is deceiving them? It does not. For what is it that they are supposed to have gotten wrong in saying that it feels to them like they are guessing? Certainly Rosenberg cannot say "It feels to them like they are guessing but in fact they are conscious of the colors and shapes," since his whole argument depends on their *not* being conscious of the colors and shapes. But then, what *is* it that they are "really" doing *rather than* guessing? Again, what is it exactly that they are wrong about?

Suppose you hit me in the back with a stone and I say that it felt like a baseball. Did introspection mislead me? Of course not. True, the object wasn't a baseball, but what introspection told me was not what the object *was*, but what it *felt like*, and it really did *feel like* a baseball. The judgment that it was in fact a baseball was not derived from introspection alone, but from introspection *together with* certain other premises – premises about what that sort of feeling has been associated with in the past, what objects people tend to throw under circumstances like the current ones, and so forth.

Similarly, when the blindsight subject says that it feels to him like he is guessing, the fact that his answers are better than what one would expect from guesses does not show that introspection is wrong. It still does *feel like* a guess, even if it turns out that it is more than that. It is the *feel* of the experience alone that introspection gives him knowledge of, not the entire reality underlying the feeling. The judgment that it is merely a guess is not derived from introspection alone, but from the introspective feel of the experience *together with* premises about what experiences that feel like this one have involved in the past, assumptions (false, as it turns out) about whether people can process visual information without consciously experiencing it, and so forth. Blindsight cases show only that the inference as a whole is mistaken, not that the introspective component by itself is mistaken.

Rosenberg might respond: "But the blindsight subject doesn't merely say it *felt like* he had guessed. He says he *did* guess. And isn't that mistaken?" But what is the difference, exactly, between *feeling like* one is guessing and *really* guessing? To guess is to propose an answer without thinking that one has sufficient evidence for it. And that is just what the blindsight subject does. True, *we* have reason to think that information is getting through his visual system in such a way that it *causes* him to answer as he does. But *he* has no access to that information, and thus it doesn't serve as *evidence* for what he says. The neuroscientific evidence suggests only that his guesses have a certain *cause*. It does *not* tell us that they weren't really *guesses* after all.

So, Rosenberg hasn't established from blindsight alone that introspection is even sometimes unreliable, let alone that it always is. But the deeper problem with his argument is that, from the fact that *some* of the information typically deriving from conscious visual experience can in *some* cases be received through the visual system without the accompanying experience, it simply does not follow that *all* such information *always does* (or even *can*) be received without conscious experience. Again, the subjects cited by Rosenberg were not always blind; they had seen colors, shapes, and the like in the past and then *became* either permanently or temporarily unable to have conscious visual experiences. There are no grounds for saying that this past experience is *irrelevant* to their ability somehow to process visual information "blindsight"-style – for denying that they can identify colors and shapes now, without visual experience of them, only because they *once did* have visual experience of them. You might as well say that, since many deaf people can read lips, it follows that perception of sounds isn't necessary for speech. Obviously, lip-reading is a non-standard way of figuring out what people are saying, and is parasitic on the normal case in which sound perception is crucial. Similarly, Rosenberg has given us no reason whatsoever to doubt that blindsight is *parasitic* on cases where conscious experience is necessary for color perception.

Here Rosenberg, like others who make sensationalistic claims in the name of neuroscience, is guilty of letting the tail wag the dog – of interpreting normal cases in light of deviant cases, rather than the other way around. Any mature and healthy dog will have four legs, and it would be absurd to suggest that examples like the occasional dog who is missing a leg because of injury or genetic defect cast any doubt on this fact. Similarly, that there are unusual cases in which people with neurological damage exhibit odd behavior casts no doubt on the commonsense understanding of what is going on in normal cases of perception.

As Bennett and Hacker note (2003, pp. 393-96), there are also problems with the way the so-called "blindsight" cases are described in the first place. For one thing, the typical cases involve patients with a scotoma – blindness in a *part* of the visual field, not all of it – who exhibit "blindsight" behavior under special experimental conditions. In ordinary contexts their visual experiences are largely normal. For another thing, how to describe the unusual behavior is by no means obvious, precisely because though in some ways it seems to indicate blindness (the subjects report that they cannot see anything in the relevant part of the visual field), in other ways it seems to indicate the presence of experience (precisely because the subject is able to discriminate phenomena in a way that would typically require visual experience). In short, the import of the cases is not *obvious*; even how one *describes* them *presupposes*, rather than establishes, crucial philosophical assumptions. It is quite ludicrous, then, glibly to proclaim that "neuroscience" has established such-and-such a philosophical conclusion. The philosophical claims are *read into* the neuroscience, not *read off* from it.

Similar errors are made by those who claim that neuroscience has shown free will to be an illusion. In Benjamin Libet's famous experiments (2004, chapter 4), subjects were asked to flex a wrist whenever they felt like doing so, and then to report on when they had become consciously aware of the urge to flex it. Their brains were wired so that the activity in the motor cortex responsible for causing their wrists to flex could be detected. While an average of 200 milliseconds passed between the conscious sense of willing and the flexing of the wrist, the activity in the motor cortex would begin an average of over *500* milliseconds before the flexing. Hence the conscious urge to flex, it is suggested, seems not to be the *cause* of the neural activity which initiates the flexing, but rather to *follow* that neural activity.

Now, Libet himself qualified his conclusions, allowing that though we don't *initiate* movements in the way we think we do, we can at least either *inhibit* or *accede to* them once initiated. But according to Rosenberg, the work done by Libet and others "shows conclusively that the conscious decisions to do things never cause the actions we introspectively think they do" and "defenders of free will have been twisting themselves into knots" trying to show otherwise (2011, p. 152). Similarly, biologist Jerry Coyne (2012) assures us that:

> "Decisions" made like that aren't conscious ones. And if our choices are unconscious, with some determined well before the moment we think we've made them, then we don't have free will in any meaningful sense.

However, as several critics have pointed out (Bennett and Hacker 2003, pp. 228-31; Tallis 2011, pp. 54-56 and 247-50; Mele 2014), this line of argument contains several fallacies. The first problem is that Libet didn't show that the kind of neural activity he measured is *invariably* followed by flexing. Given his experimental setup, only cases where the activity was actually followed by flexing were detected. He didn't check for cases where the neural activity occurred but was not followed by flexing. So we have no evidence that that kind of neural activity is *sufficient* for the flexing. For all Libet showed, it may be that the neural activity in question leads to flexing (or doesn't) depending on whether it is conjoined with a conscious free choice to flex. (Cf. Mele 2014, pp. 12-13.)

A second problem is that the sorts of actions Libet studied are highly idiosyncratic. The experimental setup required subjects to wait passively until they were struck by an urge to flex. But many of our actions don't work like that, especially those we attribute to free choice. Instead, they involve active deliberation, the weighing of considerations for and against different possible courses of action. It's hardly surprising that conscious deliberation has little influence on what we do in an experimental situation in which deliberation has been explicitly excluded. And it's a fallacy to extend conclusions derived from these artificial situations to all human action, including cases which *do* involve active deliberation. (Cf. Mele 2014, pp. 13-16.)

Third, even if the neural activity Libet identified had invariably been followed by a flexing of the wrist, that still wouldn't show that the flexing wasn't a product of free choice. For why should we assume that a choice is not free if it registers in consciousness a few hundred milliseconds after it is made? (Cf. Mele 2014, p. 16-17.) Think of making a cup of coffee. You don't explicitly think, "I will now proceed to move my hand toward the kettle; now I will pick it up; now I will pour hot water through the coffee grounds; now I will put the kettle down; now I will pick up a spoon." You simply do it. You may, after the fact, bring to consciousness the various steps you just carried out; or you may not. We take the action to be free either way. After all, you are not having a muscle spasm, or sleepwalking, or hypnotized, or under duress, or in any other way in circumstances of the sort we would normally regard as incompatible with acting of your own free will. The notion that a free action essentially involves a series of fully conscious episodes of willing, each followed by a discrete bodily movement, is a straw man.

It is also simply wrongheaded to think of voluntary actions as prompted by feelings and urges. As Bennett and Hacker point out (2003, p. 229), feeling an urge to sneeze does not make a sneeze voluntary. Since Libet is willing to allow that we might at least inhibit actions initiated by unconscious neural processes, even if we don't initiate them ourselves, Bennett and Hacker observe that:

> Strikingly, Libet's theory would in effect assimilate all human voluntary action to the status of inhibited sneezes or sneezes which one did not choose to inhibit. For, in his view, all human movements are initiated by the brain before any awareness of a

> desire to move, and all that is left for voluntary control is the inhibiting or permitting of the movement that is already under way. (2003, p. 230)

As Bennett and Hacker go on to emphasize, being moved by an urge – such as an urge to sneeze, or to vomit, or to cough – is in fact the *opposite* of a voluntary action. Once again, Libet's model of voluntary action is simply a straw man, so that his experiments have dubious relevance to the question of free will.

A fourth problem is that Libet and those who draw sensationalistic conclusions from his work fail to consider alternative interpretations of the neural activity in question. Perhaps it correlates, not with the *intention* to flex, but rather with *preparing* to flex without necessarily intending to do so, or with *imagining* or *thinking about* flexing. Or perhaps it correlates with a *general intention* to flex as opposed to a *proximal intention* to do so (Mele 2014, pp. 20-3). Think again of the coffee example. Suppose when you got up in the morning, you decided you wanted to make some coffee. You could be said to have formed a general intention to do so. But suppose also that you don't actually make it until several minutes later, after using the bathroom, getting dressed, and going to the porch to get the newspaper. Only then did you decide it was time to go to the kitchen and actually make the coffee. At that point you formed a proximal intention to make the coffee. Similarly, the participants in Libet's experiments could be said to form both a general intention that they will flex their wrists once they have a certain feeling, and then a proximal intention once the feeling actually arises. Nothing in Libet's experiment tells us that the neural activity he cites correlates with the one kind of intention rather than the other, even if we were to concede (as we should not) that there is any reason to correlate it with an intention in the first place.

As Tallis points out (2011, pp. 248-50), the nature of the intentions involved even in this simple action of flexing the wrist is actually more complex than this last point indicates. There is a sense in which the intention to perform the action could be said to have been formed many minutes before the subject flexed his wrist, when he had the experimental setup explained to him; or hours before, when he left the house to come take part in the experiment; or even days or weeks before, when he first agreed to participate. A long and complex series of psychological and physiological events played a role in what happened when the wrist was actually flexed. So why fixate on one particular bit of neural activity taken in isolation as *the* cause of the action? After all, neural activity and bodily movements do not *by themselves* entail action, free or otherwise. The spasmodic twitch of a muscle involves both neural activity and bodily movement, but it is not an action.

So, the precise significance that a bit of neural activity or a bodily movement has for a given action cannot be read off from the physiological facts alone. It is only within the larger psychological context that we can make sense of it. For it is only *the person as a whole*, and not some sub-personal part of him such as an isolated bit of neural activity, who can properly be said to intend and to act. And so it is only the person as a whole, and not the neural activity, who can be said to be the cause of his actions. In pretending otherwise, Libet and those who appeal to his research in order to cast doubt on free will are *presupposing* reductionism, and thus cannot claim that that research *supports* reductionism without begging the question.

A third sensationalistic claim sometimes alleged to have been established by neuroscience is that in perception, the brain presents us with what is largely an illusion rather than the external world as it really is. Alva Noë (2004, chapter 2; 2009, chapter 6) notes that there are two main lines of argument offered in support of this thesis. First, it is argued that the brain puts together a representation of the world that, in its detail, goes well beyond what could be gleaned from the data that actually makes it to the sensory organs. For example, since we have two eyes, the brain receives information about two retinal images. Moreover, these images are inverted. Yet what we see is only a single world, and we see it upright rather than upside-down. So, the brain must be altering the input it gets from the senses in order to generate the representation of the world we actually experience. Furthermore, the retinal images are unstable given the eyes' constant movements, the resolving power of the eye is limited insofar as there are fewer rods and cones at the periphery of the eye, each eye has a blind spot where there are no photoreceptors, it takes time for light to reach the eye, and so forth. Yet the world as we experience it seems stable, continuous, rich in detail, and immediately present. So, the brain must be filling in the gaps in the information it receives from the senses so as to create the representation that we experience.

Second, there are the phenomena of "change blindness" and "inattentional blindness," in which subjects fail to notice even dramatic things sometimes happening around them. For example, in one experiment, a person who is asked by a stranger for directions is temporarily distracted, and doesn't realize that the stranger he finishes the conversation with is not the same person as the one who initially asked for the directions. Again, the lesson some take from such examples is that the brain constructs a perceptual representation of the world that does not correspond to reality.

As Noë notes, there are in fact two kinds of skeptical conclusion that have been drawn from such considerations. The traditional skeptical lesson is the one already indicated, to the effect that the brain puts together a detailed perceptual representation of the external world that doesn't correspond to reality. But a different skeptical lesson more recently drawn by some writers (Dennett 1992; Blackmore et al. 1995) is that it only *seems* to us like the brain has constructed a detailed perceptual representation, when in fact it has not. On this view, we not only get the external world wrong, we get the internal world of experience itself wrong too.

But as Noë argues, both of these conclusions are mistaken, and certainly don't follow from the neuroscientific evidence. In fact, for all the neuroscientific evidence shows, we don't construct an internal representation of the world, and we don't seem to do so either. That is just bad phenomenology. Rather, what we *seem* to encounter in experience is precisely the *external world itself*, not some representation of it. The external world seems to us to be *directly accessible*, rather than hidden beyond some perceptual representation. What we take to be detailed is, not our perceptual representation of the world, but, again, the world itself. When we fill in the gaps in our experience of the world, we do so precisely by adverting to further experience *of the world itself*, rather than fleshing out some internal representation. We do so by virtue of *actively engaging* in the world rather than being passive spectators of an internal representation. To the extent that experience seems to us to be rich and orderly, that is simply because *the world itself* is presented to us in experience as rich and orderly. And the best explanation of why all of this seems to be the case is that it really is the case. (Recall the phenomenological and cognitive science considerations adduced in chapter 2.)

Again, nothing in the neuroscientific evidence itself shows otherwise. It seems to show otherwise only if we read into it a representationalist account of knowledge. But then it is this representationalist philosophical assumption, and not the scientific evidence itself, that is doing the work. Moreover, the assumption is highly problematic. For one thing, it appears to commit the homunculus fallacy. For example, that there are two retinal images and that they are upside down only seems problematic if we suppose that there is something like a homunculus inside the brain who is viewing the images and somehow has to figure out that there is only one external world and that it is upright. Furthermore, casting doubt on the reliability of introspection and on our knowledge of the external world has (as I have already noted) the paradoxical consequence of undermining the neuroscientific evidence that was claimed to justify these skeptical conclusions. For we need to have knowledge of the external world in order to study the brain, and we need to be able to rely on introspection in order to correlate physiological processes with perceptual states.

This dovetails with the argument of the beginning of this book, to the effect that the very possibility of science presupposes the reality and irreducibility of the conscious, thinking, embodied subject. Hence we cannot coherently eliminate that subject from our conception of the world, *especially* not in the name of science. As I also there argued, we cannot in turn make sense of this subject without deploying the fundamental concepts of Aristotelian philosophy of nature, such as actuality and potentiality, form and matter, and efficient and final causality. If science as read through the lens of philosophical naturalism seems to imply otherwise, the problem is with naturalism and not with Aristotelianism.

Thus does Aristotle have his revenge against those who claim to have overthrown him in the name of modern science. But he is a magnanimous victor, providing as he does the true metaphysical foundations for the very possibility of that science.

Bibliography

bert, David Z. 1992. *Quantum Mechanics and Experience* (Cambridge, MA: Harvard University Press).

bert, D. Z. and Loewer, B. 1988. "Interpreting the Many Worlds Interpretation," *Synthese* 77: 195-213.

en, Keith. 2016. *A Naïve Realist Theory of Colour* (Oxford: Oxford University Press).

scombe, G.E.M. 1981. "Times, Beginnings, and Causes," in G.E.M. Anscombe, *Metaphysics and the Philosophy of Mind* (Minneapolis: University of Minnesota Press).

ew, André. 2002. "Platonic and Aristotelian Roots of Teleological Arguments," in André Ariew, Robert Cummins, and Mark Perlman, eds., *Functions: New Essays in the Philosophy of Psychology and Biology* (Oxford: Oxford University Press).

ew, André. 2007. "Teleology," in David L. Hull and Michael Ruse, eds., *The Cambridge Companion to the Philosophy of Biology* (New York: Cambridge University Press).

ew, André, Robert Cummins, and Mark Perlman, eds. 2002. *Functions: New Essays in the Philosophy of Psychology and Biology* (Oxford: Oxford University Press).

stotle. 1930. *Physics*, trans. R. P. Hardie and R. K. Gaye (Oxford: Clarendon Press).

nstrong, D. M. 1983. *What is a Law of Nature?* (Cambridge: Cambridge University Press).

ley, Benedict. 1972. "Causality and Evolution," *The Thomist* 36: 199-230.

ley, Benedict. 2006. *The Way toward Wisdom* (Notre Dame, IN: University of Notre Dame Press).

Asimov, Isaac. 1993. *Understanding Physics: 3 Volumes in 1* (New York: Barnes and Noble Books).

Austin, Christopher J. 2017. "Aristotelian Essentialism: Essence in an Age of Evolution," *Synthese* 194: 2539-56.

Austin, Christopher J. 2018. *Essence in the Age of Evolution: A New Theory of Natural Kinds* (London: Routledge).

Austin, J. L. 1961. "Ifs and Cans," in J. L. Austin, *Philosophical Papers*, edited by J. O. Urmson and G. Warnock (Oxford: Clarendon Press).

Austin, J. L. 1962. *Sense and Sensibilia* (Oxford: Oxford University Press).

Ayer, Alfred Jules. 1952. *Language, Truth and Logic* (New York: Dover Publications).

Baia, Alex. 2012. "Presentism and the Grounding of Truth," *Philosophical Studies* 159: 341-56.

Baker, Lynne Rudder. 1987. *Saving Belief: A Critique of Physicalism* (Princeton, NJ: Princeton University Press).

Barbour, Julian. 1999. *The End of Time* (Oxford: Oxford University Press).

Bardon, Adrian. 2013. *A Brief History of the Philosophy of Time* (Oxford: Oxford University Press).

Barrett, Jeffrey A. 1999. *The Quantum Mechanics of Minds and Worlds* (Oxford: Oxford University Press).

Becker, Adam. 2018. *What Is Real? The Unfinished Quest for the Meaning of Quantum Physics* (New York: Basic Books).

Bedau, Mark A. 1996. "The Nature of Life," in Margaret A. Boden, ed., *The Philosophy of Artificial Life* (Oxford: Oxford University Press).

Bennett, Jeffrey. 2014. *What Is Relativity?* (New York: Columbia University Press).

Bennett, M.R. and P.M.S. Hacker. 2003. *Philosophical Foundations of Neuroscience* (Oxford: Blackwell).

Bergson, Henri. 1998. *Creative Evolution* (Mineola, NY: Dover).

ernstein, Sara. 2017. "Time Travel and the Movable Present," in J. Keller, ed., *Being, Freedom, and Method: Themes from the Philosophy of Peter van Inwagen* (Oxford: Oxford University Press).

gelow, John. 1996. "Presentism and Properties," *Philosophical Perspectives* 10: 35-52.

gelow, John. 2001. "Time Travel Fiction," in Gerhard Preyer and Frank Siebelt, eds., *Reality and Humean Supervenience: Essays on the Philosophy of David Lewis* (New York: Rowman and Littlefield).

gelow, John. 2013. "The Emergence of a New Family of Theories of Time," in H. Dyke and A. Bardon, eds., *A Companion to the Philosophy of Time* (Oxford: Wiley-Blackwell).

rd, Alexander. 1998. *Philosophy of Science* (London: UCL Press).

tle, Celestine N. 1936. *Reality and the Mind: Epistemology* (Milwaukee: Bruce Publishing Company)

tle, Celestine N. 1941. *From Aether to Cosmos: Cosmology* (Milwaukee: Bruce Publishing Company).

tle, Celestine. 1945. *The Whole Man: Psychology* (Milwaukee: Bruce Publishing Company).

ck, Max. 1962. "Review of G. J. Whitrow's 'The Natural Philosophy of Time'," *Scientific American* CCVI: 181-2.

ck, Max. 2001. "Achilles and the Tortoise," in Wesley C. Salmon, ed., *Zeno's Paradoxes* (Indianapolis: Hackett Publishing Company).

ckburn, Simon. 1990. "Filling in Space," *Analysis* 50: 62-65.

ckburn, Simon. 1991. "Losing your mind: physics, identity, and folk burglar prevention," in John D. Greenwood, ed., *The Future of Folk Psychology* (Cambridge: Cambridge University Press).

ckmore, S. J., G. Brelstaff, K. Nelson, and T. Troscianko. 1995. "Is the Richness of Our Visual World an Illusion? Transsaccadic Memory for Complex Scenes," *Perception* 24: 1075-81.

ck, Ned. 2002. "Searle's Arguments against Cognitive Science," in John Preston and Mark Bishop, eds., *Views into the Chinese Room: New Essays on Searle and Artificial Intelligence* (Oxford: Clarendon Press).

Block, Ned and Philip Kitcher. 2010. "Misunderstanding Darwin," *Boston Review* March/April, pp. 29-32.

Boden, Margaret A., ed. 1996. *The Philosophy of Artificial Life* (Oxford: Oxford University Press).

Boghossian, Paul A. 1990. "The Status of Content," *Philosophical Review* 99: 157–84.

Boghossian, Paul A. 1991. "The Status of Content Revisited," *Pacific Philosophical Quarterly* 71: 264–78.

Boorse, C. 1976. "Wright on Functions," *Philosophical Review* 85: 70-86.

Boulter, Stephen. 2013. *Metaphysics from a Biological Point of View* (Basingstoke: Palgrave Macmillan).

Bourne, Craig. 2006. *A Future for Presentism* (Oxford: Clarendon Press).

Boyd, Richard. 1989. "What Realism Implies and What it Does Not," *Dialectica* 43: 5–29.

Broad, C. D. 1923. *Scientific Thought* (London: Routledge and Kegan Paul).

Brown, James Robert. 1991. *The Laboratory of the Mind: Thought Experiments in the Natural Sciences* (London: Routledge).

Buechner, Jeff. 2008. *Gödel, Putnam, and Functionalism: A New Reading of* Representation and Reality (Cambridge, MA: The MIT Press).

Buechner, Jeff. 2011. "Not Even Computing Machines Can Follow Rules: Kripke's Critique of Functionalism," in Alan Berger, ed., *Saul Kripke* (Cambridge: Cambridge University Press).

Builder, G. 1958. "The Constancy of the Velocity of Light," *Australian Journal of Physics* 11: 457-80.

Buller, David J., ed. 1999. *Function, Selection, and Design* (Albany: State University of New York Press).

Burge, Tyler. 2010. "A Real Science of the Mind," *The New York Times*, December 19, at: https://opinionator.blogs.nytimes.com/2010/12/19/a-real-science-of-mind/ [last accessed 12.10.18]

Burtt, E. A. 1980. *The Metaphysical Foundations of Modern Physical Science* (Atlantic Highlands, N.J.: Humanities Press).

neron, Ross P. 2015. *The Moving Spotlight: An Essay on Time and Ontology* (Oxford: Oxford University Press).

npbell, John. 1994. *Past, Space, and Self* (Cambridge, MA: The MIT Press).

ıales, Jimena. 2015. *The Physicist and the Philosopher: Einstein, Bergson, and the Debate that Changed Our Understanding of Time* (Princeton: Princeton University Press).

ek, Milic. 1961. *The Philosophical Impact of Contemporary Physics* (New York: Van Nostrand Rheinhold Co.).

nap, Rudolf. 1947. *Meaning and Necessity*, Second edition (Chicago: University of Chicago Press).

nap, Rudolf. 1959a. "Psychology in Physical Language," in A. J. Ayer, ed., *Logical Positivism* (New York: The Free Press).

nap, Rudolf. 1959b. "The Elimination of Metaphysics through Logical Analysis of Language," in A. J. Ayer, ed., *Logical Positivism* (New York: The Free Press).

nap, Rudolf. 1967. *The Logical Structure of the World and Pseudoproblems in Philosophy* (Berkeley: University of California Press).

oll, Lewis. 1895. "What the Tortoise Said to Achilles," *Mind* 4: 278-280.

oll, William E. 2000. "Creation, Evolution, and Thomas Aquinas," *Revue des Questions Scientifiques* 171: 319-47.

wright, Nancy. 1983. *How the Laws of Physics Lie* (Oxford: Clarendon Press).

wright, Nancy. 1999. *The Dappled World: A Study of the Boundaries of Science* (Cambridge: Cambridge University Press).

wright, Nancy. 2005. "No God; No Laws." In S. Moriggi and E. Sindoni, eds., *Dio, la Natura e la Legge: God and the Laws of Nature* (Milan: Angelicum-Mondo X).

wright, Nancy. 2008. "Reply to Carl Hoefer," in Stephan Hartmann, Carl Hoefer, and Luc Bovens, eds., *Nancy Cartwright's Philosophy of Science* (London: Routledge).

Cartwright, Nancy. 2016. "The Dethronement of Laws in Science," in Nancy Cartwright and Keith Ward, eds., *Rethinking Order: After the Laws of Nature* (London: Bloomsbury).

Cassirer, Ernst. 1956. *Determinism and Indeterminism in Modern Physics* (New Haven: Yale University Press).

Chaberek, Fr. Michael. 2017. *Aquinas and Evolution* (The Chartwell Press).

Chakravartty, Anjan. 2007. *A Metaphysics for Scientific Realism* (Cambridge: Cambridge University Press).

Chakravartty, Anjan. 2013. "On the Prospects of Naturalized Metaphysics," in Don Ross, James Ladyman, and Harold Kincaid, eds., *Scientific Metaphysics* (Oxford: Oxford University Press).

Chalmers, David. 1996. *The Conscious Mind* (Oxford: Oxford University Press).

Chang, Hasok. 2012. *Is Water H_2O? Evidence, Realism, and Pluralism* (Dordrecht: Springer).

Chomsky, Noam. 1988. *Language and Problems of Knowledge* (Cambridge, MA: The MIT Press).

Churchland, Patricia Smith. 1995. "Take It Apart and See How It Runs," in Peter Baumgartner and Sabine Payr, eds., *Speaking Minds: Interviews with Twenty Eminent Cognitive Scientists* (Princeton: Princeton University Press).

Churchland, Patricia S. and Terrence J. Sejnowski. 1992. *The Computational Brain* (Cambridge, MA: The MIT Press).

Churchland, Paul M. 1981. "Eliminative Materialism and the Propositional Attitudes," *Journal of Philosophy* 78: 67–90.

Churchland, Paul M. 1998. "Activation Vectors vs. Propositional Attitudes: How the *Brain* Represents Reality," in Paul M. Churchland and Patricia S. Churchland, *On the Contrary: Critical Essays, 1987 - 1997* (Cambridge, MA: The MIT Press).

Clark, Andy. 1997. *Being There: Putting Brain, Body, and World Together Again* (Cambridge, MA: The MIT Press).

Clarke, W. Norris. 2001. *The One and the Many: A Contemporary Thomistic Metaphysics* (Notre Dame, IN: University of Notre Dame Press).

ıen, Yehiel. 2016. "Why Presentism Cannot Be Refuted by Special Relativity," in Yuval Dolev and Michael Roubach, eds., *Cosmological and Psychological Time* (Dordrecht: Springer).

ılter, Jeff and Wes Sharrock. 2002. "The Hinterland of the Chinese Room," in John Preston and Mark Bishop, eds., *Views into the Chinese Room: New Essays on Searle and Artificial Intelligence* (Oxford: Clarendon Press).

ne, Jerry A. 2012. "Why You Don't Really Have Free Will," *USA Today*, January 1.

ig, William Lane. 1996. "Tense and the New B-Theory of Language," *Philosophy* 71: 5–26.

ıg, William Lane. 2000a. *The Tensed Theory of Time: A Critical Examination* (Dordrecht: Kluwer Academic).

g, William Lane. 2000b. *The Tenseless Theory of Time: A Critical Examination* (Dordrecht: Kluwer Academic).

g, William Lane. 2001a. *Time and Eternity* (Wheaton, IL: Crossway Books).

g, William Lane. 2001b. *Time and the Metaphysics of Relativity* (Dordrecht: Kluwer Academic Publishers).

g, William Lane. 2008. "The metaphysics of special relativity: three views," in W. L. Craig and Q. Smith, eds., *Einstein, Relativity, and Absolute Simultaneity* (London: Routledge).

g, William Lane and Quentin Smith, eds. 2008. *Einstein, Relativity, and Absolute Simultaneity* (London: Routledge).

ιe, Tim. 2016. *The Mechanical Mind: A Philosophical Introduction to Minds, Machines and Mental Representation*, Third edition (London: Routledge).

mins, Robert. 1999. "Functional Analysis," in David J. Buller, ed., *Function, Selection, and Design* (Albany: State University of New York Press).

is, Benjamin L. and Jon Robson. 2016. *A Critical Introduction to the Metaphysics of Time* (London: Bloomsbury).

ton, Barry. 2010. *Time and Space*, Second edition (Montreal and Kingston: McGill-Queen's University Press).

ɔn, Lorraine and Peter Galison. 2007. *Objectivity* (New York: Zone Books).

Davidson, Donald. 1986. "A Coherence Theory of Truth and Knowledge," in Ernest Lepore, ed., *Truth and Interpretation: Perspectives on the Philosophy of Donald Davidson* (Oxford: Basil Blackwell).

Davidson, Donald. 2001. "Rational Animals," in *Subjective, Intersubjective, Objective* (Oxford: Clarendon Press).

Davies, Paul. 1992. *The Mind of God* (New York: Simon and Schuster).

Davies, Paul. 2010. "Universe from bit," in Paul Davies and Niels Henrik Gregersen, eds., *Information and the Nature of Reality: From Physics to Metaphysics* (Cambridge: Cambridge University Press).

Davies, Paul and Niels Henrik Gregersen, eds. 2010. *Information and the Nature of Reality: From Physics to Metaphysics* (Cambridge: Cambridge University Press).

Dawkins, Richard. 1987. *The Blind Watchmaker* (New York: W.W. Norton and Company).

Dawkins, Richard. 1989. *The Selfish Gene*, New edition (Oxford: Oxford University Press).

Deacon, Terrence W. 2010. "What is missing from theories of information?" in Paul Davies and Niels Henrik Gregersen, eds., *Information and the Nature of Reality: From Physics to Metaphysics* (Cambridge: Cambridge University Press).

Dear, Peter. 2006. *The Intelligibility of Nature: How Science Makes Sense of the World* (Chicago: University of Chicago Press).

Decaen, Christopher A. 2001. "The Viability of Aristotelian-Thomistic Color Realism," *The Thomist* 65: 179-222.

De Koninck, Charles. 1964. *The Hollow Universe* (Québec: Les Presses de l'Université Laval).

De Koninck, Charles. 2008. "The Cosmos," in *The Writings of Charles De Koninck, Volume One*, edited and translated by Ralph McInerny (Notre Dame, IN: University of Notre Dame Press).

Della Rocca, Michael. 2010. "PSR," *Philosophers' Imprint* 10: 1-13.

mbski, William A. 2002. *No Free Lunch: Why Specified Complexity Cannot Be Purchased without Intelligence* (Lanham, MD: Rowman and Littlefield Publishers, Inc.).

mbski, William A. 2004. *The Design Revolution* (Downers Grove, IL: InterVarsity Press).

mbski, William A. 2014. *Being as Communion: A Metaphysics of Information* (Farnham: Ashgate).

mbski, William A. and Jonathan Wells. 2008. *The Design of Life* (Dallas: Foundation for Thought and Ethics).

mopoulos, W. and M. Friedman. 1989. "Critical Notice: Bertrand Russell's *The Analysis of Matter*: Its Historical Context and Contemporary Interest," in C. W. Savage and C. A. Anderson, eds., *Minnesota Studies in the Philosophy of Science, Volume XII: Rereading Russell: Essays on Bertrand Russell's Metaphysics and Epistemology* (Minneapolis: University of Minnesota Press).

nnett, Daniel C. 1981. "Intentional Systems," in *Brainstorms: Philosophical Essays on Mind and Psychology* (Cambridge, MA: The MIT Press).

nnett, Daniel C. 1992. *Consciousness Explained* (Boston: Little, Brown).

nnett, Daniel C. 1995. *Darwin's Dangerous Idea* (New York: Simon and Schuster).

nnett, Daniel C. 1998. "Real Patterns," in *Brainchildren: Essays on Designing Minds* (Cambridge, MA: The MIT Press).

pew, David J. 2015. "Accident, Adaptation, and Teleology in Aristotle and Darwinism," in Phillip R. Sloan, Gerald McKenny, and Kathleen Eggleson, eds., *Darwin in the Twenty-First Century* (Notre Dame, IN: University of Notre Dame Press).

s Chene, Dennis. 1996. *Physiologia: Natural Philosophy in Late Aristotelian and Cartesian Thought* (Ithaca: Cornell University Press).

spagnat, Bernard. 2006. *On Physics and Philosophy* (Princeton: Princeton University Press).

utsch, David and Michael Lockwood. 1994. "The Quantum Physics of Time Travel," *Scientific American* 270: 50-56.

vitt, Michael. 2008. "Resurrecting Biological Essentialism," *Philosophy of Science* 75: 344-82.

DeWitt, Richard. 2004. *Worldviews: An Introduction to the History and Philosophy of Science* (Oxford: Blackwell).

Dicker, Georges. 1998. *Hume's Epistemology and Metaphysics* (London: Routledge).

Dodds, Michael J. 2012. *Unlocking Divine Action: Contemporary Science and Thomas Aquinas* (Washington, D.C.: The Catholic University of America Press).

Donceel, J. F. 1961. *Philosophical Psychology*, Second edition (New York: Sheed and Ward).

Dowe, Phil. 2000. *Physical Causation* (Cambridge: Cambridge University Press).

Drake, Stillman, ed. 1957. *Discoveries and Opinions of Galileo* (New York: Anchor Books).

Dreyfus, Hubert L. 1992. *What Computers Still Can't Do: A Critique of Artificial Reason* (Cambridge, MA: The MIT Press).

Dreyfus, Hubert L. 1993. "Heidegger's Critique of the Husserl/Searle Account of Intentionality," *Social Research* 60: 17-38.

Dreyfus, Hubert and Charles Taylor. 2015. *Retrieving Realism* (Cambridge, MA: Harvard University Press).

Duhem, Pierre. 1991. *The Aim and Structure of Physical Theory* (Princeton, NJ: Princeton University Press).

Dummett, Michael. 1960. "A Defense of McTaggart's Proof of the Unreality of Time," *Philosophical Review* 69: 497-504.

Dumsday, Travis. 2012. "A New Argument for Intrinsic Biological Essentialism," *The Philosophical Quarterly* 62: 486-504.

Dupré, John. 1993. *The Disorder of Things: Metaphysical Foundations of the Disunity of Science* (Cambridge, MA: Harvard University Press).

Dupré, John. 2012. *Processes of Life: Essays in the Philosophy of Biology* (Oxford: Oxford University Press).

Eames, Elizabeth R. 1989. "Cause in the Later Russell," in C. Wade Savage and C. Anthony Anderson, eds., *Rereading Russell: Essays on Bertrand Russell's Metaphysics and Epistemology* (Minneapolis: University of Minnesota Press).

arman, J. and M. Friedman. 1973. "The Meaning and Status of Newton's Law of Inertia and the Nature of Gravitational Forces," *Philosophy of Science* 40: 329-59.

ddington, Arthur. 1929. *Science and the Unseen World* (New York: Macmillan).

ddington, Arthur. 1958. *The Nature of the Physical World* (Ann Arbor: University of Michigan Press).

nstein, Albert. 1970. "Remarks Concerning the Essays Brought Together in this Co-operative Volume," in Paul Arthur Schilpp, ed., *Albert Einstein: Philosopher-Scientist, Volume Two*, Third edition (La Salle, IL: Open Court).

nstein, Albert. 1988. *The Meaning of Relativity*, Fifth edition (Princeton, NJ: Princeton University Press).

der, Crawford L. 2004. *Real Natures and Familiar Objects* (Cambridge, MA: The MIT Press).

der, Crawford L. 2008. "Biological Species are Natural Kinds," *Southern Journal of Philosophy* 46: 339-62.

der, Crawford L. 2011. *Familiar Objects and their Shadows* (Cambridge: Cambridge University Press).

is, Brian. 1965. "The Origin and Nature of Newton's Laws of Motion." In Robert G. Colodny, ed., *Beyond the Edge of Certainty: Essays in Contemporary Science and Philosophy* (Englewood Cliffs, NJ: Prentice-Hall).

is, Brian. 2002. *The Philosophy of Nature: A Guide to the New Essentialism* (Chesham: Acumen).

s, Brian. 2009. *The Metaphysics of Scientific Realism* (Montreal and Kingston: McGill-Queen's University Press).

s, George. 2014. "Time really exists! The evolving block universe," *Euresis Journal* 7: 11-26.

;, Berent. 2002. "Indeterminacy of Functional Attributions," in André Ariew, Robert Cummins, and Mark Perlman, eds., *Functions: New Essays in the Philosophy of Psychology and Biology* (Oxford: Oxford University Press).

dicott, Ronald P. 1996. "Searle, Syntax, and Observer Relativity," *Canadian Journal of Philosophy* 26:101-22.

Ereshefsky, Marc, ed. 1992. *The Units of Evolution: Essays on the Nature of Species* (Cambridge, MA: The MIT Press).

Ewing, A. C. 1937. "Meaninglessness." *Mind* 46: 347-64.

Feigl, Herbert. 1981. "Physicalism, Unity of Science and the Foundations of Psychology," in Herbert Feigl, *Inquiries and Provocations: Selected Writings 1929-1974*, edited by Robert S. Cohen (Dordrecht: D. Reidel Publishing Company).

Feser, Edward. 2006. *Philosophy of Mind* (Oxford: Oneworld Publications).

Feser, Edward. 2007. *Locke* (Oxford: Oneworld Publications).

Feser, Edward. 2008. *The Last Superstition: A Refutation of the New Atheism* (South Bend, IN: St. Augustine's Press).

Feser, Edward. 2009. *Aquinas* (Oxford: Oneworld Publications).

Feser, Edward. 2010. "Teleology: A Shopper's Guide" *Philosophia Christi* 12: 142-59. Reprinted in Feser 2015.

Feser, Edward. 2011a. "Existential Inertia and the Five Ways," *American Catholic Philosophical Quarterly*, Vol. 85, No. 2.

Feser, Edward. 2011b. "Hayek, Popper, and the Causal Theory of the Mind," in Leslie Marsh, ed., *Hayek in Mind: Hayek's Philosophical Psychology*, special issue of *Advances in Austrian Economics*, Vol. 15. Reprinted in Feser 2015.

Feser, Edward. 2013a. "Kripke, Ross, and the Immaterial Aspects of Thought" *American Catholic Philosophical Quarterly* 87: 1-32. Reprinted in Feser 2015.

Feser, Edward. 2013b. "Between Aristotle and William Paley: Aquinas's Fifth Way" *Nova et Vetera* (English edition) 11: 707-49. Reprinted in Feser 2015.

Feser, Edward, ed. 2013c. *Aristotle on Method and Metaphysics* (Basingstoke: Palgrave Macmillan).

Feser, Edward. 2014a. "Being, the Good, and the Guise of the Good," in Daniel D. Novotny and Lukas Novak, eds., *Neo-Aristotelian Perspectives in Metaphysics* (London: Routledge). Reprinted in Feser 2015.

Feser, Edward. 2014b. *Scholastic Metaphysics: A Contemporary Introduction* (Heusenstamm: Editiones Scholasticae/Transaction Publishers).

Feser, Edward. 2015a. *Neo-Scholastic Essays* (South Bend, IN: St. Augustine's Press).

ser, Edward. 2015b. "Why Searle Is a Property Dualist," in *Neo-Scholastic Essays* (South Bend, IN: St. Augustine's Press).

ser, Edward. 2017. *Five Proofs of the Existence of God* (San Francisco: Ignatius Press).

ser, Edward. 2018. "Aquinas on the Human Soul," in Jonathan Loose, Angus Menuge, and J. P. Moreland, eds., *The Blackwell Companion to Substance Dualism* (Oxford: Wiley-Blackwell).

yerabend, Paul. 1993. *Against Method*, Third edition (New York: Verso).

yerabend, Paul. 1999. *Conquest of Abundance* (Chicago: University of Chicago Press).

ynman, Richard. 1994. *The Character of Physical Law* (New York: Modern Library).

h, William. 2010. *Philosophy of Perception: A Contemporary Introduction* (London: Routledge).

ridi, Luciano ed. 2008. *Philosophy of Computing and Information: 5 Questions* (Automatic Press).

lor, Jerry A. 1987. *Psychosemantics: The Problem of Meaning in the Philosophy of Mind* (Cambridge, MA: The MIT Press).

lor, Jerry A. 1990. *A Theory of Content and Other Essays* (Cambridge, MA: The MIT Press).

lor, Jerry. 1992. "The big idea: Can there be a science of the mind?," *Times Literary Supplement*, July 3, pp. 5-7.

lor, Jerry. 1998. *In Critical Condition: Polemical Essays on Cognitive Science and the Philosophy of Mind* (Cambridge, MA: The MIT Press).

or, Jerry. 2000. *The Mind Doesn't Work That Way* (Cambridge, MA: The MIT Press).

or, Jerry. 2007. "Why Pigs Don't Have Wings," *London Review of Books*, October 18, pp. 19-22.

or, Jerry and Massimo Piattelli-Palmarini. 2011. *What Darwin Got Wrong*, Updated edition (New York: Picador).

Ford, Kenneth M., Clark Glymour, and Patrick J. Hayes, eds. 2006. *Thinking about Android Epistemology* (Cambridge, MA: The MIT Press).

Forrest, Peter. 2004. "The Real but Dead Past: A Reply to Braddon-Mitchell," *Analysis* 64: 358-62.

Foster, John. 1982. *The Case for Idealism* (London: Routledge and Kegan Paul).

French, Steven. 2014. *The Structure of the World: Metaphysics and Representation* (Oxford: Oxford University Press).

Gale, Richard M. 1968. *The Language of Time* (London: Routledge and Kegan Paul).

Gale, Richard M. 1991. *On the Nature and Existence of God* (Cambridge: Cambridge University Press).

Garber, Daniel. 1992. *Descartes' Metaphysical Physics* (Chicago: University of Chicago Press).

Gardeil, H. D. 1956. *Introduction to the Philosophy of St. Thomas Aquinas, Volume III: Psychology* (St. Louis: B. Herder Book Co.).

Gardeil, H. D. 1958. *Introduction to the Philosophy of St. Thomas Aquinas, Volume II: Cosmology* (St. Louis: B. Herder Book Co.).

Garrigou-Lagrange, Reginald. 1939. *God: His Existence and His Nature*, Volume I (St. Louis: B. Herder Book Co.).

Gascoigne, Neil and Tim Thornton. 2013. *Tacit Knowledge* (Durham: Acumen).

Gibson, James J. 1979. *The Ecological Approach to Visual Perception* (Boston: Houghton Mifflin).

Gilson, Etienne. 1984. *From Aristotle to Darwin and Back Again* (London: Sheed and Ward).

Gleick, James. 2016. *Time Travel: A History* (New York: Pantheon Books).

Gödel, Kurt. 1970. "A Remark About the Relationship Between Relativity Theory and Idealistic Philosophy," in Paul Arthur Schilpp, ed., *Albert Einstein: Philosopher-Scientist, Volume Two*, Third edition (La Salle, IL: Open Court).

Godfrey-Smith, Peter. 2003. *Theory and Reality: An Introduction to the Philosophy of Science* (Chicago: University of Chicago Press).

ɔdfrey-Smith, Peter. 2007. "Information in Biology," in David L. Hull and Michael Ruse, eds., *The Cambridge Companion to the Philosophy of Biology* (Cambridge: Cambridge University Press).

ɔdfrey-Smith, Peter. 2010. "It Got Eaten," *London Review of Books* Vol. 32, No. 13, pp. 29-30.

dfrey-Smith, Peter. 2014. *Philosophy of Biology* (Princeton: Princeton University Press).

odman, Nelson. 1983. *Fact, Fiction, and Forecast*, Fourth edition (Cambridge, MA: Harvard University Press).

uld, S. J. and R. C. Lewontin. 1979. "The Spandrels of San Marco and the Panglossian Paradigm: A Critique of the Adaptationist Programme," *Proceedings of the Royal Society of London. Series B, Biological Sciences* 205: 581-98.

eene, Merrill F. 1952. "Hylemorphism: Dead or Alive," in Joseph T. Clark, ed., *Hylemorphism and Contemporary Physics* (Woodstock, MD: Woodstock College Press).

ene, Marjorie. 1974. "Biology and Teleology," in *The Understanding of Nature: Essays in the Philosophy of Biology* (Dordrecht: D. Reidel).

enier, Henri. 1948. *Thomistic Philosophy, Volume I* (Charlottetown, Canada: St. Dunstan's University).

bbin, John. 1984. *In Search of Schrödinger's Cat* (New York: Bantam Books).

ce, H.P. 1961. "The Causal Theory of Perception," *Proceedings of the Aristotelian Society* (Supplementary Volume) 35: 121-52.

ff, Ruth and John Greco, eds. 2013. *Powers and Capacities in Philosophy: The New Aristotelianism* (London: Routledge).

ve, Stanley F. 2008. *Quantum Theory and Aquinas's Doctrine on Matter*. Ph.D. dissertation, Catholic University of America.

nbaum, Adolf. 2001. "Modern Science and Refutation of the Paradoxes of Zeno," in Wesley C. Salmon, ed., *Zeno's Paradoxes* (Indianapolis: Hackett Publishing Company).

dane, John and Patrick Lee. 2003. "Aquinas on Human Ensoulment, Abortion and the Value of Life," *Philosophy* 78: 255-78.

Hanson, Norwood Russell. 1958. *Patterns of Discovery* (Cambridge: Cambridge University Press).

Hanson, Norwood Russell. 1963. "The Law of Inertia: A Philosophers' Touchstone," *Philosophy of Science* 30: 107-21.

Hanson, Norwood Russell. 1965a. "Newton's First Law: A Philosopher's Door into Natural Philosophy." In Robert G. Colodny, ed., *Beyond the Edge of Certainty: Essays in Contemporary Science and Philosophy* (Englewood Cliffs, NJ: Prentice-Hall).

Hanson, Norwood Russell. 1965b. "A Response to Ellis's Conception of Newton's First Law." In Robert G. Colodny, ed., *Beyond the Edge of Certainty: Essays in Contemporary Science and Philosophy* (Englewood Cliffs, NJ: Prentice-Hall).

Harrington, James. 2015. *Time: A Philosophical Introduction* (London: Bloomsbury).

Hasker, William. 1999. *The Emergent Self* (Ithaca: Cornell University Press).

Hattab, Helen. 2009. *Descartes on Forms and Mechanisms* (Cambridge: Cambridge University Press).

Haugeland, John. 2002. "Syntax, Semantics, Physics," in John Preston and Mark Bishop, eds., *Views into the Chinese Room: New Essays on Searle and Artificial Intelligence* (Oxford: Clarendon Press).

Hawking, Stephen W. 1988. *A Brief History of Time* (New York: Bantam Books).

Hayek, F. A. 1952. *The Sensory Order: An Inquiry into the Foundations of Theoretical Psychology* (Chicago: University of Chicago Press).

Healey, Richard. 2002. "Can Physics Coherently Deny the Reality of Time?" in Craig Callender, ed., *Time, Reality, and Experience* (Cambridge: Cambridge University Press).

Heathwood, Chris. 2005. "The Real Price of the Dead Past: A Reply to Forrest and to Braddon-Mitchell," *Analysis* 65: 249-51.

Heidegger, Martin. 1962. *Being and Time*, translated by John Macquarrie and Edward Robinson (New York: Harper and Row).

Heil, John. 2003. *From an Ontological Point of View* (Oxford: Clarendon Press).

Heisenberg, Werner. 2007. *Physics and Philosophy* (New York: HarperCollins).

npel, Carl. 1962. "Explanation in Science and in History," in R. G. Colodny, ed., *Frontiers of Science and Philosophy* (Pittsburgh, PA: University of Pittsburgh Press).

ıdry, Robin Findlay. 2006. "Is There Downwards Causation in Chemistry?," in Davis Baird, Eric Scerri, and Lee McIntyre, eds., *Philosophy of Chemistry: Synthesis of a New Discipline* (Dordrecht: Springer).

ıdry, Robin Findlay. 2010a. "Ontological Reduction and Molecular Structure," *Studies in History and Philosophy of Modern Physics* 41: 183-91.

ıdry, Robin Findlay. 2010b. "Science and Everyday Life: 'Water' vs 'H_2O,'" *Insights* 3

ıdry, Robin F. 2017. "Prospects for Strong Emergence in Chemistry," in Michele Paolini Paoletti and Francesco Orilia, eds., *Philosophical and Scientific Perspectives on Downward Causation* (London: Routledge).

:hliff, Mark. 2000. "A Defense of Presentism in a Relativistic Setting," *Philosophy of Science* 67: S585-S576.

fer, Carl. 2008. "For Fundamentalism," in Stephan Hartmann, Carl Hoefer, and Luc Bovens, eds., *Nancy Cartwright's Philosophy of Science* (London: Routledge).

ıen, Peter. 1955. *The Philosophical Nature of Physical Bodies* (West Baden Springs, IN: West Baden College).

ıen, Peter. 1958. "Simultaneity and the Principle of Neo-Positivism," in Henry J. Koren, ed., *Readings in the Philosophy of Nature* (Westminster, MD: The Newman Press).

ıen, Peter. 1960. *The Philosophy of Inorganic Compounds* (West Baden Springs, IN: West Baden College).

nan, Paul. 2009. "Does Efficient Causation Presuppose Final Causation? Aquinas vs. Early Modern Mechanism," in Samuel Newlands and Larry M. Jorgensen, eds., *Metaphysics and the Good: Themes from the Philosophy of Robert Merrihew Adams* (Oxford: Oxford University Press).

nann, Roald. 1995. *The Same and Not the Same* (New York: Columbia University Press).

Hoffmeyer, Jesper. 2010. “Semiotic freedom: an emerging force,” in Paul Davies and Niels Henrik Gregersen, eds., *Information and the Nature of Reality: From Physics to Metaphysics* (Cambridge: Cambridge University Press).

Horgan, John. 2014. “Physicist George Ellis Knocks Physicists for Knocking Philosophy, Falsification, Free Will,” *ScientificAmerican.com*, at https://blogs.scientificamerican.com/cross-check/physicist-george-ellis-knocks-physicists-for-knocking-philosophy-falsification-free-will/ [last accessed 23.4.17]

Horwich, Paul. 1987. *Asymmetries in Time: Problems in the Philosophy of Science* (Cambridge, MA: The MIT Press).

Huggett, Nick. 2010. *Everywhere and Everywhen: Adventures in Physics and Philosophy* (Oxford: Oxford University Press).

Hugon, Edouard. 2013. *Cosmology*, translated, with notes, by Françisco J. Romero Carrasquillo (Heusenstamm: Editiones Scholasticae).

Hull David L. 1974. *Philosophy of Biological Science* (Englewood Cliffs, NJ: Prentice-Hall, Inc.).

Hull, David L. 1992. “The Effect of Essentialism on Taxonomy: Two Thousand Years of Stasis,” in Marc Ereshefsky, ed., *The Units of Evolution: Essays on the Nature of Species* (Cambridge, MA: The MIT Press).

Hull, David L. 1995. “Mechanistic explanation,” in Robert Audi, general editor, *The Cambridge Dictionary of Philosophy* (Cambridge: Cambridge University Press).

Humphreys, Paul. 2008. “How Properties Emerge,” in Mark A. Bedau and Paul Humphreys, eds., *Emergence: Contemporary Readings in Philosophy and Science* (Cambridge, MA: The MIT Press).

Husserl, Edmund. 1970. *The Crisis of European Sciences and Transcendental Phenomenology*, trans. D. Carr (Evanston, IL: Northwestern University Press).

Husserl, Edmund. 1991. *On the Phenomenology of the Consciousness of Internal Time*, trans. J. Brough (Dordrecht: Kluwer Academic Publishers).

Husserl, Edmund. 2002. *Ideas: General Introduction to Pure Phenomenology* (New York: Routledge).

r, Ferris. 2014. “Why Nothing Is Truly Alive,” *The New York Times*, March 12, at: https://www.nytimes.com/2014/03/13/opinion/why-nothing-is-truly-alive.html [last accessed 8.9.18]

nes, William. 1890. *The Principles of Psychology* (New York: Dover).

ns, Sir James. 1931. *The Mysterious Universe* (Cambridge: Cambridge University Press).

nson, Monte Ransome. 2005. *Aristotle on Teleology* (Oxford: Clarendon Press).

nston, Mark. 1997. “How to Speak of the Colors,” in Alex Byrne and David R. Hilbert, eds., *Readings on Color, Volume 1: The Philosophy of Color* (Cambridge, MA: The MIT Press).

ce, George Hayward. 1924. *Principles of Natural Theology*, Second edition (London: Longmans, Green and Co.).

k, John W. 2007. “The Natural Motion of Matter in Newtonian and Post-Newtonian Physics,” *The Thomist* 71: 529-54.

k, John W. 2011. “The Messiness of Matter and the Problem of Inertia.” Paper presented at the Society for Aristotelian Studies Meeting, June 17, 2011, Santa Paula, California.

nedy, J. B. 2003. *Space, Time and Einstein: An Introduction* (Montreal and Kingston: McGill-Queen’s University Press).

aid, Harold. 2013. “Introduction: Pursuing a Naturalist Metaphysics,” in Don Ross, James Ladyman, and Harold Kincaid, eds., *Scientific Metaphysics* (Oxford: Oxford University Press).

ner, Philip. 1984. “Species,” *Philosophy of Science* 51: 308-33.

ner, Philip. 1993. *The Advancement of Science* (Oxford: Oxford University Press).

ertanz, George P. 1953. *The Philosophy of Human Nature* (New York: Appleton-Century-Crofts, Inc.).

as, John F. X. 2003. *Being and Some Twentieth-Century Thomists* (New York: Fordham University Press).

s, Robert C. 2000. *Realism Regained: An Exact Theory of Causation, Teleology, and the Mind* (Oxford: Oxford University Press).

Koons, Robert C. 2018a. “Hylomorphic Escalation: An Aristotelian Interpretation of Quantum Thermodynamics and Chemistry,” *American Catholic Philosophical Quarterly* 92:159-78.

Koons, Robert C. 2018b. “Knowing Nature: Aristotle, God, and the Quantum,” in Andrew B. Torrance and Thomas H. McCall, eds., *Knowing Creation: Perspectives from Theology, Philosophy, and Science, Volume 1* (Grand Rapids: Zondervan).

Koons, Robert C. 2018c. “The Many Worlds Interpretation of QM: A Hylomorphic Critique and Alternative,” in William M. R. Simpson, Robert C. Koons, and Nicholas J. Teh, eds., *Neo-Aristotelian Perspectives on Contemporary Science* (London: Routledge).

Koons, Robert C. Unpublished. “Quantum Hylomorphism, Part I: An Aristotelian Interpretation of Quantum Mechanics.”

Koons, Robert C. and Logan Paul Gage. 2012. “St. Thomas Aquinas on Intelligent Design,” *Proceedings of the American Catholic Philosophical Association* 85: 79-97.

Koons, Robert C. and Timothy H. Pickavance. 2015. *Metaphysics: The Fundamentals* (Oxford: Wiley Blackwell).

Koperski, Jeffrey. 2015. *The Physics of Theism* (Oxford: Wiley Blackwell).

Koren, Henry J. 1955. *An Introduction to the Philosophy of Animate Nature* (St. Louis: B. Herder Book Co.).

Koren, Henry J. 1962. *An Introduction to the Philosophy of Nature* (Pittsburgh: Duquesne University Press).

Korzybski, Alfred. 1933. *Science and Sanity* (Lancaster: International Non-Aristotelian Library Publishing Company).

Kosso, Peter. 1998. *Appearance and Reality: An Introduction to the Philosophy of Physics* (Oxford: Oxford University Press).

Koyré, Alexandre. 1965. “The Significance of the Newtonian Synthesis,” in *Newtonian Studies* (London: Chapman and Hall).

Kripke, Saul A. 1980. *Naming and Necessity* (Cambridge, MA: Harvard University Press).

ripke, Saul A. 1982. *Wittgenstein on Rules and Private Language* (Cambridge, MA: Harvard University Press).

ihn, Thomas S. 1962. *The Structure of Scientific Revolutions* (Chicago: University of Chicago Press).

ippers, Bernd-Olaf. 2010. "Information and communication in living matter," in Paul Davies and Niels Henrik Gregersen, eds., *Information and the Nature of Reality: From Physics to Metaphysics* (Cambridge: Cambridge University Press).

dyman, James. 2002. *Understanding Philosophy of Science* (London: Routledge).

dyman, James. 2014. "Structural Realism," *Stanford Encyclopedia of Philosophy*, at https://plato.stanford.edu/entries/structural-realism/ [last accessed 23.4.2017]

dyman, James and Don Ross with David Spurrett and John Collier. 2007. *Every Thing Must Go: Metaphysics Naturalized* (Oxford: Oxford University Press).

ndini, Gregory. 2011. *Russell* (London: Routledge).

udan, Larry. 1984. "A Confutation of Convergent Realism," in Jarrett Leplin, ed., *Scientific Realism* (Berkeley and Los Angeles: University of California Press).

wrence, Nathaniel. 1971. "Time Represented as Space," in Eugene Freeman and Wilfrid Sellars, eds., *Basic Issues in the Philosophy of Time* (La Salle, IL: Open Court).

clerc, Ivor. 1972. *The Nature of Physical Existence* (London: George Allen and Unwin).

nnox, James G. 1993. "Darwin *was* a Teleologist," *Biology and Philosophy* 8: 409-421.

Poidevin, Robin. 2003. *Travels in Four Dimensions: The Enigmas of Space and Time* (Oxford: Oxford University Press).

Poidevin, Robin. 2007. *The Images of Time: An Essay on Temporal Representation* (Oxford: Oxford University Press).

vine, Joseph. 2001. *Purple Haze: The Puzzle of Consciousness* (Oxford: Oxford University Press).

Lewens, Tim. 2016. *The Meaning of Science: An Introduction to the Philosophy of Science* (Oxford: Oxford University Press).

Lewis, David. 1973. *Counterfactuals* (Oxford: Blackwell).

Lewis, David. 1976. "The Paradoxes of Time Travel," *American Philosophical Quarterly* 13: 145-52.

Lewis, David. 1980. "Veridical Hallucination and Prosthetic Vision," *Australasian Journal of Philosophy* 58: 239-49.

Lewis, Peter J. 2016. *Quantum Ontology: A Guide to the Metaphysics of Quantum Mechanics* (New York: Oxford University Press).

Libet, Benjamin. 2004. *Mind Time: The Temporal Factor in Consciousness* (Cambridge, MA: Harvard University Press).

Lilla, Mark. 2007. *The Stillborn God: Religion, Politics, and the Modern West* (New York: Alfred A. Knopf).

Lloyd, Seth. 2006. *Programming the Universe* (New York: Vintage Books).

Lloyd, Seth. 2010. "The computational universe," in Paul Davies and Niels Henrik Gregersen, eds., *Information and the Nature of Reality: From Physics to Metaphysics* (Cambridge: Cambridge University Press).

Lockwood, Michael. 1989. *Mind, Brain, and the Quantum* (Oxford: Basil Blackwell).

Lockwood, Michael. 2005. *The Labyrinth of Time* (Oxford: Oxford University Press).

Lonergan, Bernard. 1992. *Insight: A Study of Human Understanding*, edited by Frederick E. Crowe and Robert M. Doran (Toronto: University of Toronto Press).

Lowe, E. J. 1995. *Locke on Human Understanding* (London: Routledge).

Lowe, E. J. 1999. *The Possibility of Metaphysics: Substance, Identity, and Time* (Oxford: Clarendon Press).

Lowe, E. J. 2002. *A Survey of Metaphysics* (Oxford: Oxford University Press).

Lucas, J. R. 1973. *A Treatise on Time and Space* (London: Methuen).

Mach, Ernst. 1984. *The Analysis of Sensations and the Relation of the Physical to the Psychical*, trans. by C. M. Williams (La Salle: Open Court).

ackie, J. L. 1976. *Problems from Locke* (Oxford: Clarendon Press).

ackie, J. L. 1982. *The Miracle of Theism* (Oxford: Clarendon Press).

adden, James D. 2013. *Mind, Matter, and Nature: A Thomistic Proposal for the Philosophy of Mind* (Washington, D.C.: Catholic University of America Press).

agnus, P. D. and Craig Callender. 2004. "Realist Ennui and the Base Rate Fallacy," *Philosophy of Science* 71: 320-38.

anent, Pierre. 1995. *An Intellectual History of Liberalism* (Princeton, N.J.: Princeton University Press).

anent, Pierre. 1998. *The City of Man* (Princeton, N.J.: Princeton University Press).

aritain, Jacques. 1951. *Philosophy of Nature* (New York: Philosophical Library).

aritain, Jacques. 1977. "Toward a Thomist Idea of Evolution," in *Untrammeled Approaches: The Collected Works of Jacques Maritain, Volume 20* (South Bend: University of Notre Dame Press).

aritain, Jacques. 1995. *The Degrees of Knowledge* (Notre Dame: University of Notre Dame Press).

arkosian, Ned. 2004. "A Defense of Presentism," in Dean Zimmerman, ed., *Oxford Studies in Metaphysics, Volume 1* (Oxford: Oxford University Press).

arshall, Richard. 2014. "Jerry Fodor: Meaningful Words Without Sense, and Other Revolutions," in *Philosophy at 3:AM: Questions and Answers with 25 Top Philosophers* (Oxford: Oxford University Press).

rtin, C. B. 2008. *The Mind in Nature* (Oxford: Clarendon Press).

xwell, Grover. 1962. "The Ontological Status of Theoretical Entities," in Herbert Feigl and Grover Maxwell, eds., *Minnesota Studies in the Philosophy of Science, Volume 3* (Minneapolis: University of Minnesota Press).

xwell, Grover. 1970a. "Structural Realism and the Meaning of Theoretical Terms," in S. Winokur and M. Radner, eds., *Minnesota Studies in the Philosophy of Science, Volume 4: Analyses of Theories and Methods of Physics and Psychology* (Minneapolis: University of Minnesota Press).

xwell, Grover. 1970b. "Theories, Perception and Structural Realism," in R. Colodny, ed., *University of Pittsburgh Series in the Philosophy of Science, Volume*

4: *The Nature and Function of Scientific Theories* (Pittsburgh: University of Pittsburgh Press).

Maxwell, Grover. 1972. "Scientific Methodology and the Causal Theory of Perception," in H. Feigl, W. Sellars and K. Lehrer, eds., *New Readings in Philosophical Analysis* (New York: Appleton-Century Crofts).

Mayfield, John E. 2013. *The Engine of Complexity: Evolution as Computation* (New York: Columbia University Press).

Mayr, Ernst. 1961. "Cause and Effect in Biology," *Science* 134: 1501-1506.

Mayr, Ernst. 2004. *What Makes Biology Unique?* (Cambridge: Cambridge University Press).

Mazur, Joseph. 2008. *Zeno's Paradox* (New York: Plume).

McCormick, John F. 1940. *Scholastic Metaphysics, Part I: Being, Its Division and Causes* (Chicago: Loyola University Press).

McGinn, Colin. 1983. *The Subjective View: Secondary Qualities and Indexical Thoughts* (Oxford: Clarendon Press).

McGinn, Colin. 1991. *The Problem of Consciousness* (Oxford: Blackwell).

McGinn, Colin. 1993. *Problems in Philosophy: The Limits of Inquiry* (Oxford: Blackwell).

McGinn, Colin. 2011. *Basic Structures of Reality: Essays in Meta-Physics* (Oxford: Oxford University Press).

McLaughlin, Thomas. 2004. "Local Motion and the Principle of Inertia: Aquinas, Newtonian Physics, and Relativity," *International Philosophical Quarterly*, Vo. 44, No. 1.

McLaughlin, Thomas J. 2008. "Nature and Inertia," *Review of Metaphysics*, Vol. 62, No. 2.

McInerny, D.Q. 2001. *The Philosophy of Nature* (Lincoln, NE: The Alquin Press).

McMullin, Ernan, ed. 1963. *The Concept of Matter in Greek and Medieval Philosophy* (Notre Dame: University of Notre Dame Press).

McMullin, Ernan, ed. 1978. *The Concept of Matter in Modern Philosophy*, Revised edition (Notre Dame: University of Notre Dame Press).

:Mullin, Ernan. 1984. "A Case for Scientific Realism," in Jarrett Leplin, ed., *Scientific Realism* (Berkeley and Los Angeles: University of California Press).

:Taggart, J. M. E. 1908. "The Unreality of Time," *Mind* 17: 457-74.

:Taggart, J. M. E. 1927. *The Nature of Existence, Volume II* (Cambridge: Cambridge University Press).

Williams, James A. 1950. *Cosmology*, Second revised edition (New York: Macmillan).

le, Alfred R. 2014. *Free: Why Science Hasn't Disproved Free Will* (Oxford: Oxford University Press).

llor, D. H. 1981. *Real Time* (Cambridge: Cambridge University Press).

llor, D. H. 1998. *Real Time II* (London: Routledge).

llor, D. H. 2005. "Time," in F. Jackson and M. Smith, eds., *The Oxford Handbook of Contemporary Philosophy* (Oxford: Oxford University Press).

nuge, Angus. 2004. *Agents Under Fire* (Lanham: Rowman and Littlefield).

rleau-Ponty, Maurice. 1967. *The Structure of Behavior*, translated by Alden L. Fisher (Boston: Beacon).

rleau-Ponty, Maurice. 2012. *Phenomenology of Perception*, translated by Donald A. Landes (London: Routledge).

rricks, Trenton. 2001. *Objects and Persons* (Oxford: Oxford University Press).

yer, Stephen C. 2009. *Signature in the Cell: DNA and the Evidence for Intelligent Design* (New York: HarperOne).

ikan, Ruth Garrett. 1984. *Language, Thought, and Other Biological Categories* (Cambridge, MA: The MIT Press).

ak, C. J. 1995. *Verificationism: Its History and Prospects* (London: Routledge).

nar, George. 2003. *Powers: A Study in Metaphysics* (Oxford: Oxford University Press).

nton, Bradley. 2006. "Presentism and Quantum Gravity," in D. Dieks, ed., *The Ontology of Spacetime* (Amsterdam: Elvesier).

Moreno, Antonio. 1974. "The Law of Inertia and the Principle '*Quidquid movetur ab alio movetur*,'" *The Thomist*, Vol. 38.

Moural, Josef. 2003. "The Chinese Room Argument," in Barry Smith, ed., *John Searle* (Cambridge: Cambridge University Press).

Mullahy, Bernard. 1946. *Thomism and Mathematical Physics*. Ph.D. dissertation, Laval University.

Mumford, Stephen. 2004. *Laws in Nature* (London: Routledge).

Mumford, Stephen. 2009. "Causal Powers and Capacities," in Helen Beebee, Christopher Hitchcock, and Peter Menzies, eds., *The Oxford Handbook of Causation* (Oxford: Oxford University Press).

Mundle, C. W. K. 1967. "The Space-Time World" *Mind* 76: 264-69.

Nagel, Thomas. 1979. "What Is It Like to Be a Bat?" in *Mortal Questions* (Cambridge: Cambridge University Press).

Nagel, Thomas. 1986. *The View from Nowhere* (Oxford: Oxford University Press).

Nagel, Thomas. 1997. *The Last Word* (Oxford: Oxford University Press).

Nagel, Thomas. 2012. *Mind and Cosmos* (Oxford: Oxford University Press).

Neurath, Otto. 1983. "Physicalism: The Philosophy of the Vienna Circle," in *Philosophical Papers 1913 - 1946*, edited by R.S. Cohen and M. Neurath (Dordrecht: D. Reidel Publishing Company).

Newman, M. H. A. 1928. "Mr. Russell's Causal Theory of Perception," *Mind* 37: 137-48.

Newton-Smith, W. H. 1980. *The Structure of Time* (London: Routledge and Kegan Paul).

Noë, Alva. 2004. *Action in Perception* (Cambridge, MA: The MIT Press).

Noë, Alva. 2005. "Against Intellectualism," *Analysis* 65: 278-90.

Noë, Alva. 2009. *Out Of Our Heads: Why You Are Not Your Brain, and Other Lessons from the Biology of Consciousness* (New York: Hill and Wang).

ɔë, Alva. 2012. "Are the Mind and Life Natural?," *13.7 Cosmos and Culture*, October 12, at: https://www.npr.org/sections/13.7/2012/10/12/162725315/are-the-mind-and-life-natural?ft=1&f=114424647 [last accessed 15.9.2018]

ɔë, Alva. 2013. "Do We Know How Life Began? Not Really," *13.7 Cosmos and Culture*, February 8, at: https://www.npr.org/sections/13.7/2013/02/08/171463418/do-we-know-how-life-began-not-really [last accessed 15.9.2018]

orton, John D. 2007. "Causation as Folk Science," in Huw Price and Richard Corry, eds., *Causation, Physics, and the Constitution of Reality: Russell's Republic Revisited* (Oxford: Clarendon Press).

erberg, David S. 1993. *The Metaphysics of Identity over Time* (London: Macmillan).

erberg, David S. 2004. "Temporal Parts and the Possibility of Change," *Philosophy and Phenomenological Research* 69: 686-708.

erberg, David S. 2006. "Instantaneous Change without Instants," in Craig Paterson and Matthew S. Pugh, eds., *Analytical Thomism: Traditions in Dialogue* (Aldershot: Ashgate).

erberg, David S. 2007. *Real Essentialism* (London: Routledge).

erberg, David S. 2008. "Teleology: Inorganic and Organic," in Ana Marta González, ed., *Contemporary Perspectives in Natural Law* (Aldershot: Ashgate).

erberg, David S. 2009. "Persistence," in J. Kim, E. Sosa, and G. Rosenkrantz, eds., *A Companion to Metaphysics*, Second edition (Oxford: Wiley-Blackwell).

erberg, David S. 2013. "Synthetic Life and the Bruteness of Immanent Causation," in Edward Feser, ed., *Aristotle on Method and Metaphysics* (Basingstoke: Palgrave Macmillan).

ısha, Samir. 2002. "Darwinian Metaphysics: Species and the Question of Essentialism," *Synthese* 131: 191-213.

fson, Frederick A. 2001. *Naturalism and the Human Condition: Against Scientism* (London: Routledge).

Oldroyd, David. 1989. *The Arch of Knowledge: An Introductory Study of the History of the Philosophy and Methodology of Science* (Kensington: New South Wales University Press).

O'Neill, John. 1923. *Cosmology: An Introduction to the Philosophy of Matter*, Volume I (London: Longmans, Green and Co.).

O'Rourke, Fran. 2004. "Aristotle and the Metaphysics of Evolution," *The Review of Metaphysics* 58: 3-59.

Osler, Margaret. 1996. "From Immanent Natures to Nature as Artifice: The Reinterpretation of Final Causes in Seventeenth Century Natural Philosophy," *The Monist* 79: 389-90.

Ott, Walter. 2009. *Causation and Laws of Nature in Early Modern Philosophy* (Oxford: Oxford University Press).

Parker, DeWitt H. 1970. *Experience and Substance: An Essay in Metaphysics* (Westport, CT: Greenwood Press).

Pasnau, Robert and Christopher Shields. 2004. *The Philosophy of Aquinas* (Boulder, CO: Westview Press).

Perlman, Mark. 2002. "Pagan Teleology: Adaptational Role and the Philosophy of Mind," in André Ariew, Robert Cummins, and Mark Perlman, eds., *Functions: New Essays in the Philosophy of Psychology and Biology* (Oxford: Oxford University Press).

Perry, John. 1979. "The Problem of the Essential Indexical," *Nous* 13: 3-29.

Phillips, R.P. 1950. *Modern Thomistic Philosophy, Volume I: The Philosophy of Nature* (Westminster, MD: The Newman Press).

Place, U.T. 1996. "Dispositions as Intentional States," in D.M. Armstrong, C.B. Martin, and U.T. Place, *Dispositions: A Debate*, ed. Tim Crane (London: Routledge).

Poincaré, Henri. 1952. *Science and Hypothesis* (New York: Dover Publications).

Polanyi, Michael. 1962. *Personal Knowledge* (Chicago: University of Chicago Press).

Polanyi, Michael. 1966. *The Tacit Dimension* (Garden City, NY: Doubleday).

Popper, Karl. 1968. *Conjectures and Refutations: The Growth of Scientific Knowledge* (New York: Harper and Row).

›pper, Karl. 1992. *The Logic of Scientific Discovery* (London: Routledge).

›pper, Karl. 1998. "Beyond the Search for Invariants." In Karl Popper, *The World of Parmenides* (London: Routledge).

:e, Huw. 1996. *Time's Arrow and Archimedes' Point* (Oxford: Oxford University Press).

›r, Arthur. 1970. "The Notion of the Present," *Studium Generale* 33: 245-48.

›r, Arthur. 1996. "Some Free Thinking about Time," in B. J. Copeland, ed., *Logic and Reality: Essays on the Legacy of Arthur Prior* (Oxford: Oxford University Press).

ser, Simon. 2016. *Experiencing Time* (Oxford: Oxford University Press).

ss, Alexander R. 2006. *The Principle of Sufficient Reason: A Reassessment* (Cambridge: Cambridge University Press).

ss, Alexander R. 2009. "The Leibnizian Cosmological Argument," in William Lane Craig and J. P. Moreland, eds., *The Blackwell Companion to Natural Theology* (Oxford: Wiley-Blackwell).

ss, Alexander R. 2018. "A Traveling Forms Interpretation of Quantum Mechanics," in William M. R. Simpson, Robert C. Koons, and Nicholas J. Teh, eds., *Neo-Aristotelian Perspectives on Contemporary Science* (London: Routledge).

os, Stathis. 1999. *Scientific Realism: How Science Tracks Truth* (London: Routledge).

am, Hilary. 1967. "Time and Physical Geometry," *Journal of Philosophy* 64: 240-47.

am, Hilary. 1979. "What is Mathematical Truth?" in Hilary Putnam, *Mathematics, Matter and Method: Philosophical Papers, Volume 1*, Second edition (Cambridge, MA: Cambridge University Press).

am, Hilary. 1987. *The Many Faces of Realism* (LaSalle, Illinois: Open Court).

am, Hilary. 1988. *Representation and Reality* (Cambridge, MA: The MIT Press).

am, Hilary. 1992. *Renewing Philosophy* (Cambridge, MA: Harvard University Press).

Putnam, Hilary. 1999. *The Threefold Cord: Mind, Body, and World* (New York: Columbia University Press).

Putnam, Hilary. 2012a. "A Philosopher Looks at Quantum Mechanics (Again)," in Hilary Putnam, *Philosophy in an Age of Science* (Cambridge, MA: Harvard University Press).

Putnam, Hilary. 2012b. "Quantum Mechanics and Ontology," in Hilary Putnam, *Philosophy in an Age of Science* (Cambridge, MA: Harvard University Press)

Quine, W. V. 1960. *Word and Object* (Cambridge, MA: The MIT Press).

Quine, W. V. 1980. "Two Dogmas of Empiricism," in W. V. Quine, *From a Logical Point of View*, Second edition, Revised (Cambridge, MA: Harvard University Press).

Ray, Christopher. 1991. *Time, Space, and Philosophy* (London: Routledge).

Reichenbach, Hans. 1957. *The Philosophy of Space and Time* (New York: Dover Publications).

Reith, Herman. 1956. *An Introduction to Philosophical Psychology* (Englewood Cliffs, NJ: Prentice-Hall, Inc.).

Reppert, Victor. 1992. "Eliminative Materialism, Cognitive Suicide, and Begging the Question," *Metaphilosophy* 23: 378–92.

Rey, Georges. 2002. "Searle's Misunderstandings of Functionalism and Strong AI," in John Preston and Mark Bishop, eds., *Views into the Chinese Room: New Essays on Searle and Artificial Intelligence* (Oxford: Clarendon Press).

Rickles, Dean. 2016. *The Philosophy of Physics* (Cambridge: Polity Press).

Rizzi, Anthony. 2004. *The Science Before Science* (Baton Rouge: IAP Press).

Robinson, Howard. 1982. *Matter and Sense* (Cambridge: Cambridge University Press).

Rorty, Richard. 1979. *Philosophy and the Mirror of Nature* (Princeton, NJ: Princeton University Press).

Rosenberg, Alexander. 1994. *Instrumental Biology or The Disunity of Science* (Chicago: The University of Chicago Press).

Rosenberg, Alex. 2006. *Darwinian Reductionism* (Chicago: University of Chicago Press).

senberg, Alex. 2011. *The Atheist's Guide to Reality* (New York: W.W. Norton and Co.).

senberg, Alex. 2013. "How Jerry Fodor Slid Down the Slippery Slope to Anti-Darwinism, and How We Can Avoid the Same Fate," *European Journal for Philosophy of Science* 3: 1-17.

enberg, Alex and Daniel W. McShea. 2008. *Philosophy of Biology: A Contemporary Introduction* (London: Routledge).

s, Don, James Ladyman, and Harold Kincaid, eds. 2013. *Scientific Metaphysics* (Oxford: Oxford University Press).

s, James F. 1990. "The Fate of the Analysts: Aristotle's Revenge," *Proceedings of the American Catholic Philosophical Association* 64: 51-74.

s, James. 2008. *Thought and World: The Hidden Necessities* (Notre Dame, IN: University of Notre Dame Press).

man, Stephen. 2015. *The Paradox of Evolution: The Strange Relationship Between Natural Selection and Reproduction* (Amherst, NY: Prometheus Books).

ce, James E. 1961. *Man and His Nature* (New York: McGraw-Hill Book Company).

sell, Bertrand. 1915. "On the Experience of Time," *The Monist* 25: 212-33.

sell, Bertrand. 1927. *The Analysis of Matter* (New York: Harcourt, Brace).

ell, Bertrand. 1931. *The Scientific Outlook* (New York: W. W. Norton and Company).

ell, Bertrand. 1948. *Human Knowledge: Its Scope and Limits* (New York: Simon and Schuster).

ell, Bertrand. 1963. "Mathematics and the Metaphysicians," in *Mysticism and Logic and Other Essays* (London: Unwin Books).

ell, Bertrand. 1985. *My Philosophical Development* (London: Unwin Paperbacks).

ell, Bertrand. 1995. "Mind and Matter," in *Portraits from Memory and Other Essays* (Nottingham: Spokesman).

ell, Bertrand. 2003. "On the Notion of Cause," in Bertrand Russell, *Russell on Metaphysics*, ed. Stephen Mumford (London: Routledge).

Russell, Bertrand. 2009. *ABC of Relativity* (London: Routledge).

Russell, Bertrand and F. C. Copleston. 1964. "A Debate on the Existence of God," in John Hick, ed., *The Existence of God* (New York: Macmillan).

Ryckman, Thomas. 2005. *The Reign of Relativity: Philosophy in Physics 1915-1925* (Oxford: Oxford University Press).

Ryle, Gilbert. 1945-46. "Knowing How and Knowing That." *Proceedings of the Aristotelian Society* 46: 1-16.

Ryle, Gilbert. 1949. *The Concept of Mind* (New York: Barnes and Noble).

Sachs, Joe. 1995. *Aristotle's* Physics: *A Guided Study* (New Brunswick: Rutgers University Press).

Salmon, Wesley C., ed. 2001. *Zeno's Paradoxes* (Indianapolis: Hackett Publishing Company).

Sanson, David and Ben Caplan. 2010. "The Way Things Were," *Philosophy and Phenomenological Research* 81: 24-39.

Saunders, Simon. 2002. "How Relativity Contradicts Presentism," in Craig Callender, ed., *Time, Reality, and Experience* (Cambridge: Cambridge University Press).

Scerri, Eric R. 1994. "Has Chemistry Been at Least Approximately Reduced to Quantum Mechanics?" *PSA* 1: 160-70.

Scerri, Eric R. 2000. "Realism, Reduction, and the 'Intermediate Position,'" in Nalini Bhushan and Stuart Rosenfeld, eds., *Of Minds and Molecules: New Philosophical Perspectives on Chemistry* (Oxford: Oxford University Press).

Scerri, Eric R. 2007a. "The Ambiguity of Reduction," *HYLE* 13: 67-81.

Scerri, Eric. R. 2007b. *The Periodic Table: Its Story and Its Significance* (Oxford: Oxford University Press).

Scerri, Eric R. 2016. "The Changing Views of a Philosopher of Chemistry on the Question of Reduction," in Eric Scerri and Grant Fisher, eds., *Essays in the Philosophy of Chemistry* (Oxford: Oxford University Press).

Schaffer, Jonathan. 2007. "The Metaphysics of Causation," *Stanford Encyclopedia of Philosophy*, at: https://plato.stanford.edu/entries/causation-metaphysics/ [last accessed 18.3.2017] hlick, Moritz. 1985. *General Theory of Knowledge*, translated by Albert E. Blumberg (La Salle, IL: Open Court).

hrödinger, Erwin. 1956. "On the Peculiarity of the Scientific World-View," in *What is Life? and Other Scientific Essays* (New York: Doubleday).

hrödinger, Erwin. 1992. "Mind and Matter." In Erwin Schrödinger, *What is Life? with Mind and Matter and Autobiographical Sketches* (Cambridge: Cambridge University Press).

hulman, Adam L. 1989. *Quantum and Aristotelian Physics*. Ph.D. dissertation, Harvard University.

arle, John R. 1980. "Minds, Brains and Programs," *Behavioral and Brain Sciences* 3: 417-57.

arle, John R. 1983. *Intentionality: An Essay in the Philosophy of Mind* (Cambridge: Cambridge University Press).

arle, John R. 1984. *Minds, Brains, and Science* (Cambridge, MA: Harvard University Press).

arle, John R. 1992. *The Rediscovery of the Mind* (Cambridge, MA: The MIT Press).

arle, John R. 1997. *The Mystery of Consciousness* (New York: The New York Review of Books).

arle, John R. 2008a. "Is the Brain a Digital Computer?" in John R. Searle, *Philosophy in a New Century: Selected Essays* (Cambridge: Cambridge University Press).

arle, John R. 2008b. "Why I Am Not a Property Dualist," in John R. Searle, *Philosophy in a New Century: Selected Essays* (Cambridge: Cambridge University Press).

hon, Scott. 2005. *Teleological Realism: Mind, Agency, and Explanation* (Cambridge, MA: The MIT Press).

llars, Wilfrid. 1956. "Empiricism and the Philosophy of Mind," in H. Feigl and M. Scriven, eds., *Minnesota Studies in the Philosophy of Science*, Volume I (Minneapolis, MN: University of Minnesota Press).

llars, Wilfrid. 1963. "Philosophy and the Scientific Image of Man," in *Science, Perception and Reality* (London: Routledge and Kegan Paul).

Shannon, Claude E. and Warren Weaver. 1949. *The Mathematical Theory of Communication* (Urbana, IL: University of Illinois Press).

Shields, Christopher. 2007. *Aristotle* (London: Routledge).

Shoemaker, Sydney. 1969. "Time Without Change," *Journal of Philosophy* 66: 363-81.

Sider, Theodore. 2001. *Four-Dimensionalism* (Oxford: Clarendon Press).

Simon, Yves R. 2001. *The Great Dialogue of Nature and Space* (South Bend, Indiana: St. Augustine's Press).

Simpson, William M. R., Robert C. Koons, and Nicholas J. Teh, eds. 2018. *Neo-Aristotelian Perspectives on Contemporary Science* (London: Routledge).

Sklar, Lawrence. 1985a. "Inertia, Gravitation, and Metaphysics," in Lawrence Sklar, *Philosophy and Spacetime Physics* (Berkeley and Los Angles: University of California Press).

Sklar, Lawrence. 1985b. "Time, Reality, and Relativity," in Lawrence Sklar, *Philosophy and Spacetime Physics* (Berkeley and Los Angeles: University of California Press).

Sklar, Lawrence. 1992. *Philosophy of Physics* (Boulder: Westview Press).

Sklar, Lawrence. 1993. *Physics and Chance: Philosophical Issues in the Foundations of Statistical Mechanics* (Cambridge: Cambridge University Press).

Skow, Bradford. 2015. *Objective Becoming* (Oxford: Oxford University Press).

Smart, J. J. C. 1963. *Philosophy and Scientific Realism* (London: Routledge and Kegan Paul).

Smart, J. J. C. 1978. "Spatialising Time," in Richard M. Gale, ed., *The Philosophy of Time* (New Jersey: Humanities Press).

Smith, John Maynard. 2010. "The concept of information in biology," in Paul Davies and Niels Henrik Gregersen, eds., *Information and the Nature of Reality: From Physics to Metaphysics* (Cambridge: Cambridge University Press).

Smith, Quentin. 1993. *Language and Time* (Oxford: Oxford University Press).

Smith, Quentin. 2008. "A radical rethinking of quantum gravity: Rejecting Einstein's relativity and unifying Bohmian quantum mechanics with a Bell-neo-Lorentzian absolute time, space and gravity," in W. L. Craig and Q.

Smith, eds., *Einstein, Relativity, and Absolute Simultaneity* (London: Routledge).

nith, Vincent Edward. 1950. *Philosophical Physics* (New York: Harper).

nith, Vincent Edward. 1958. *The General Science of Nature* (Milwaukee: Bruce Publishing Company).

nith, Wolfgang. 1999. "From Schrödinger's Cat to Thomistic Ontology," *The Thomist* 63: 49-63.

nith, Wolfgang. 2005. *The Quantum Enigma*, Third edition (Hillsdale, NY: Sophia Perennis).

nolin, Lee. 2007. *The Trouble with Physics* (New York: Mariner Books).

nolin, Lee. 2013. *Time Reborn* (New York: Houghton Mifflin Harcourt).

owdon, P.F. 1980–81. "Perception, Vision and Causation," *Proceedings of the Aristotelian Society* (New Series) 81: 175-92.

owdon, Paul. 2004. "Knowing How and Knowing That: A Distinction Reconsidered," *Proceedings of the Aristotelian Society* 104:1-29.

ber, Elliott. 1993a. *The Nature of Selection* (Chicago: University of Chicago Press).

ber, Elliott. 1993b. *Philosophy of Biology* (Boulder: Westview Press).

ber, Elliott. 2010. "Natural Selection, Causality, and Laws: What Fodor and Piatelli-Palmarini Got Wrong," *Philosophy of Science* 77: 594-607.

mos, David N. 2003. *The Species Problem: Biological Species, Ontology, and the Metaphysics of Biology* (Lanham, MD: Lexington Books).

nley, Jason and Timothy Williamson. 2001. "Knowing How," *Journal of Philosophy* 98: 411-444.

bbing, L. Susan. 1958. *Philosophy and the Physicists* (New York: Dover Publications).

in, Howard. 1968. "On Einstein-Minkowski Space-time," *Journal of Philosophy* 65: 5-23.

relny, Kim. 2007. *Dawkins vs. Gould: Survival of the Fittest*, New edition (Cambridge: Icon Books).

Sterelny, Kim and Paul E. Griffiths. 1999. *Sex and Death: An Introduction to the Philosophy of Biology* (Chicago: The University of Chicago Press).

Stich, Stephen P. and Stephen Laurence. 1996. "Intentionality and Naturalism." In Stephen P. Stich, *Deconstructing the Mind* (Oxford: Oxford University Press).

Stove, David. 2006. *Darwinian Fairytales* (New York: Encounter Books).

Strawson, Galen. 1994. *Mental Reality* (Cambridge, MA: The MIT Press).

Strawson, Galen. 2008. "Real Materialism," in *Real Materialism and Other Essays* (Oxford: Clarendon Press).

Strawson, P. F. 1959. *Individuals: An Essay in Descriptive Metaphysics* (London: Methuen).

Strawson, P. F. 1979. "Perception and its Objects," in G. F. Macdonald, ed., *Perception and Identity: Essays Presented to A.J. Ayer with His Replies* (Ithaca: Cornell University Press).

Strawson, P. F. 1989. *The Bounds of Sense: An Essay on Kant's* Critique of Pure Reason (London: Routledge).

Stroud, Barry. 2000. *The Quest for Reality: Subjectivism and the Metaphysics of Color* (Oxford: Oxford University Press).

Stroud, Barry. 2011. *Engagement and Metaphysical Dissatisfaction: Modality and Value* (Oxford: Oxford University Press).

Stump, Eleonore. 2003. *Aquinas* (London: Routledge).

Stump, Eleonore. 2006. "Substance and Artifact in Aquinas's Metaphysics," in Thomas M. Crisp, Matthew Davidson, and David Vanderlaan, eds., *Knowledge and Reality: Essays in Honor of Alvin Plantinga* (Dordrecht: Springer).

Tahko, Tuomas E., ed. 2012. *Contemporary Aristotelian Metaphysics* (Cambridge: Cambridge University Press).

Tallant, Jonathan. 2009. "Ontological Cheats Might Just Prosper," *Analysis* 69: 422-30.

Tallis, Raymond. 2009. "Neurotrash," *New Humanist* 124: 18-21.

llis, Raymond. 2011. *Aping Mankind: Neuromania, Darwinitis, and the Misrepresentation of Humanity* (Durham: Acumen).

llis, Raymond. 2017. *Of Time and Lamentation* (Newcastle upon Tyne: Agenda Publishing).

gmark, Max. 2014. *Our Mathematical Universe* (New York: Alfred A. Knopf).

omas Aquinas. 1948. *Summa Theologica.* Translated by the Fathers of the English Dominican Province (New York: Benziger Bros.).

omas Aquinas. 1952. *On the Power of God*, trans. English Dominican Fathers (Westminster, MD: The Newman Press).

omas Aquinas. 1964. *Exposition of Aristotle's Treatise On the Heavens*, trans. Fabian R. Larcher and Pierre H. Conway (Columbus: College of St. Mary of the Springs).

omas Aquinas. 1975. *Summa Contra Gentiles*, translated by Anton C. Pegis, James F. Anderson, Vernon J. Bourke, and Charles J. O'Neil (Notre Dame: University of Notre Dame Press).

omas Aquinas. 1986. *The Divisions and Methods of the Sciences*, Fourth revised edition, trans. Armand Maurer (Toronto: Pontifical Institute of Mediaeval Studies).

omasson, Amie L. 2007. *Ordinary Objects* (Oxford: Oxford University Press).

ompson, Ian J. 2010. *Philosophy of Nature and Quantum Reality* (Pleasanton, CA: Eagle Pearl Press).

ompson, Michael. 1995. "The Representation of Life," in R. Hursthouse, G. Lawrence, and W. Quinn, eds., *Virtues and Reasons* (Oxford: Clarendon Press).

orne, Kip S. 1994. *Black Holes and Time Warps: Einstein's Outrageous Legacy* (New York: W. W. Norton).

ley, Michael. 1997. *Time, Tense, and Causation* (Oxford: Clarendon Press).

ner, J. Scott. 2007. *The Tinkerer's Accomplice: How Design Emerges from Life Itself* (Cambridge, MA: Harvard University Press).

, Michael. 2017. "Qualia," *Stanford Encyclopedia of Philosophy*, at: https://plato.stanford.edu/entries/qualia/ [last accessed 15.9.2018]

Van Brakel, J. 2000. *Philosophy of Chemistry* (Leuven: Leuven University Press).

Van Fraassen, Bas C. 1980. *The Scientific Image* (Oxford: Clarendon Press).

Van Fraassen, Bas C. 2008. *Scientific Representation* (Oxford: Clarendon Press).

Van Inwagen, Peter. 1990. *Material Beings* (Ithaca, NY: Cornell University Press).

Van Melsen, Andrew G. 1954. *The Philosophy of Nature*, Second edition (Pittsburgh: Duquesne University Press).

Van Melsen, Andrew G. 1960. *From Atomos to Atom: The History of the Concept* Atom (New York: Harper and Brothers).

Van Melsen, Andrew G. 1965. *Evolution and Philosophy* (Pittsburgh: Duquesne University Press).

Wallace, David Foster. 2010. *Everything and More: A Compact History of Infinity* (New York: W. W. Norton and Company).

Wallace, W. A. 1956. "Newtonian Antinomies Against the *Prima Via*," *The Thomist* 19: 151-92.

Wallace, William A. 1982. "St. Thomas's Conception of Natural Philosophy and Its Method," in Leo Elders, ed., *La Philosophie de la Nature de Saint Thomas d'Aquin* (Rome: Pontifical Academy of St. Thomas).

Wallace, William A. 1983. *From a Realist Point of View: Essays on the Philosophy of Science*, Second edition (New York: University Press of America).

Wallace, William A. 1996. *The Modeling of Nature: Philosophy of Science and Philosophy of Nature in Synthesis* (Washington, D.C.: Catholic University of America Press).

Wallace, William A. 1997. "Thomism and the Quantum Enigma," *The Thomist* 61: 455-68.

Walsh, D. 2002. "Brentano's Chestnuts," in André Ariew, Robert Cummins, and Mark Perlman, eds., *Functions: New Essays in the Philosophy of Psychology and Biology* (Oxford: Oxford University Press).

Walsh, D. 2006. "Evolutionary Essentialism," *British Journal of the Philosophy of Science* 57: 425-48.

Wasserman, Ryan. 2018. *Paradoxes of Time Travel* (Oxford: Oxford University Press).

eatherall, James Owen. 2016. *Void: The Strange Physics of Nothing* (New Haven: Yale University Press).

eisberg, Michael, Paul Needham, and Robin Hendry. 2011. "Philosophy of Chemistry," *Stanford Encyclopedia of Philosophy*, at https://plato.stanford.edu/entries/chemistry/ [last accessed 13.7.2018]

eisheipl, James A. 1955. *Nature and Gravitation* (River Forest, IL: Albertus Magnus Lyceum).

eisheipl, James A. 1985. *Nature and Motion in the Middle Ages*, ed. William E. Carroll (Washington, D. C.: Catholic University of America Press).

eiskrantz, Lawrence. 2009. *Blindsight: A Case Study Spanning 35 Years and New Developments* (Oxford: Oxford University Press).

eyl, Hermann. 1934. *Mind and Nature* (Philadelphia: University of Pennsylvania Press).

eyl, Hermann. 1949. *Philosophy of Mathematics and Natural Science* (Princeton: Princeton University Press).

heeler, John Archibald with Kenneth Ford. 1998. *Geons, Black Holes, and Quantum Foam: A Life in Physics* (New York: W.W. Norton and Company).

hitehead, Alfred North. 1948. *Essays in Science and Philosophy* (New York: Philosophical Library).

hitehead, Alfred North. 1967. *Science and the Modern World* (New York: The Free Press).

kins, John. 2014. "Information is the new Aristotelianism (and Dawkins is a hylomorphist)," *Scientia Salon*, at: http://scientiasalon.wordpress.com/2014/05/01/information-is-the-new-aristotelianism-and-dawkins-is-a-hylomorphist/ [last accessed 2.5.2014]

lliams, Bernard. 1990. *Descartes: The Project of Pure Inquiry* (Harmondsworth: Penguin).

ppel, John F. 2000. *The Metaphysical Thought of Thomas Aquinas* (Washington, D.C.: Catholic University of America Press).

ttgenstein, Ludwig. 1961. *Tractatus Logico-Philosophicus,* trans. D. F. Pears and B. F. McGuinness (London: Routledge and Kegan Paul).

Wittgenstein, Ludwig. 1968. *Philosophical Investigations*, Third edition (New York: Macmillan).

Wittgenstein, Ludwig. 1972. *On Certainty* (New York: Harper and Row).

Woodruff, David M. 2011. "Presentism and the Problem of Special Relativity," in W. Hasker, T. J. Oord, and D. Zimmerman, eds., *God in an Open Universe: Science, Metaphysics, and Open Theism* (Eugene, OR: Pickwick Publications).

Worrall, John. 1996. "Structural Realism: The Best of Both Worlds?" in David Papineau, ed., *The Philosophy of Science* (Oxford: Oxford University Press).

Worrall, John. 2007. "Miracles and Models: Why Reports of the Death of Structural Realism May Be Exaggerated," in Anthony O'Hear, ed., *Philosophy of Science* (Cambridge: Cambridge University Press).

Wright, Larry. 1999. "Functions," in David J. Buller, ed., *Function, Selection, and Design* (Albany: State University of New York Press).

Wuellner, Bernard. 1956. *Summary of Scholastic Principles* (Chicago: Loyola University Press).

Yourgrau, Palle. 1991. *The Disappearance of Time: Kurt Gödel and the Idealistic Tradition in Philosophy* (Cambridge: Cambridge University Press).

Yourgrau, Palle. 1999. *Gödel Meets Einstein: Time Travel in the Gödel Universe* (Chicago and La Salle, IL: Open Court).

Yourgrau, Palle. 2005. *A World Without Time: The Forgotten Legacy of Gödel and Einstein* (New York: Basic Books).

Zahar, Elie. 2001. *Poincaré's Philosophy: From Conventionalism to Phenomenology* (Chicago and La Salle, IL: Open Court).

Zahar, Elie. 2007. *Why Science Needs Metaphysics: A Plea for Structural Realism* (Chicago and La Salle, IL: Open Court).

Zimmerman, Dean. 2008. "The Privileged Present: Defending an 'A-Theory' of Time," in T. Sider, J. Hawthorne, and D. W. Zimmerman, eds., *Contemporary Debates in Metaphysics* (Oxford: Blackwell).

Zimmerman, Dean. 2011a. "Open Theism and the Metaphysics of the Space-Time Manifold," in W. Hasker, T. J. Oord, and D. Zimmerman, eds., *God in an Open Universe: Science, Metaphysics, and Open Theism* (Eugene, OR: Pickwick Publications).

ımerman, Dean. 2011b. “Presentism and the Space-Time Manifold,” in C. Callender, ed., *The Oxford Handbook of Philosophy of Time* (Oxford: Oxford University Press).

ırt, P. J. 1975. *About Time* (Amsterdam: North-Holland).

Index